THE
BIOLOGY
BOOK

Books by Michael C. Gerald

Nursing Pharmacology and Therapeutics

Pharmacology: An Introduction to Drugs

The Complete Idiot's Guide to Prescription Drugs

The Drug Book: From Arsenic to Xanax, 250 Milestones in the History of Drugs

The Poisonous Pen of Agatha Christie

THE BIOLOGY BOOK

FROM THE ORIGIN OF LIFE TO EPIGENETICS,
250 MILESTONES IN THE HISTORY OF BIOLOGY

Michael C. Gerald, *with* Gloria E. Gerald

STERLING
New York

STERLING
New York

An Imprint of Sterling Publishing
387 Park Avenue South
New York, NY 10016

ISBN 978-1-4549-1068-8

Distributed in Canada by Sterling Publishing
⅌ Canadian Manda Group, 165 Dufferin Street
Toronto, Ontario, Canada M6K 3H6
Distributed in the United Kingdom by GMC Distribution Services
Castle Place, 166 High Street, Lewes, East Sussex, England BN7 1XU
Distributed in Australia by Capricorn Link (Australia) Pty. Ltd.
P.O. Box 704, Windsor, NSW 2756, Australia

For information about custom editions, special sales, and premium and corporate purchases, please contact
Sterling Special Sales at 800-805-5489 or specialsales@sterlingpublishing.com.

Manufactured in China

2 4 6 8 10 9 7 5 3 1

www.sterlingpublishing.com

This book is dedicated to our fantastic children,
Marc Jonathan Gerald and Melissa Suzanne Gerald,
with gratitude and love and with the tremendous
joy and pride we have in their accomplishments.

To the memory of my brother, Steven Gerald, and our parents,
Tobias and Ruby Gerald and Hyman and Esther Gruber,
for their love, encouragement, and inspiration,
we also dedicate this book.

Contents

Introduction

The first page in the history of biology was undoubtedly "written" when our preliterate forebears became consciously aware of the distinction between living and nonliving objects. Later they recognized, at least at a superficial level, the similarities among, and differences between, living organisms encountered in their local environment while hunting for game and gathering food. During the course of preparing animals for a meal, their internal structures were revealed, but there is little reason to suspect that the differences between and among such animals aroused their intellectual hunters' curiosity. Of far greater importance in the lives of our ancestors were the supernatural forces that were responsible for their existence and who rewarded them with good fortune and children, while punishing them by withholding food sources and inflicting them with disease. They hoped those decisions could be influenced by human and animal sacrifices. Some 12,000 years ago, humans began assuming greater control of their living environment by cultivating plants to provide food and domesticating animals, notably dogs, to assist and accompany them.

The earliest students of biology were healers—variously referred to as witch doctors, medicine men/women, or shamans, who were the resident experts in treating disease. Their "therapy" combined plant-based medicines and prayers and supplications to supernatural forces with healing practices they acquired by experience—not systematic study. Among the greatest and earliest scholars of living organisms was Aristotle (384–322 BCE), who systematically examined animals and plants and their characteristics, and categorized them based on meticulous observation, reasoning, and interpretation, devoid of supernatural explanations, sharing this knowledge in no fewer than four books.

In the late seventeenth century, Leeuwenhoek—trained as a linen merchant and functioning as a self-taught amateur lens grinder, and whose letters to European scientific societies were limited to his native Dutch—uncovered a previously unknown microscopic world that was inhabited by living beings that were neither plants nor animals. Using a microscope, it was possible for Schneider and Schwann in the 1830s to identify the cell as the basic structural and functional unit of life—all life, both plant and animal—just as the atom is the basic unit in chemistry.

Prior to the nineteenth century, the study of living things, then called *natural history*, focused primarily on the diversity and classification of plants and animals as well as on the anatomy and physiology of animals. These naturalists employed observational rather than experimental methods of study. This dramatically changed in the nineteenth century, which hosted the explosive emergence of the systematic study of living things and descriptions of how living organisms functioned. *Natural science* was replaced by the newly coined designation *biology*. Advances in organic chemistry were applied to studying the chemistry of living organisms by such biochemical pioneers as Claude Bernard; those studies have continued to this present day with increasing sophistication.

Perhaps some of the most momentous findings in biology occurred in the decade between 1859 and 1868. In 1859, Charles Darwin advanced his theory of natural selection, the basis for evolution. Evolution, now the central theme of biology, has been used to explain both the unity and diversity of *all* living organisms. The scientific world reverberated in response to the appearance of Darwin's *On the Origin of Species by Means of Natural Selection* but barely took notice when Gregor Mendel, an obscure Czech priest, published the results of his studies on the height of garden peas conducted in a monastery garden. Over three decades later, Mendel's paper was rediscovered and served as the foundation for the new science of genetics. It also provided the basis for explaining mutations that lead to natural selection, an explanation that perplexed Darwin and his acolytes and that had challenged his theory of evolution. Since ancient times, living beings were believed to have arisen from nonliving sources, namely, by spontaneous generation. In a simple but elegant experiment, Louis Pasteur provided convincing evidence that living beings arose from earlier life forms. But the question still remains, what was the original source of life?

Among the most significant studies in the twentieth century, and continuing to the present, are those that have sought to understand the role of the individual cell components and their unique contributions to cell function. James Watson and Frances Crick's 1953 determination of the structure of DNA generated a revolution in biological research and a popular interest in science that continues to excite. Subsequent research focused upon explaining how genes, working through DNA, serve as the molecular basis for inheritance, direct the synthesis of proteins, and influence our health. Manipulation of DNA and biotechnology have been valuable contemporary tools in the development and modification of innovative medicines and the plants and animals we consume.

Unlike Aristotle, who was interested in and attempted to understand *all* knowledge of his day, the study of biology became increasingly detailed, diversified, and specialized in the late nineteenth century, giving rise to subdisciplines and professionally trained specialists who placed an increasing emphasis on active experimentation. General biology or zoology/botany courses and departments have fragmented into departments of biochemistry, molecular and cellular biology, anatomy and physiology, microbiology, evolutionary biology, genetics, and ecology. The milestones in each of these specialty disciplines will be found in *The Biology Book*.

Our goal in writing *The Biology Book* is to provide insights into the 250 most significant events in biology in a readable and enjoyable manner. Each entry is intended to be readily comprehensible to all readers, while providing new information and insights for the scientifically well-prepared. In the available space of several paragraphs, we will set the stage for the milestone by providing essential background information, without subjecting the reader to the task of slogging through highly technical discussions and explanations. In short, each entry, arranged chronologically, is intended to be scientifically sound, yet accessible and engaging, and each can stand alone and be read in any order. We have provided cross-references to other entries related to the topic, as well as to sources of more detailed information. As we are sure you will appreciate, the dates assigned to some entries vary in precision: experts are not always of a single mind regarding a date or, for that matter, even which researcher should be most credited with the milestone.

Since most leading college-level biology textbooks are over 1,000 pages long, what was our rationale for selecting only 250 milestones? First and foremost, each milestone must have represented a significant scientific advance in its day, for hundreds of years, and perhaps even today. Some of these milestones incrementally built upon and extended earlier discoveries, and were readily accepted by their contemporaries. Others, particularly those that were revolutionary in nature—a paradigm shift as described by the philosopher of science Thomas Kuhn—and that significantly departed from the then-accepted "truth," were all too often greeted by naysayers with a firestorm of derision, criticism, and even outright hostility. Although contrary to the self-image of scientists as being rational and objective, over time, some scholars have resisted and rejected new and novel ideas for political, philosophical, economic, or religious reasons, because such ideas ran counter to traditional time-honored, venerated beliefs they embraced, or because of simple, unvarnished ignorance. However, as irrefutable evidence was presented, the scientific community embraced Andreas Vesalius's correction of Galen's erroneous description of the human body, which had been taught without question to medical students for almost 1,500 years. Robert Koch's demonstration that germs—and not supernatural forces or "bad air"—were the causes of infectious diseases was another victory for the scientific method that revolutionized medicine.

Some of the greatest of all scientists, regardless of discipline, made monumental discoveries in the biological sciences. *The Biology Book* highlights their particular discovery and, when appropriate and interesting, profiles these scientists. Consider the internationally acclaimed physiologist Nobel laureate Ivan Pavlov, who linked psychic stimulation and digestive function, and who was tolerated by the Soviet regime during the 1920s and 1930s, notwithstanding his outspoken negative views of Communism; or Otto Loewi, who provided convincing proof of chemical neurotransmission and used his Nobel Prize money to buy his way out of Austria after the Nazi invasion. Finally, we must confess that a few milestone entries are included mainly because they provide an interesting story and, we suspect, all of us like a good story.

The milestones described in *The Biology Book* illustrate Isaac Newton's saying, "If I have seen further it is by standing on the shoulders of giants." We shall attempt to explain the importance of the discovery or concept in biology from a historical perspective, and underline the influence the finding has had on subsequent researchers and on our contemporary thinking. It is our hope that readers of *The Biology Book* will come away with a vastly enriched appreciation of the living world around them.

Acknowledgments

We wish to express great thanks to our daughter, Melissa Gerald, whose advice and suggestions as a biological anthropologist were of utmost assistance in preparing all aspects of this book. Our son, Marc Gerald, was instrumental in introducing us to Sterling Publishing and providing encouragement and invaluable professional support throughout this project. Christina Gerald is thanked for her loving support during the writing of this book, and we also wish to acknowledge Jon Ivans for his suggestions for book-entry inclusions. The excellent support of the editorial and production staff of Sterling Publishing—in particular, Melanie Madden, our editor—and Scott Calamar at LightSpeed Publishing helped bring this book to fruition, and all are also most greatly appreciated.

Origin of Life

Louis Pasteur (1822–1895), **J.B.S. Haldane** (1892–1964), **Alexander Oparin** (1894–1980)

The remains of fossils of microorganisms reveal that life on Earth began perhaps as early as 4–4.2 billion years ago. But how did life begin? The notion that life could arise from nonliving matter (spontaneous generation) can be traced back to ancient Greece, and it wasn't until 1859 that Louis Pasteur conducted a series of experiments that apparently disproved this concept. But in the mid-1920s, spontaneous generation reappeared, now relabeled as *abiogenesis*. The Russian biochemist Alexander Oparin and British evolutionary biologist J.B.S. Haldane, working independently, suggested that conditions on the primordial Earth were very different from those that now exist and favored chemical reactions leading to the synthesis of organic molecules from inorganic starting materials. The scientific literature abounds with theories postulating how the origin of life occurred, and while none have gained universal acceptance, most use some variation of the Oparin-Haldane hypothesis as their basis.

The process of abiogenesis (or biopoiesis), by which life arises from self-replicating abiotic (nonliving) simple organic molecules, occurs in several stages: Small organic molecules, such as amino acids and nitrogen-containing bases, are synthesized from atmospheric carbon dioxide and nitrogen, with energy provided by intense sunlight or UV (ultraviolet) radiation. These small organic molecules are linked to form macromolecules, such as proteins and nucleic acids. The macromolecules are brought in proximity within protocells, vesicles that are precursors of living cells and surrounded by a membrane controlling the cell's internal chemical contents. Under these conditions, reproduction and energy-producing and energy-utilizing chemical reactions can occur. In the final stage, a self-reproducing ribonucleic acid (RNA) is formed that is required for protein synthesis and that can perform enzyme functions needed for RNA replication. The unique chemistry of these newly formed RNA molecules makes them most successful in self-replication, allowing them to pass their favorable traits to daughter RNA molecules. This may represent the earliest example of natural selection.

SEE ALSO: Prokaryotes (c. 3.9 Billion BCE), Eukaryotes (c. 2 Billion BCE), Metabolism (1614), Refuting Spontaneous Generation (1668), Fossil Record and Evolution (1836), Darwin's Theory of Natural Selection (1859), Enzymes (1878), Miller-Urey Experiment (1953), Domains of Life (1990).

The question of how life first began on Earth has challenged scholars and philosophers for thousands of years. Conditions on our planet some one billion years after its formation were very different and conducive to the formation of simple organic molecules from the raw materials in the primordial atmosphere.

Last Universal Common Ancestor

Charles Darwin (1809–1882)

Charles Darwin's theory of evolution postulated that all life on earth descended from a common ancestor. In *Origin of Species*, Darwin wrote, "Therefore I should infer from analogy that all the organic beings which have ever lived on this earth have descended from the same one primordial form, into which life was first breathed." This last universal common ancestor (LUCA), also called the last common ancestor (LCA), was not necessarily the first living being. Rather, it was the most recent ancestor from which all other life that now exists had evolved approximately 3.9 billion years ago, and which now shares its genetic characteristics. There are three major branches of living organisms: the **eukaryotes**, including plants, animals, protozoa, and all others that have a nucleus; and bacteria and archaea, two branches of organisms that do not have a nucleus.

This search for characteristics that would enable us to anoint LUCA has proven to be both elusive and controversial. When the search began, it was assumed that LUCA was a crude, simple mass, but it is now believed that this was a gross oversimplification. In 2010, a formal test was proposed to evaluate the common features a LUCA candidate would have to meet.

LUCA was a single-celled organism, with a lipid membrane surrounding that cell. Some of the other features are in the general areas of genetics, biochemistry, energy source, and reproduction. In all living forms, genetic information is encoded in **deoxyribonucleic acid (DNA)**, and the genetic code by which DNA is translated into the production of **enzymes** and other proteins is almost identical from bacteria to humans. This translation of genetic information provides support for the concept of LUCA, and it appears far less likely that life evolved from multiple ancestors.

One of the greatest complications involved in the identification of LUCA is related to gene swapping. Genes have been shown to be able to jump from organism to organism, making it difficult to determine whether the characteristics we are viewing are universal or those characteristics that have moved.

SEE ALSO: Origin of Life (c. 4 Billion BCE), Prokaryotes (c. 3.9 Billion BCE), Eukaryotes (c. 2 Billion BCE), Darwin's Theory of Natural Selection (1859), Deoxyribonucleic Acid (DNA) (1869), Enzymes (1878), Miller-Urey Experiment (1953).

The tree of life, an allusion to our common origin, is a metaphor used in theologies and mythologies worldwide. This painting of the tree of life from the Palace of the Shaki Khans (c. 1797) is now on display in the Azerbaijan National Art Museum.

Prokaryotes

Carl Woese (1928–2012), **George E. Fox** (b. 1945)

Life first began on earth approximately four billion years ago, 600 million years after the formation of the earth. Prokaryotes are the most primitive and most abundant life forms. Their successful survival has been attributed to a number of factors: Most prokaryotes have a cell wall, which protects the cell and serves to maintain its shape. Many exhibit taxis, the innate ability to move toward nutrients and oxygen and away from noxious stimuli. Most significantly, prokaryotes can rapidly reproduce asexually by binary fission (dividing in half) and rapidly adapt to unfavorable environmental conditions.

Based on Woese and Fox's **domain classification of life**, two of the three domains—Archaea and Bacteria—are prokaryotic, lacking membranes surrounding their nucleus and organelles, structures (such as **ribosomes** and **mitochondria**) that have specific functions within the cell. The interior of the prokaryotic cell is a gelatinous fluid, the cytosol, in which subcellular materials are suspended; **deoxyribonucleic acid (DNA)** is contained in the nucleoid region of the cytosol.

Archaea have the remarkable ability to adapt, survive, and thrive in extreme environments in which few other living forms can exist. Referred to as extremophiles, some Arachaea live in volcanic hot springs, while others reside in the Great Salt Lake in Utah, where the salt concentration is tenfold higher than seawater.

By far, the majority of prokaryotes are bacteria, some with which animals have symbiotic or mutualistic (mutually beneficial) relationships. Bacteria are far better known because of their role in disease, and it has been estimated that one half of all human diseases have a bacterial cause. When examined under a microscope, bacteria can assume a number of shapes, the most common being sphere, rod, spiral, or comma shaped. Based on the chemical composition of their cell walls, and their response to a dye (**Gram stain**), bacteria are classified as gram-positive or gram-negative, which has important implications in clinical medicine for the diagnosis and treatment of infectious disorders with **antibiotics**.

SEE ALSO: Origin of Life (c. 4 Billion BCE), Eukaryotes (c. 2 Billion BCE), Leeuwenhoek's Microscopic World (1674), Cell Theory (1838), Darwin's Theory of Natural Selection (1859), Deoxyribonucleic Acid (DNA) (1869), Gram Stain (1884), Germ Theory of Disease (1890), Probiotics (1907), Antibiotics (1928), Ribosomes (1955), Domains of Life (1990).

Prokaryotes are the most populous living organisms and were the earliest to exist. Bacteria, which represent one of the three domains of life and are the most familiar prokaryotes, have a cell wall and assume four major shapes: rod (shown), sphere, spiral, and comma.

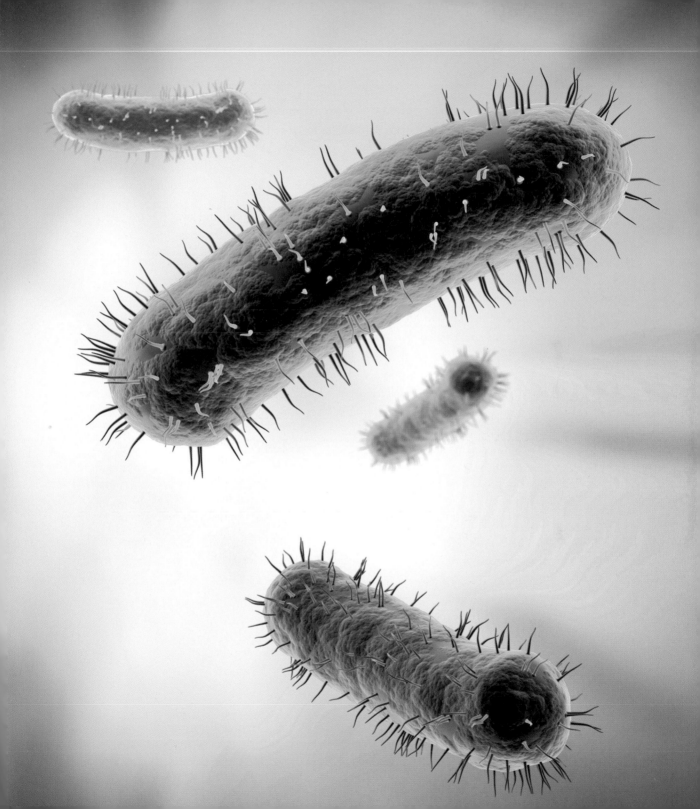

Algae

BASE OF THE FOOD CHAIN. Algae range in complexity from a single simple cell to millions of cells. In size, they span over seven orders of magnitude—from tiny *Micromonas* (1 μmeter in diameter) to giant kelp (200 feet or 60 meters). By the process of **photosynthesis**, algae form organic food molecules from carbon dioxide and water; these represent the base of the food chain upon which the existence of all marine life is dependent. Oxygen is a by-product of photosynthesis, and algae produce 30–50 percent of the global oxygen required by land animals for respiration. Crude oil and natural gas are photosynthetic products of ancient algae.

The heterogeneous nature of algae defies a universally agreed-upon biological classification. Some share features with protozoa and **fungi**, which diverged from algae over one billion years ago. As a group, algae are not closely related, nor do they form a single evolutionary linkage. A massive spike in atmospheric oxygen levels around 2.3 Billion BCE, believed to result from photosynthesis by cyanobacteria (blue-green algae), suggests that their evolutionary history began 2.5 billion years ago. Red algae and green algae evolved from a common ancient ancestor more than one billion years ago, with the oldest red fossil dating back some 1.5 billion years. The lineage that produced green algae were the forebears of **land plants**, and some biologists have proposed including green algae in the plant kingdom.

Some classifications of algae are based on whether they have a **cell nucleus** (eukaryotic) or lack one (prokaryotic) or are ecologically grouped by their habitat. Since the 1830s, algae have been classified into major groups predicated upon their color (red, green, brown), a photosynthetic accessory pigment that masks the green of chlorophyll. There are some 6,000 known species of red algae, with their different shades dependent on the depth of the sea, and they are most abundant in the warm coastal waters of tropical oceans. Most red algae are multicellular, the largest of which is referred to as "seaweed." There are over 7,000 species of green algae of the chlorophyte group, most found in fresh water.

SEE ALSO: Prokaryotes (c. 3.9 Billion BCE), Eukaryotes (c. 2 Billion BCE), Fungi (c. 1.4 Billion BCE), Land Plants (c. 450 Million BCE), Cell Nucleus (1831), Photosynthesis (1845), Food Webs (1927).

All life depends upon photosynthesis in algae. By this process, organic molecules are formed, upon which marine life is dependent. As a by-product of photosynthesis, oxygen is produced, which is essential for the survival of land life.

Eukaryotes

All advanced life forms have eukaryotic cells. Some 1.6–2.1 billion years ago, living organisms with eukaryotic cells, which are thought to have evolved from a prokaryotic ancestor by the process of endosymbiosis, appeared. Eukaryotic cells are ten times larger and more organizationally complex than prokaryotic cells. Eukaryotes (also spelled eucaryotes) exhibit staggering differences in shape and size that vary from amoebas to whales and from red **algae** (among the first eukaryotes) to **dinosaurs**.

The principal distinguishing difference between prokaryotic and eukaryotic cell types is the presence of membranes surrounding the eukaryotic nucleus and other intracellular organelles. These walled-off compartments permit the organelles to perform their specific functions (such as energy transformation, nutrient digestion, protein synthesis) far more efficiently, unaffected by other processes that are occurring simultaneously elsewhere in the cell.

The largest of these organelles is the **cell nucleus**, which contains DNA packaged in chromosomes that carry genetic information. Eukaryotic reproduction involves two processes: **mitosis**, in which one cell gives rise to two genetically identical daughter cells; and **meiosis**, where a pair of chromosomes divide, with each possessing half the number of chromosomes of the original cell.

The Eukarya, a third domain of life, includes the animal, plant, and **fungi** kingdoms that are all multicellular, and the protist kingdom, which is mostly unicellular; protists are by far the most diverse and numerous eukaryotes. One approach to differentiate members of each of these kingdoms is to consider how they meet their nutritional needs. Plants produce their own food by the process of photosynthesis, fungi absorb dissolved nutrients (decomposed organisms, dead wastes) from their environment, and animals eat and digest other organisms. No generalizations can be drawn with respect to protists and how they obtain their nutrition: in this respect, algae are plantlike, slime molds resemble fungi, and amoeba are more animal-like. The appropriate classification and evolutionary history of the protists has been subject to revision in recent years, when genetic analysis revealed that some protists are more closely related to animals and fungi than to fellow protists.

SEE ALSO: Origin of Life (c. 4 Billion BCE), Prokaryotes (c. 3.9 Billion BCE), Algae (c. 2.5 Billion BCE), Fungi (c. 1.4 Billion BCE), Leeuwenhoek's Microscopic World (1674), Cell Nucleus (1831), Cell Theory (1838), Darwin's Theory of Natural Selection (1859), Meiosis (1876), Mitosis (1882), Gram Stain (1884), Germ Theory of Disease (1890), Endosymbiont Theory (1967), Domains of Life (1990), Protist Taxonomy (2005).

Eukaryotes include multicellular plants, animals, and fungi, as well as unicellular protists. In nature, there are some 600 species of the fungus Amanita, *which are responsible for 95 percent of all fatal mushroom poisonings.*

Fungi

Apart from molds and mushrooms, fungi don't spring to mind as being important life forms, and yet they have a great impact on us. Fungi assist in the breakdown and recycling of dead and decaying organic matter in our environment. In addition to including mushrooms, morels, and truffles, some fungi are used to ripen cheese, and yeasts are used in the production of bread, alcoholic beverages, and industrial chemicals. Fungi are the source of the most important of all drugs: penicillin, and cyclosporine, a drug that prevents the rejection of transplanted organs. But 30 percent of the 100,000 species of fungi are parasites or pathogens. Plants are their favorite target; they've devastated fruit harvests and caused the **American chestnut tree blight**, Dutch elm disease, and **ergotism**, which killed 40,000 people in France in year 944 and has been implicated in causing hallucinations in those accused in the Salem witch trials. Fungi cause infections of the skin (athlete's foot), yeast infections (candidiasis), and life-threatening systemic infections.

Fungi were formerly classified as plants. They grow in the soil, are sessile (immotile), and have cell walls. But molecular evidence reveals that they are more closely related to animals, having evolved from a common aquatic unicellular ancestor at least 1.4 billion years ago. The oldest **fossil** of a land fungus is 460 million years old.

With the exception of single-celled fungi, as yeasts, fungi are composed of hyphae (thread-like tubular filaments), which are surrounded by cell walls composed of chitin (as in the external skeletons of insects) and not cellulose, found in plants. Hyphae tips branch out into mycelium (interwoven mats), which are above ground and produce fruiting bodies containing spores that serve a reproductive function.

Unlike animals that ingest food, and plants that manufacture food, fungi obtain their nutrients using several different approaches: as heterotrophs, they absorb nutrients from the environment; as saprophytes, they secrete **enzymes** that break down large organic molecules in living and dead cells (fallen logs, animal corpses) into small molecules that fungi can absorb; and as parasites, they secrete other enzymes that penetrate the walls of cells and absorb nutrients into their own cells.

SEE ALSO: Eukaryotes (c. 2 Billion BCE), Ergotism and Witchcraft (1670), Enzymes (1878), Antibiotics (1928), Domains of Life (1990), Protist Taxonomy (2005), American Chestnut Tree Blight (2013).

Yeasts, molds, and mushrooms are fungi that have been eaten as foods (mushrooms, truffles) and used to make bread and alcoholic beverages by fermentation (using yeast). Inoculation of milk curds with selected fungi imparts unique flavors and textures to cheese.

Arthropods

Arthropods are the most successful animals on the planet, having populated land, sea, and air, from the highest mountain to the deepest sea, from the poles to the tropics. They make up over three-quarters of all living and fossil animals, and it has been estimated that one billion billion (10^{18}) are currently living on earth, representing over one million species that have been described, with many millions more, living in tropical rainforests, yet to be identified. Their size ranges from microscopic insects and crustaceans to blue king crabs in the Bering Sea, with a leg span extending beyond 6 feet (1.8 meters) and often weighing more than 18 pounds (8 kilograms).

The origin and evolution of arthropods are clouded in controversy because many of the earliest members did not leave fossil remains. It is generally believed that all arthropods evolved from a common annelid ancestor—a marine worm—some 550–600 million years ago. Scientists are not of a single mind whether all arthropods evolved only once or multiple times from this common antecedent. The earliest fossil remains are of the now-extinct marine trilobites, dating back over 530 million years. The first land animals were myriapod arthropods (centipede-related) appearing some 450 million years ago.

Arthropods, the most diverse phylum, are invertebrates that are categorized into five major groups—insects, spiders, scorpions, crustaceans, and centipedes—all of which have common characteristics: They are bilaterally symmetrical (as are humans), that is, the left half of the body is the mirror image of the right half. They are surrounded by a cuticle, an exoskeleton (external skeleton) composed of chitin (a carbohydrate polymer), which provides protection, points of attachment for muscles that move their appendages, and prevents water loss from the body. Insect bodies are segmented and their appendages are jointed (arthropod means "jointed feet"), permitting them to move their legs, claws, and mouthparts, although their body is encased in the inflexible exoskeleton. Appendages have evolved to become fewer in number and more specialized in function, such as for locomotion (walking or swimming), feeding, defense, sensory perception (which is well developed), and reproduction.

SEE ALSO: Insects (c. 400 Million BCE).

Three-quarters of all living and fossil animals are classified as arthropods, including such crustaceans as the lobster. This watercolor, Hawaiian Lobster, *was painted in 1819 by a sixteen-year-old French artist named Adrien Taunay the Younger.*

Iles Sandwich

Voyage de l'Uranie, crustacés, pl. N° 3

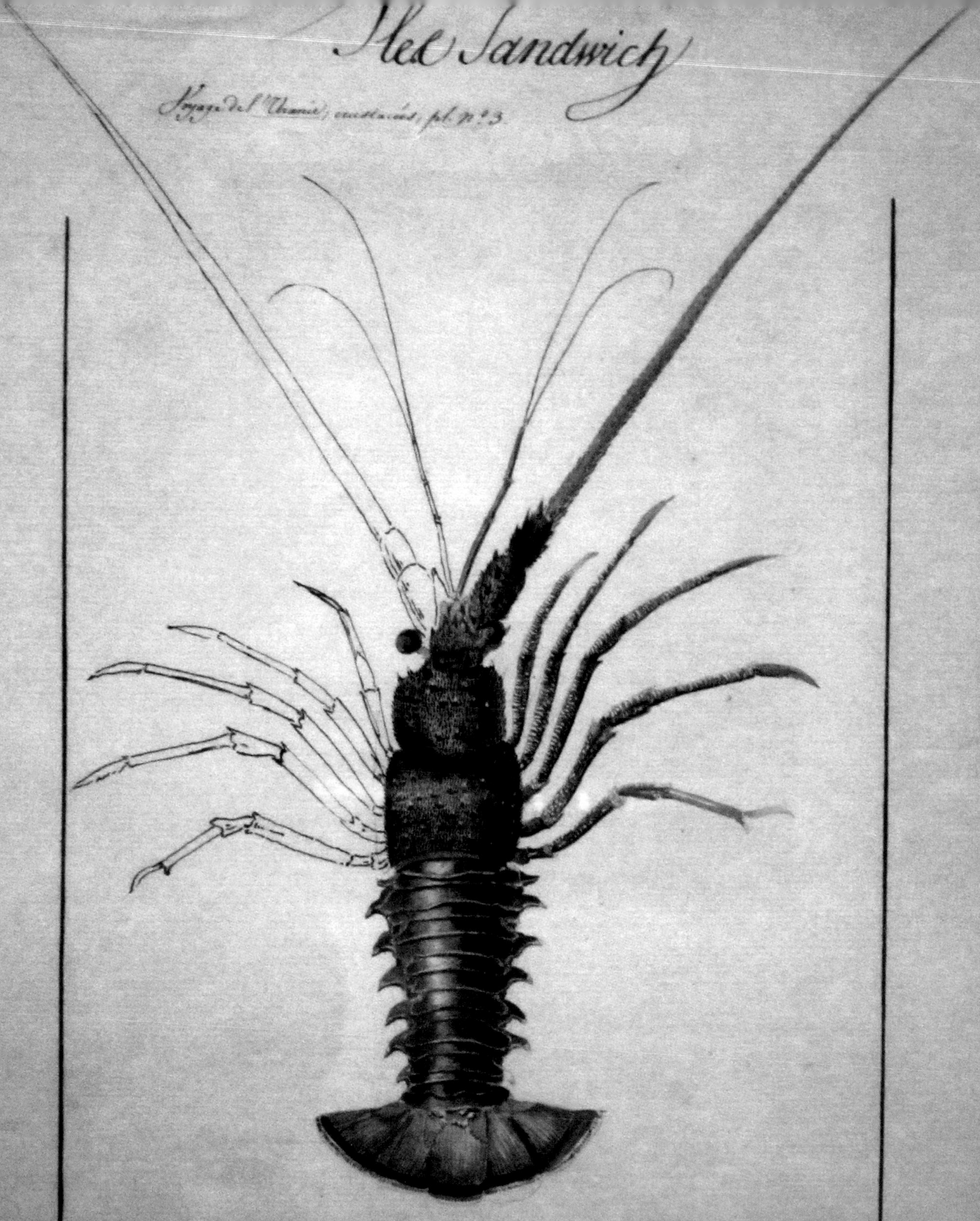

Medulla: The Vital Brain

When thoughts turn to the brain, we undoubtedly think about reasoning, emotions, and, of course, thinking—activities that are controlled at the highest levels. However, far more basic are those vital functions that are essential for survival, and these are regulated by the medulla oblongata, likely the first brain part to have evolved. Some authorities argue that the medulla is the most important brain component.

Animals possessing a bilaterally symmetrical body—bilaterians, the common ancestor of all vertebrates—first appeared some 555–558 million years ago. Among their characteristics was a hollow gut tube that ran from the mouth to the anus and contained a nerve cord, a precursor of the spinal cord. More than 500 million years ago, the first vertebrates, which are thought to have resembled the modern hagfish, emerged. Their anatomy had developed three swellings at the mouth end of their nerve cord: the forebrain, midbrain, and hindbrain.

The medulla, a structure in the hindbrain, developed from the top of the spinal cord. It is the lowermost portion of the brain and the most primitive portion of the vertebrate brain. The medulla regulates those functions upon which life is most dependent and which occur without our voluntary action: control of breathing, heart rate, and **blood pressure**. Chemoreceptors, located in the medulla, monitor oxygen and carbon dioxide levels in the blood and orchestrate appropriate changes in the rate of breathing. Its destruction causes instant death from respiratory failure. Baroreceptors, located in the aorta and carotid artery, detect changes in arterial blood pressure and, via nerve impulses, transmit messages to the cardiovascular center in the medulla, which in turn trigger changes that restore blood pressure and heart rate to normal levels.

The medulla is also the site of a number of reflex centers that respond without delay, in the absence of cognitive processing, when required to initiate vomiting, coughing, and swallowing. Additionally, it provides a pathway for nerves entering and leaving the brain and transmitting messages between the brain and spinal cord.

SEE ALSO: Fish (c. 530 Million BCE), Blood Pressure (1733), Nervous System Communication (1791), Neuron Doctrine (1891).

The medulla, the most primitive brain structure, controls such essential functions as breathing, heart rate, and blood pressure, as well as such reflex responses as coughing and sneezing. This poster, warning US soldiers to cover their coughs and sneezes in order to prevent the spread of germs, was issued by the Office of War Information during World War II.

COVER COUGHS

COVER SNEEZES

NEVER GIVE A GERM A BREAK!

Fish

The earliest ancestors of vertebrates appeared in oceans over some 550 million years ago. Later, the **Devonian Period** (417 to 359 million years ago)—dubbed the Age of Fishes—saw the remarkable evolution of fish, which are now more diverse than any other vertebrates. Of the 52,000 vertebrate species, 32,000 are fish, which may be simply and collectively characterized as gill-bearing vertebrates that lack limbs with digits.

Agnathans, which appeared during the Cambrian Explosion around 530 million years ago, were the first fish. Jawless with plated armored heads, their round mouthparts were used for sucking or filter-feeding; lampreys and hagfish are their only surviving descendants. The development of jaws, seen in the present-day cartilaginous fish and bony fish, permitted them to ingest a wider range of foods and become active hunters. The cartilaginous fish (Chondrichthyes) lack true bone but have instead rather light, flexible skeletons of cartilage, which make fish like rays, sharks, and skates very agile predators.

There are 19,000 species of true bony fish (Osteichthyes), which are as disparate as eels, horse fish, trout, and tuna. Most species have a swim bladder, which is a gas-filled bag that enables fish to effortlessly maintain their buoyancy at a desired depth. Sharks and rays lack swim bladders and, since they are heavier than water (as are all fish), they will sink. They may choose to rest on the sea floor or remain in constant motion, which involves a large expenditure of energy. Water contains a fraction of the oxygen found in air. Multiple pairs of gills, through which water continuously passes, are highly efficient at extracting oxygen and removing carbon dioxide, the end-product of metabolism.

There are two major groups of bony fish: the far more common ray-finned fish, which are named for the bony rays that support their fins, and the lobe-finned fish, such as **coelacanth, the "living fossil,"** which have rod-shaped bones surrounded by muscles in their pectoral and pelvic fins. These lobed fins evolved into the limbs and feet of tetrapods, four-legged land animals, which include humans.

SEE ALSO: Devonian Period (c. 417 Million BCE), Amphibians (c. 360 Million BCE), Paleontology (1796), Coelacanth: "The Living Fossil" (1938).

The most primitive fish, such as this sea lamprey, were jawless and used their round mouths to filter-feed. Lampreys are considered an invasive species in the North American Great Lakes, being fortunate to have no natural enemies there and feeding on such commercially valuable fish as lake trout.

Land Plants

Aristotle (384–322 BCE), Charles Darwin (1809–1882)

Aristotle was first to divide the living world into animals and plants, a division that still enjoys popular usage: animals moved and plants did not. There are 300,000 to 315,000 species of land plants (embryophytes), which include flowering plants, conifers, ferns, and mosses, but not algae and fungi.

All land plants arose from a single type of green **algae**, the carophytes, which made their first appearance on land 1.2 million years ago. Many green algae live on the edge of ponds and lakes, which are subject to drying. **Darwin's theory of natural selection** suggests that those algae that adapted and lived above the water line survived when the water receded. Such early land plants, which appeared some 450 million years ago, enjoyed the benefits of greater access to bright sunlight and carbon dioxide— conditions that favored the generation of organic molecules to feed themselves by **photosynthesis**—as well as nutrient-rich soil. Seed plants, which represent 85–90 percent of all land plants, appeared 360 million years ago, followed by flowering plants, 140 million years later, and the most recent major group, the grasses, some 40 million years ago.

Green plants vary from small weeds to giant redwood trees. All have eukaryotic (nucleus-containing) cells and cell walls composed of cellulose, and the vast majority derive their energy by photosynthesis. The first seed plants were the now-extinct seed ferns. Seeds consist of an embryo—a sperm-fertilized egg—and its food supply surrounded by a protective coat; such seeds can remain dormant for years after their release from their parent plant.

Some 12,000 years ago, humans in many parts of the world began to cultivate wild seed plants, transforming themselves from hunter-gatherers to farmers. Seed plants are the major sources of our food, as well as fuel, wood products (absent from seedless plants), and medicines. Seedless plants, which include ferns, mosses, liverworts, and horsetails, do not grow flowers or grow from seeds.

SEE ALSO: Algae (c. 2.5 Billion BCE), Eukaryotes (c. 2 Billion BCE), Plant Defenses against Herbivores (c. 400 Million BCE), Seeds of Success (c. 350 Million BCE), Gymnosperms (c. 300 Million BCE), Angiosperms (c. 125 Million BCE), Plant-Derived Medicines (c. 60,000 BCE), Agriculture (c. 10,000 BCE), Photosynthesis (1845), Darwin's Theory of Natural Selection (1859).

Land plants vary in size from inconspicuous weeds to giant redwood trees. Since the moss depicted here have no system for transporting water through the plant, they require a damp environment in which to grow and must be surrounded by liquid water in which to reproduce. These roof tiles, which retain moisture in a rainy climate, are ideal.

Devonian Period

Adam Sedgwick (1785–1873), Roderick Murchison (1792–1871)

The Devonian Period was so-named in 1839 by geologists Adam Sedgwick and Roderick Murchison, after the county of Devon, England, where the rocks of this period were first studied. The Devonian, from 417–359 million years ago, witnessed major changes in plant life and **fish**, including some fish that ventured on land. Oceans covered 85 percent of the globe, and there were two supercontinents: Laurasia in the Northern Hemisphere and Gondwanaland in the Southern Hemisphere.

The first **land plants** appeared some 450 million years ago, and the oldest vascular plants appeared at the beginning of the Devonian. Unlike the earlier nonvascular plants, such as liverworts, ferns, and moss, vascular plants have an extensive system of tubes that transport water and nutrients throughout the plant. During this time, vegetation was simple, limited to the water's edge, and consisting of small plants, the tallest a mere 3 feet (1 meter) in height. Wood appeared, increasing axial strength, permitting trees to grow taller to reach open sunlight and support the greater weight of branches and leaves. Changes in the soil promoted the development of plant rooting systems; some 385 million years ago, the first tree-like organisms growing in forests were in evidence.

A Revolution in the Biotic World. During the Middle Devonian, the agnathans (jawless fish, with plate-like armor) started to decline in number. Jawed fish, such as the shark-like cartilaginous fish and the vast majority of extant bony fish, increased in number and variety and became the dominant marine and freshwater predators; this period has been justifiably called the Age of Fishes. From the lobe-finned jawed fish evolved the earliest tetrapods, which were able to crawl from the mud and move about out of water to feed on terrestrial invertebrates.

Toward the end of the Devonian, some 70 percent of the invertebrates disappeared, with the greatest losses among marine and, to a lesser degree, among freshwater species; **coral reefs** disappeared entirely. Estimates of the duration of Late Devonian extinction range from 500,000 to 25 million years. Although its cause remains unknown, it ranks as one of the five major extinction events in biota history.

SEE ALSO: Fish (c. 530 Million BCE), Land Plants (c. 450 Million BCE), Amphibians (c. 360 Million BCE), Gymnosperms (c. 300 Million BCE), Coral Reefs (c. 8000 BCE), Paleontology (1796), Fossil Record and Evolution (1836), Continental Drift (1912).

This image from Ernst Haeckel's Kunsiformen der Natur (Art Forms of Nature) *of 1904 shows various sea anemones, classified as* Actiniae, *which are water-dwelling, predatory animals that may have flourished in coral reefs during the Devonian period. Since most Actiniaria lack hard parts, however, fossil records are sparse.*

Insects

Edward O. Wilson (b. 1929)

THE MOST POPULOUS ANIMAL. Some one million species of insects have been described, and scientists speculate that another six to ten million are waiting to be discovered. As such, insects are the largest group in the animal kingdom and are more numerous than all other animal life forms combined. Members of the invertebrate phylum Arthropoda, they have been divided into at least thirty insect orders, each with an external skeleton and the same common characteristics: three pairs of legs; a head, thorax, and abdomen that are the insect's three distinct body segments; a pair of antennae, which detects sounds, vibrations, and chemical signals (semiochemicals, including **pheromones**), and external mouthpieces specialized for feeding.

Insects appeared some 400 million years ago, with the oldest fossils resembling modern-day silverfish. Scientists uncovered the full-body fossil impression of a flying insect that is thought to have lived 300 million years ago. Insects were the first animals to fly and the only invertebrates to do so. Flying provided insects with a major competitive advantage, permitting them to escape from predators, find food and mates, and move to new habitats.

Many insects undergo metamorphosis as an integral part of their lifecycle, and this may be incomplete metamorphosis, as with grasshoppers, where the young (nymph) resemble the adult but miniaturized. By contrast, the butterfly undergoes complete metamorphosis, in which the young (larva) goes through four distinct stages and is completely transformed in appearance as an adult. The American biologist E. O. Wilson has studied ants and their social behavior, which he described in his co-authored book *The Ants* (1990). Eusocial ants live in groups, cooperatively care for the brood, have overlapping generations and reproductive division of labor.

Insects have a multifaceted relationship with their environment. Many of us perceive them to be pests, feeding on animal hosts (as mosquitoes), transmitting disease (malaria), and destroying crops (locusts) and structures (termites). But they also pollinate flowering plants, are useful for genetics research, and serve as food for animals.

SEE ALSO: Arthropods (c. 570 Million BCE), Plant Defenses against Herbivores (c. 400 Million BCE), Angiosperms (c. 125 Million BCE), Paleontology (1796), Insect Dance Language (1927), Pheromones (1959), Sociobiology (1975).

Wenceslaus Hollar (1607–1677) was a Bohemian-born etcher whose works included a broad spectrum of subjects, including this 1646 etching, Forty-One Insects.

Plant Defenses against Herbivores

Some 400 million years ago, 50 million years after the appearance of the first **land plants**, fossil evidence reveals that **insects** were feasting on plants. The earliest terrestrial vertebrates, the **amphibians**, dating back some 360 million years, initially fed on fish and insects, and later expanded their diets to include plants. Herbivores are animals that have adapted anatomically and physiologically to eat plant material as a major component of their diet, which provides a rich source of carbohydrates. To thwart the efforts of these herbivores, and to enhance their own survival and reproductive fitness, plants have evolved physical and chemical defense mechanisms that can deter, injure, or even kill their predators. But just as plants evolved, the herbivores co-evolved, too, to overcome or reduce the effectiveness of these defense mechanisms and enable themselves to continue their plant diets.

Physical defenses or mechanical barriers, such as the thorns on the stems of roses and the spines on cacti, are intended to deter or injure herbivores. Trichomes are hairs that cover the leaves and stems of plants and represent an effective deterrent to most insect herbivores, although some insects have evolved counter-defensive mechanisms. Waxes or resins that coat plant parts serve to modify their texture and make cell walls difficult to eat and digest.

The chemicals produced for defensive purposes are by-products of plant **metabolism** and are secondary metabolites that do not participate in such fundamental functions as growth, development, and reproduction. Rather, they promote the long-term survival of the plant by acting as herbivore repellants or toxins. Among these chemical types are those classified as alkaloids and cyanogenic glycosides, and both contain nitrogen. Alkaloids are derivatives of amino acid metabolism and include such familiar chemicals as cocaine, strychnine, morphine, and nicotine, the latter long used as an insecticide in the garden and agricultural fields. Alkaloids adversely affect herbivores by modifying the activity of their **enzymes**, inhibiting protein synthesis and DNA repair mechanisms, and interfering with nerve function. When herbivores feed on plants containing cyanogenic glycosides, hydrogen cyanide is released, which poisons the cellular respiration of the would-be predator.

SEE ALSO: Land Plants (c. 450 Million BCE), Insects (c. 400 Million BCE), Amphibians (c. 360 Million BCE), Metabolism (1614), Nitrogen Cycle and Plant Chemistry (1837), Coevolution (1873), Enzymes (1878), Mitochondria and Cellular Respiration (1925).

Thorns, spines, and prickles are plant structures with sharp, stiff ends, which are used to mechanically deter herbivores. Although these terms are commonly used interchangeably, botanists distinguish them based upon where on the plant they originate. These prickles were found on a rose.

Amphibians

Some 360 million years ago, the fins of lobe-finned **fish** evolved into limbs and feet with digits. This descendent of the early tetrapods was able to leave the water, which may have proved advantageous to avoid aquatic competition and predation, and allowing pursuit of prey through the thick vegetation at the water's edge. These tetrapods evolved into amphibians ("living a double life"), with many members inhabiting both aquatic and terrestrial habitats during the course of their lifecycle.

There are 5000–6000 amphibian species divided into three groups, each with its specific characteristics: Salamanders and newts (Urodela = "tailed ones") have long tails and two pairs of limbs. Caecilians (Apoda = "legless ones") are legless, nearly blind, worm-like creatures found in tropical habitats.

Some 90 percent of all living amphibians are classified in the third group, frogs and toads (Anura = "tailless ones") that are aquatic as juveniles and land-based in damp habitats as adults. The female lays her eggs in water where they are externally fertilized by the male's sperm. The tadpole, the larval stage of the frog, has gills to extract oxygen from water, a long tail, and a lateral line system, which is a sensory system permitting it to detect movements and pressure changes in water. After metamorphosis, the tadpole develops powerful muscular hind limbs, a large head and eyes, a pair of external eardrums, and a digestive system suited for a carnivorous diet; it loses its tail, lateral line system, and gills. The exchange of respiratory gases—oxygen and carbon dioxide— occur through the skin, a common characteristics in all amphibians. Many, but not all, amphibians undergo metamorphosis.

A Bellwether for Biodiversity. Since the 1980s, there has been an alarming worldwide decline in the number of amphibians and frogs, which has led to the extinction of some species. This represents a major threat to global biodiversity. Amphibians feed on algae and zooplankton, and are active predators of **insects**, which reduces the threat of many insect-borne diseases; they, in turn, are sources of food for other vertebrates. The causes of this decline have not been established but may involve habitat destruction or modification, pollution, and fungal infections.

SEE ALSO: Fish (c. 530 Million BCE), Insects (c. 400 Million BCE), Plant Defenses against Herbivores (c. 400 Million BCE), Gas Exchange (1789), Thyroid Gland and Metamorphosis (1912), Food Webs (1927).

The red-eyed tree frog is an inhabitant of the Central American rainforest. When startled, the frog's bulging red eyes flash open and its bright sides are displayed—a defensive mechanism called startle coloration—*which surprises a predator bird or snake, allowing the frog to make its escape.*

Seeds of Success

The first plants, related to modern moss and ferns, arrived on land some 450 million years ago and ruled the vegetative world for over 100 million years. These early ferns were seedless and dependent upon water for their sexual reproductive success. The male gametophyte released sperm that were obliged to swim through a watery film to rendezvous with an egg for fertilization, forming a zygote. Seed plants, among the most important organisms on earth, made their first appearance some 350 million years ago and since then have been and continue to be the dominant and most familiar form of vegetation. Seeds and pollen have permitted plants to succeed on land, freeing them from dependence on water for reproduction and enabling them to adapt to drought and the harmful ultraviolet radiation in sunlight.

Seedless plants produce only a single kind of spore, which gives rise to a bisexual gametophyte. Over time, some of these plants evolved into seed plants that generate two types of spores: a microspore, giving rise to multiple male gametophytes, and a megaspore that produces a single female gametophyte. The female gametophyte and its surrounding protective coat is an immature seed called an *ovule*. Pollen grains are male gametophytes containing sperm enclosed in a protective coat, which prevents sperm from drying out, helps them withstand mechanical damage, and enables them to travel long distances and spread their genes. Unlike sperm in seedless plants that must travel to the ovule, seed plant sperm are passively carried in air currents.

Transfer of the pollen grain to the ovule-containing part of the plant is called *pollination*. After the egg cell, inside an ovule, is fertilized by sperm, the ovule produces an embryo that develops into a seed. Seeds provide the embryo with protection and nourishment, and enable the embryo to remain in a dormant stage for decades, if need be, awaiting favorable climatic conditions for germination.

Seed plants (spermatophytes) are of two major types: **gymnosperms** ("naked" seeds), including the conifers, and **angiosperms** (flowering plants), of which there are some 250,000 species—about 90 percent of the plant world.

SEE ALSO: Land Plants (c. 450 Million BCE), Gymnosperms (c. 300 Million BCE), Angiosperms (c. 125 Million BCE), Depletion of the Ozone Layer (1987).

Seed plants permit pollen grains to carry male sperm long distances, under adverse climatic conditions, to fertilize ovules. These seeds are from "lucky bean" trees (Erythrina lysistemon), members of the pea family that are cultivated in parks and gardens and believed to have magical and medicinal properties.

Reptiles

Some 320 million years ago, the earliest reptiles appeared, arising from **amphibians**, with lungs, more muscular legs, and laying hard-shelled external eggs that could better withstand land than the amphibian's water-bound eggs. Reptiles would enjoy their prominence during the Mesozoic era (250–265 million years ago), fittingly called the Age of Reptiles, when they were the most numerous and dominate vertebrates. After this period of pre-eminence, only sea turtles, crocodiles, snakes, and lizards were left, with lizards representing over 95 percent of extant reptiles.

Reptiles are amniote (tetrapods that lay eggs on land, excluding birds and mammals), which have scales or bony exterior plates, and are ectothermic, relying on external sources to provide their body heat. The oldest reptilian fossil remains, dating back 315 million years, were found in Nova Scotia, and consist of a series of footprints, made by a lizard-like animal, 8–12 inches (20–30 centimeters) in length. Among the earliest groups of reptiles were the diapsids, characterized by a pair of holes on each side of their skull. The diapsids gave rise to two branches: the lepidosaurs and the archosaurs (saur = lizard).

The lepidosaurs include the lizards, snakes, and tuataras, the latter, a lizard attaining a length of about 39 inches (1 meter), which once inhabited many parts of the world, but now solely resides on coastal New Zealand islands. The most impressive lepidosaur was the mosasaur, an extinct marine reptile resembling the monitor lizard that could attain a length of 57 feet (17.5 meters). It was quick, agile, and the dominant marine predator for almost 20 million years.

Two prominent archosaurs were the pterosaurs and **dinosaurs**. The pterosaurs (formerly called pterodactyls) were the earliest vertebrates that had mastered powered flight. They were the largest flying animals of all time, having a wingspan of 40 feet (12 meters); the smallest of the 120 species of pterosaurs was the size of a sparrow. Their bones were hollow, like that of **birds**, and their extremely long fourth digit provided support for a wing that was unlike those of bats and birds. Pterosaurs first appeared some 215 million years ago and thrived for 150 million years before becoming extinct.

SEE ALSO: Amphibians (c. 360 Million BCE), Dinosaurs (c. 230 Million BCE), Birds (c. 150 million BCE), Paleontology (1796), Thermoreception (c.1882).

The green tree python is found in New Guinea, Indonesia, and Cape York Peninsula in northern Australia. This python can reach more than 6 feet (180 cm) in length and resides in a distinctive coiled position around tree branches.

Gymnosperms

IF TREES COULD TALK. Some of the oldest, tallest, and thickest living organisms are gymnosperms, many of them residing in California. Redwoods live for thousands of years, and the Methuselah, a bristlecone pine, over 4,600 years old, is thought to be the world's oldest tree. Coast redwoods sometimes exceed 360 feet (110 meters) in height, and the world's tallest tree, the Stratosphere Giant, is 370 feet (113 meters). Fossil records reveal that the first gymnosperms appeared some 300 million years ago, 50 million years after the first seed plants, and provided nourishment for giant plant-eating dinosaurs. Although gymnosperms have been displaced for primacy in the plant world by **angiosperms** (flowering plants), which appeared 125 million years ago, conifers are more dominant and thrive at higher altitudes, in the colder climes of North America and northern Eurasia to the edge of the Arctic tundra, and under dry conditions. There are about six hundred conifer species, by far the largest group of gymnosperms, most of which are evergreens.

Gymnosperms ("naked" seeds) are plants that reproduce by means of an exposed seed (ovule) usually found on modified leaves that form cones; by contrast, the seeds of angiosperms are enclosed in mature ovaries (fruits). As is typical of spermatophytes (seed plants), the plant body of gymnosperms has a stem, roots, leaves, and a vascular system with two conducting pathways: the xylem channels water and minerals from the roots to the shoots, and the phloem conducts organic materials manufactured in the leaves to the nonphotosynthetic parts of the plant.

The gymnosperms are plants of great economic importance. Most of the commercial lumber in the Northern Hemisphere comes from the trunks of conifers such as pine, spruce, and Douglas fir, so-called softwoods, as does most plywood. Conifers are the source of essential oils, and their resins include derived turpentine, rosin, wood alcohol, and balsam. Some non-conifer gymnosperms are used as medicines including ephedra (the source of ephedrine), employed in China for thousands of years for respiratory disorders; Ginkgo biloba, claimed to be effective for the treatment of Alzheimer's disease, high blood pressure, and menopause; and the anticancer drug Taxol, extracted from the yew bark.

SEE ALSO: Land Plants (c. 450 Million BCE), Seeds of Success (c. 350 Million BCE), Dinosaurs (c. 230 Million BCE), Angiosperms (c. 125 Million BCE), Plant-Derived Medicines (c. 60,000 BCE), Paleontology (1796).

Ancient bristlecone pines, such as this one in the Inyo National Forest in California's Sierra Nevada mountain range, are thought to be the oldest living trees. Their longevity, measured in thousands of years, has been attributed to their very dense, durable, and resinous wood that is resistant to insects and fungi.

Dinosaurs

William Buckland (1784–1856), Richard Owen (1804–1892)

THE JURASSIC WORLD. From around 230 million years ago, and for the next 135 million years, dinosaurs were the dominant land vertebrates. William Buckland first described their fossils in the scientific literature in 1824, followed by Richard Owen who, in 1842, coined the name dinosaur ("terrible lizard"). They were not lizards.

Dinosaurs, classified as **reptiles**, were an extremely varied group, with over 1,000 species, making it impractical to list meaningful all-encompassing distinguishing characteristics apart from laying eggs and exhibiting nesting behavior. Some were herbivores, others carnivores; some stood erect and were bipedal, while others, quadrupedal. It was long believed that dinosaurs were lumbering creatures but, in recent decades, evidence suggests that some (e.g., the velociraptor) were agile and fast moving, as well as highly sociable, congregating in flocks. Dinosaurs varied in size from that of a pigeon to the largest land animals ever known. The herbivorous Apatosaurus (Brontosaurus) had a very long neck and a relatively small head and measured some 75 feet (23 meters) in length. The most familiar dinosaur was Tyrannosaurus rex, a bipedal carnivore 40 feet (12 meters) in length that shared a common ancestor with **birds**.

There is general agreement that birds descended from dinosaurs, and that Archaeopteryx, which lived some 150 million years ago and was first found in Bavaria in 1861, may be the missing link. Although its fossil remains failed to disclose feathers, other feathered dinosaurs have been uncovered since the 1990s, further supporting their relationship to birds.

Some 66 million years ago, all non-avian dinosaurs became extinct, as did 95 percent of all life on earth. The event causing this mass extinction has been the subject of considerable speculation and theorizing. The prevailing theory favors an impact event producing a toxic atmosphere and blocking out sunlight for an extended time period, extinguishing plant and animal life. Although dinosaurs are gone, they have not been forgotten. They have remained a popular culture staple in children's stuffed toys and books, and movies including A. Conan Doyle's *The Lost World* (1925), *King Kong* (1933), and *Jurassic Park* (1990s–2000s).

SEE ALSO: Reptiles (c. 320 Million BCE), Birds (c. 150 Million BCE), Paleontology (1796).

The Majungasaurus was a bipedal dinosaur that lived in Madagascar some 66–70 million years ago. Typically measuring 20–23 feet (6–7 meters) in length and weighing 2,400 lbs (1,130 kg), these carnivores were the alpha-predators in their environment.

Mammals

Carl Linnaeus (1707–1778)

For the past 65 million years, mammals have been the planet's paramount terrestrial animals and, with the exception of insects and arachnids (spiders), have the widest worldwide distribution. Every terrestrial and aquatic biome is inhabited by mammals, and their ecological success has been largely attributed to the ability to control body temperature. The 5,500–5,700 mammalian species range in size from the bumblebee bat (1.2–1.6 inches or 30–40 millimeters) to the largest living animal, the blue whale (greater than 100 feet or 30 meters).

The first true mammals made their appearance some 200 million years ago and, over the course of tens of millions of years, diverged into three branches: The monotremes are egg-laying mammals, such as the duckbilled platypus, found only in Australia and New Guinea. Kangaroos and opossums are marsupials, found in Australia and the Americas, in which the newborn continue to develop outside the womb in a marsupium (pouch). Ninety percent of all mammals are placental (eutherians), in which the fetus is carried in the womb until birth at which time it is at an advanced stage of development. In 2013, a shrew-size Chinese fossil was found, named *Juramaia sinensis*, which, at 160 million years, is thought to be the oldest placental mammal. Humans are placental mammals classified as **primates**.

Mammals have a number of unique characteristics not shared by other vertebrates: Mammary glands are modified sweat glands that allow females to nourish their young with milk, the offspring's primary source of nutrition. (In 1758, Linnaeus named these animals *mammals*, from the Latin = "breast.") Hair or fur, present at some time during life, protects mammals against extreme cold. The middle ear contains three bones that transmit sound vibrations into nerve impulses. The lower mammalian jaw consists of only a single bone on each side. Other characteristics, although not necessarily unique to mammals, include being warm-blooded (endothermic) and having specialized or differentiated teeth, a larger brain (in particular, the neocortex, the most advanced brain area), a diaphragm (a muscular sheet that separates the heart and lungs from the abdominal cavity), and an efficient four-chambered heart.

SEE ALSO: Primates (c. 65 Million BCE), Placenta (1651), Paleontology (1796).

This 1937 poster promoted infant nursing. Breastfeeding was common since ancient times, but declined significantly from 1900 to 1960 because of negative social attitudes and the increased popularity of infant formulas. Since then, the practice has increased, and experts recommend it for at least the first six months of infant life.

NURSE THE BABY
YOUR PROTECTION AGAINST TROUBLE
INFORM YOURSELF THROUGH THE HEALTH BUREAU PUBLICATIONS
AND CONSULT YOUR DOCTOR

WPA FEDERAL ART PROJELT

Birds

Charles Darwin (1809–1882)

There are some 10,000 species of living birds on all continents that have the common characteristics of being feathered, winged, bipedal, warm-blooded, and egg-laying. However, their distinguishing characteristic, when compared with almost all other vertebrates, is the ability of most of their species to fly. The power of flight endowed birds with many benefits, which descended from the winged Archaeopteryx or a related theropod (bipedal **dinosaur**) some 150 million years ago. In addition to serving as their primary mode of locomotion, flying enhanced the ability of birds to hunt, forage, breed, escape from grounded predators, travel to more fruitful feeding areas, and to migrate.

To facilitate their power of flight, birds evolved a number of characteristics: Their bodies are streamlined to minimize air resistance. Their body weight is reduced by hollow bones, the loss or adaptation of nonessential bones, and the lack of a urinary bladder and teeth. To accommodate their need for large volumes of oxygen, their respiratory system is modified. But, by far, the most important adaptations are their feathers and wings, the latter modified from forelimbs. The size and shape of wings and feathers of different bird species have aerodynamically evolved for greater speed, lower energy expenditures, and to better soar, glide, and maneuver. In addition to serving as an aid in flying, feathers also provide insulation against the cold and rain, assist in maintaining body temperature, and are used to attract mates in courtship.

Among the most distinctive features seen in birds are the size and shape of their beaks (also called bills), a feature noted by Charles Darwin during his stay on the Galápagos Islands in 1835. This observation provided one of the critical keys in his formulation of the theory of evolution based on natural selection. Darwin noted that the beaks of the dozen finches he observed were adapted to the specific food available to them. While birds use their beaks primarily for eating, beaks are also used to probe for food, kill prey, manipulate objects, groom, feed their young, and during courtship.

SEE ALSO: Reptiles (c. 320 Million BCE), Dinosaurs (c. 230 Million BCE), Animal Migration (c. 330 BCE), Darwin and the Voyages of the *Beagle* (1831), Darwin's Theory of Natural Selection (1859).

The Griffon Vulture is a bird of prey with an impressive wingspan that can exceed 9 feet (2.7 meters). Like other Old World vultures, it feeds mostly on carrion but lacks the keen sense of smell found in New World vultures.

Angiosperms

THE FLOWERING PLANTS. Angiosperms made their first appearance 125 million years ago and, 25 million years later, thanks to climatic changes, rapidly diversified and have since been the most dominant and familiar spermatophytes—the seed plants. Evidence suggests that angiosperms did not evolve from the more primitive **gymnosperms** (which include conifer evergreens) but developed separately. Angiosperms are more adaptable than gymnosperms and can grow in different soils and under disparate climatic conditions. When compared with gymnosperms, the angiosperms have a more efficient reproductive apparatus. Their seeds are protected inside fruits and are pollinated by **insects** and other animals, and so the seeds are more efficiently carried than just by wind currents. Some three-quarters of all angiosperms are eudicots (formerly called dicots), which include such diverse flowering plants as carnations, roses, **tobacco**, legumes, potatoes, grains, maples, and sycamores.

Representing 90 percent of all plants, with 250,000 species, angiosperms are second only to insects in number. They exhibit a wide variety of sizes, shapes, colors, smells, and arrangements, features which account for their highly specialized partnerships with mutualistic pollinators, a classic case of co-evolution of two distinctive organisms; wind-pollinated flowers lack colorful parts.

The flowers of angiosperms are often not only beautiful but also serve as highly efficient and distinctive reproductive systems that contain male or female reproductive structures, or both, within the same flower. Gametes are produced in separate organs within the flower, and fertilization and development of the embryo occur internally, thus providing protection against vagaries of the weather. Pollen, which contains sperm, is formed by stamen, while the ovule is produced by the pistil. Pollen is transferred to a receptive surface on the pistil where it germinates. The sperm is then transported to the ovule, through the pollen tube, and fertilizes the ovule in the pistil. Tissues grow around the embryo, which thickens, and the seeds develop into fruit. Like pollen and seeds, fruits are not only dispersed by winds but also by animals that eat the fruit; the seeds pass through the digestive tract undigested and are deposited in the feces at another location.

SEE ALSO: Land Plants (c. 450 Million BCE), Insects (c. 400 Million BCE), Seeds of Success (c. 350 Million BCE), Gymnosperms (c. 300 Million BCE), Ecological Interactions (1859).

There are over 250,000 species of angiosperms, the flowering plants, likenesses of which appear on these two Iranian panels of earthenware tiles dating from the first half of nineteenth century.

Primates

There is a lack of agreement regarding the scientific names and classification of the some 350 species of primates, and most contentiously, whether humans should be classified separately or be included with great apes. In a simple and traditional classification, prosimians—the earliest primates—include lemurs, lorises, and tarsiers. Anthropoids (simians) include monkeys, apes (gibbons, orangutans, gorillas, chimpanzees, bonobos), and humans. Biologists do agree, however, that humans did not evolve from apes, but rather, humans and apes had a common ancestor and diverged some 5–8 million years ago.

Fossil records vary in age from 85–55 million years as to when our early ancestors made their debut, but the consensus hovers about 65 million years. These forebears were followed some 35–55 million years ago by species of lemurs and lorises, which had large eyes and brains, small snouts, and more erect body positions. The first monkeys appeared 35 million years ago, and 10–15 million years later, apes diverged from Old World monkeys.

There are a number of characteristics seen to varying degrees among primates, arising from their arboreal history, but these are not necessarily unique to primates. (Most primates still live in trees, in the subtropics as well as in tropical rain forests of Africa, Asia, and the Americas.) Primate hands and feet have adapted for grasping, with specialized nerve endings that confer greater tactile sensitivity; flat nails are found on these digits and not claws. Apes and some monkeys have opposable thumbs that, as exemplified in humans, permit them to manipulate tools, including computer keyboards. The eyes are forward-looking and close together, which permits stereoscopic vision, with depth perception possible, an asset when swinging through trees. Monkeys and apes place greater reliance on vision as their primary sense rather than olfaction, as in other **mammals**.

The most distinguishing characteristic of primates is their high degree of sociability and advanced cognitive skills. In ascending rank order of intelligence are New World monkeys, Old World monkeys, apes, and their cousins, humans. Primates have a slower rate of development than other mammals, with an extended period of juvenile and adolescent life, presumably a period in which to learn from elders.

SEE ALSO: Mammals (c. 200 Million BCE), Neanderthals (c. 350,000 BCE), Anatomically Modern Humans (c. 200,000 BCE).

Mandrills, the most colorful primates, are Old World monkeys closely related to baboons. They live mostly in the tropical rainforests and forest-savanna mosaics of western Africa and can survive up to 31 years in captivity.

Amazon Rainforest

The biodiversity of the Amazon rainforest (known in English as "Amazonia") is unparalleled anywhere in the world: one in ten biological species on the planet resides here, and it contains the world's largest living collection of plants and animals. Although the inventory is far from being completed and described, there are over one million insect species, 40,000 plants, 2,200 fish, and 2,000 birds and mammals. More than 20 percent of the world's oxygen is produced in the rainforest—"the lungs of the planet."

The Amazon River, 4,000 miles (6,400 kilometers) long, and the second longest river in the world, flows from its origin in the Andes to its mouth in the Atlantic Ocean. Amazonia is the drainage basin of the river and its 1,100 tributaries, and stretches across nine countries, with 60 percent located in Brazil. This river basin is the largest in the world and encompasses 2.7 million square miles (7,000 square kilometers). The rainforest has been in existence some 55 million years, and its existence has been attributed to high annual rainfall and high humidity, and a high year-round temperature.

Since the 1960s, the unique biodiversity of Amazonia has been threatened by deforestation, which has continued until the early years of this century when its rate slowed; some sources claim that one-fifth of the rainforest has been lost. In its place, forests have been cleared for their excellent hardwood timber and to provide pastureland for cattle and crop farms.

Environmentalists have expressed alarm that the loss of the rainforest will have profound effects on its plant and animal inhabitants, known, and yet to be discovered; these include potential medicinal plants that have been used by indigenous healers for countless generations, which have yet to be critically evaluated for their effectiveness. Moreover, the rainforest serves as a major sink for carbon dioxide, which, if permitted to accumulate, will increase climate change, in particular, **global warming**.

SEE ALSO: Land Plants (c. 450 Million BCE), Insects (c. 400 Million BCE), Seeds of Success (c. 350 Million BCE), Gymnosperms (c. 300 Million BCE), Angiosperms (c. 125 Million BCE), Plant-Derived Medicines (c. 60,000 BCE), Global Warming (1896), Green Revolution (1945).

The poison dart frog is native to Central and South America, including the Amazon rainforest. Many of these frogs secrete an alkaloid poison through their skin that serves as a chemical defense against predators. Indigenous Amerindians have used these secretions to poison the tips of blowdarts.

Neanderthals

Philippe-Charles Schmerling (1790–1836)

Based on reconstructions appearing in museums, book illustrations, and movies, we have long envisioned Neanderthals to be stooped-over and inarticulate brutes, with hair enveloping their ape-like features. However, a growing body of evidence accumulating in recent years supports the notion that they walked upright, communicated by speech, used tools, buried their dead, had a cranial capacity that was equivalent to or even larger than our own, and had no more facial hair than modern man. In 2013, a 120,000-year old Neanderthal fossil was found, with clues that its owner had fibrous dysplasia, a cancer seen in modern humans. Contemporary museum reconstructions of Neanderthals now more closely resemble modern Europeans but with larger skulls, low foreheads, no chin, heavily built bones, and much stronger hands and arms.

In 1829, Philippe-Charles Schmerling discovered the first fossil remains of a Neanderthal—that of a small child—in a cave in present-day Belgium, although it was not so identified until 1936. The first recognized fossil human form was uncovered in 1856 in Neander's Valley in Germany and was called *Neanderthal* (thal = "valley"), and since then, other remains have been found in Western Europe, the Near East, and Siberia. The Neanderthals lived some 600,000 to 350,000 years ago, numbering, at their peak, about 70,000 in Europe. Then, around 30,000 to 45,000 years ago, they became extinct from causes that have yet to be more than speculation and theory—which also seems the case for what we "know" or think we know about them.

DNA findings suggest that Neanderthals and *Homo sapiens* diverged from a common ancestor some 400,000 to 500,000 years ago, and it continues to be debated whether Neanderthals are a subspecies of *Homo sapiens* (the weight of opinion points to Neanderthals as a separate species). For many thousands of years, Neanderthals lived in the same geographic regions as modern humans, and it appears that they interbred with humans. Some 99.7 percent of Neanderthal DNA is identical to present-day human DNA, and Europeans and Asians derive 1–4 percent of their genes from Neanderthals.

SEE ALSO: Primates (c. 65 Million BCE), Anatomically Modern Humans (c. 200,000 BCE), Fossil Record and Evolution (1836), Oldest DNA and Human Evolution (2013).

In 1899, Dragutin Gorjanovic-Kramberger (1856–1936), a Croatian geologist, paleontologist, and archeologist, found over 800 Neanderthal fossil remains in Krapina, a town located in northern Croatia. These remains, along with the statues shown here, are found at the Krapina Neanderthal Museum.

Anatomically Modern Humans

Édouard Lartet (1801–1871)

Fossil records provide evidence that some 150,000 to 200,000 years ago a group of early humans first appeared in Ethiopia. According to the prevailing "out of Africa" view, some 50,000 years ago, they reappeared in Europe. Alternatively, other scientists have favored a "multiregional theory," that modern humans originated independently in different parts of the world. When compared to contemporary humans, they were of approximately the same height, walked upright, but had a more robust build, and had less prominent eye ridges, the protruding layers of bone above the eye socket. In short, they closely resembled us in appearance and, therefore, have been designated Anatomically Modern Humans (AMH) or Early Modern Humans.

Experts are currently debating whether the AMH achieved anatomical modernity first (150,000–200,000 years ago) and later adopted modern behavior—such as developing a modern language, the capacity for abstract thought and symbolism, making more sophisticated tools—or whether anatomical and behavioral modernity were achieved simultaneously. The earliest AMH to appear in Western Europe have been commonly referred to as *Cro-Magnon*. During the past twenty years, experts have argued that since they were not that different from modern humans, they are undeserving of a separate designation and should be called European Early Modern Humans. Their first skeletal remains were discovered in 1868 by the French geologist Édouard Lartet in a cave that now bears his name. Later-discovered remains suggest that they inhabited Europe 45,000 to up to 10,000 years ago.

The Cro-Magnons were nomadic hunter-gathers who wove clothing and practiced elaborate rituals. Their remains of carved statuettes of animals and people provide evidence that they produced the earliest human art. Paleolithic cave paintings have been found in Spain and France, and the most famous such cave—which contains some 600 multicolored paintings and drawings of animals and symbols, dated to c. 15,000 BCE—was discovered at Lascaux, France, in 1940. Evidence suggests that the Cro-Magnons coexisted with the earlier **Neanderthals** for at least 10,000 years prior to the Cro-Magnon becoming extinct. The Neanderthals disappeared over 30,000 years ago.

SEE ALSO: Primates (c. 65 Million BCE), Neanderthals (c. 350,000 BCE), Radiometric Dating (1907), Lucy (1974), Mitochondrial Eve (1987), Oldest DNA and Human Evolution (2013).

This 13,000-year-old skull of a Cro-Magnon man, said to be Europe's first anatomically modern human, was found at the Bichon Cave in the western Swiss canton of Neuchâtel.

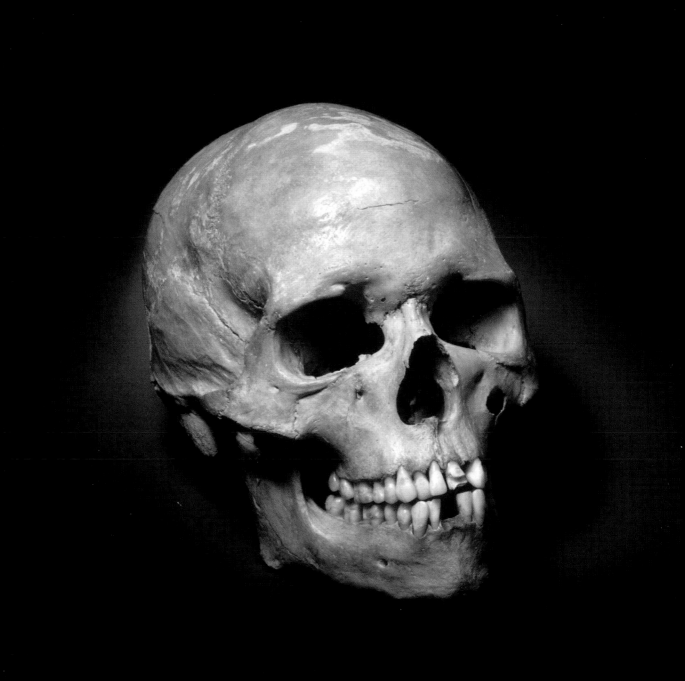

Plant-Derived Medicines

During their search for sources of nutrition in new environments, our earliest human ancestors sampled the local plants and, undoubtedly, emulated the example of the local animals and birds. This sampling was a trial-and-error process: some plants satisfied their hunger, while others produced unexpected effects—some bad, some good. The herb may have produced frightening thoughts, severe toxic effects, or even death, at least at the quantity taken; perhaps fewer leaves, berries, or roots would have been safer.

On the other hand, the gatherer's feelings of hunger, pain, fever, or constipation might have been relieved but also the herb might have made her sufficiently drowsy to catch up on much-needed sleep, or the application of leaves or the juice from a fruit might have quieted an itchy rash. In time, experience revealed that ingesting only the leaves, roots, seeds, berries, or their juices produced the effects sought with fewer side effects. Such experiences were passed on to subsequent generations of healers.

Use of white willow bark to relieve pain and reduce fever, and yarrow to promote sweating, are examples of plants with a long history of being cultivated in Europe and China. They were excavated from Shanidar IV, a Neanderthal burial site in Iraq that dates back some 60,000 years. Birch polypore, an edible mushroom used as a laxative, was carried by Ötzi the Iceman, a 5,000 year-old mummy preserved in ice where he perished in western Austria and was discovered in 1991.

The World Health Organization estimates that 75–80 percent of the earth's population now uses plant medicines either exclusively or in part. Plant-derived medicines continue to be used in homeopathic treatments and health food products. With advances in the isolation and purification of the active constituents of plants starting in the nineteenth century, these derived chemicals, of known composition, purity, and dose, have largely displaced plants in Western medicine. Nevertheless, many very important drugs in modern medicine are plants or derivatives of plant products and include morphine, codeine, aspirin (pain), atropine (eye exams), digoxin (heart failure), quinine (malaria), cocaine (local anesthetic), warfarin (anticoagulant), colchicine (gout), and Taxol and vinblastine (cancer).

SEE ALSO: Land Plants (c. 450 Million BCE), Gymnosperms (c. 300,000 Million BCE), Amazon Rainforest (c. 55 Million BCE), Neanderthals (c. 350,000 BCE), Anatomically Modern Humans (c. 200,000 BCE), Agriculture (c. 10,000 BCE), Mummification (c. 2600 BCE).

In 1775, the English physician William Withering (1741–1799) was called upon to assess the complex secret recipe of an old woman in Shropshire for the treatment of "dropsy" (fluid buildup from heart failure). Withering identified the active ingredient as Digitalis purpurea *(foxglove, shown) and, ten years later, after meticulous study, introduced one of the most important drugs in the history of medicine.*

Wheat: The Staff of Life

Wheat was one of the first crops to be cultivated and stored on a large-scale basis, transforming hunter-gathers into farmers, and it was instrumental in the establishment of city-states leading to the Babylonian and Assyrian empires. Wheat originally grew wild in the Fertile Crescent of the Middle East and in southwestern Asia. The archeological evidence traces the origins of wheat to wild grasses, such as wild emmer (*Triticum dicoccum*), which was gathered for food in Iraq in 11,000 BCE, and einkorn (*T. monococcum*), grown in Syria 7800–7500 BCE. Wheat was farmed in the Nile Valley of Egypt before 5000 BCE, where Joseph of the Hebrew Bible was overseeing grain stores in 1800 BCE.

A natural hybrid, wheat was derived from cross-pollination of grains. Over thousands of years, farmers and breeders have cross-hybridized grains to maximize the qualities they deemed most desirable. During the nineteenth century, single genetic strains were selectively produced that possessed the traits they were seeking. With a growing understanding of **Mendelian inheritance**, two lines were crossbred, and the progeny inbred for ten or more generations to obtain and maximize specific characteristics. The twentieth century saw the development and planting of hybrids selected on such desirable characteristics as large kernels, short straw, hardiness to cold, and resistance to insects and to fungal, bacterial, and viral diseases.

In recent decades, bacteria have been used to transfer genetic information to produce transgenic wheat. Such **genetically modified crops** (GMC) have been engineered to produce greater yields, require less nitrogen to grow, and offer greater nutritional value. In 2012, the whole genome of bread wheat was completed and found to have 96,000 genes. This marks an important step in continuing the production of genetically modified wheat, in which more specific desirable characteristics can be inserted in specific loci on the wheat chromosomes.

As rice is a dietary staple in Asia, so is wheat in Europe, North America, and western Asia. Wheat is the most widely consumed cereal grain in the world, and world trade in wheat is greater than all other crops combined.

SEE ALSO: Angiosperms (c. 125 Million BCE), Agriculture (c. 10,000 BCE), Rice Cultivation (c. 7000 BCE), Artificial Selection (Selective Breeding) (1760), Mendelian Inheritance (1866), Genetically Modified Crops (1982).

This Chinese farmer is carrying bushels of dry wheat, as did his ancestors for thousands of years.

Agriculture

From small groups of hunter-gathers living off the land and foraging berries and other edible plants, agriculture, a type of applied biology, evolved to domestication and cultivation of crops. This active involvement originated at different times and places, and to various extents based on environmental conditions: archeological evidence suggests its origin dated from the end of the Ice Age, as early as 14,500 to 12,000 years ago. The earliest agricultural successes coexisting with the rise of great ancient civilizations appeared in major river valleys where the annual river flooding not only provided water but also a consistent source of silt, a natural fertilizer. These included the birthplace of agriculture in the Fertile Crescent between the Tigris and Euphrates Rivers in Mesopotamia and the Nile in Egypt; Indus in India; and the Huang in China.

Explanations for the adoption of agriculture and its consequences vary: Some experts contend that it was intended to meet the increasing food needs in ever-burgeoning populations, needs that could not be satisfied by food gathering or hunting. Alternatively, agriculture may not have originated in response to food scarcity but rather that the population in a given area increased significantly only after stable sources of food had been established. Evidence supporting each has been adduced. Whereas in the Americas, villages sprang up after the development of crops, villages and towns in Europe appeared earlier than or at the same time as agricultural advances.

Agricultural success depended not only upon the whims of nature providing favorable climatic conditions but also upon the ability of early farmers to utilize irrigation, crop rotation, fertilizers, and domestication—the conscious selection of developing plants whose characteristics increased their utility. Tools intended for the simple acquisition of wild foods were replaced by those for production, such as the plow and those powered by animals. The earliest domesticated crops include rye, **wheat**, and figs in the Middle East; rice and millet in China; wheat and some legumes in the Indus Valley; maize, potatoes, tomatoes, pepper, squash, and beans in the Americas; and wheat and barley in Europe.

SEE ALSO: Wheat: The Staff of Life (c. 11,000 BCE), Domestication of Animals (c. 10,000 BCE), Rice Cultivation (c. 7000 BCE), Botany (c. 320 BCE), Artificial Selection (Selective Breeding) (1760), Green Revolution (1945), Genetically Modified Crops (1982).

The National Grange of the Order of Patrons of Husbandry, an association of farmers, was founded in the United States in 1867 to promote community wellness and agriculture. This 1873 poster, "Gift for the Grangers," promotes the organization through idyllic scenes of farm life. In 1870, 70–80 percent of the US population was employed in agriculture; by 2008, this number had dwindled to only 2–3 percent.

GIFT FOR THE GRANGERS

"I PAY FOR ALL"

FAITH HOPE CHARITY FIDELITY

Domestication of Animals

Domesticated animals were initially developed from species that were social in the wild and could breed in captivity, thus allowing genetic modifications to increase those traits that are advantageous to humans. Depending upon the species, such desirable traits might include: being docile and easy to control; having the ability to produce more meat, wool, or fur; and suitability for traction, transportation, pest control, assistance, companionship, or as a form of currency.

The most familiar domesticated animal, the dog (*Canis lupus familiaris*), is a subspecies of the gray wolf (*Canis lupis*), with the oldest fossil remains showing a split in their lineage some 35,000 years ago. Dogs were the first animals to be domesticated with the earliest evidence being a jawbone found in a cave in Iraq and dating back some 12,000 years. Images on Egyptian paintings, Assyrian sculpture, and Roman mosaics show that even in ancient times, domestic dogs were of many sizes and shapes. The first dogs were domesticated by hunter-gatherers but their job description has since been expanded beyond hunting to include herding, protection, pulling loads, aiding police and the military, assisting handicapped individuals, serving as human food, and providing loyal companionship. The American Kennel Club now lists 175 breeds, with most only several hundred years old.

Around 10,000 years ago, sheep and goats were domesticated in southwest Asia. While alive, they served as a source of manure for crop fertilization and, when dead, as a regular supply of food, leather, and wool. Researchers have long been puzzled about the origins and evolution of the domestic horse (*Equus ferus caballus*), whose wild ancestor first appeared 160,000 years ago and is now extinct. Based on archeological and genetic evidence, including bit wear on horse teeth that were found at sites associated with the ancient Botai culture, in 2012 researchers concluded that their domestication dates back some 6,000 years in the western Eurasian Steppe (Kazakhstan). As they were domesticated, these early horses were regularly bred with wild horses to provide meat and skin and later to play an essential role in war, transportation, and sport.

SEE ALSO: Agriculture (c. 10,000 BCE), Artificial Selection (Selective Breeding) (1760), Fossil Record and Evolution (1836), Darwin's Theory of Natural Selection (1859).

Dogs, which have all evolved from the gray wolf, were the first domesticated animals and have been the working partner and loyal companion of humans for some 12,000 years. They are now commonly functionally categorized as companion, guarding, hunting, herding, and working dogs.

Coral Reefs

Coral reefs are one of the most diverse ecosystems on the planet, with estimates of reef inhabitants ranging from 600,000 to nine million. In addition to coral, reefs are home to **algae**, **fungi**, sponges, mollusks, crustaceans, many species of **fish**, and sea birds. In short, they are the "rainforests of the oceans."

Corals, members of the invertebrate *cnidaris* group—which also includes sea anemone, jellyfish, and hydras—are sessile, that is they are not mobile but stay fixed in place. Living in colonies, with each individual called a *polyp*, they secrete calcium carbonate at the base of the polyp, which serves as the skeletal foundation of the colony. Calcium carbonate is continually deposited by a living colony, adding to the size of the structure. Corals reside on the surface of the structure, completely covering it, with the size and shape of reefs differing depending upon the species of coral and the color from algae.

Corals feed by extending their tentacles and capturing prey such as small fish and plankton. In addition, coral maintain a symbiotic relationship with algae (*zooxanthellae*) that reside inside coral polyps where the algae carry out **photosynthesis**. The algae provide the coral with energy and nutrients in exchange for protection and the light required for photosynthesis to occur. Coral reefs are located in shallow, clear water where light can reach the polyps, and in tropical or subtropical climates. Although reefs were in existence hundreds of millions of years ago during the Cambrian and Devonian periods (among others), they have been subject to numerous devastating extinction events. Most reefs existing today were established less than 10,000 years ago after glacial melting caused a rise in sea level and widespread flooding of continental shelves.

A number of ecological challenges threaten the survival of coral reefs. Natural stresses, such as hurricanes, are generally of short-term duration. Far more serious are a variety of human-induced stresses, such as agricultural runoffs of herbicides, pesticides, and fertilizers; industrial runoffs; pollution from human sewage and toxic discharges; destructive fishing practices; and coral mining. By 2030, 90 percent of reefs are estimated to be at risk for extinction unless aggressive steps are taken to reverse human-induced climate changes and rises in sea temperature leading to coral bleaching, acidification, and pollution of the oceans.

SEE ALSO: Algae (c. 2.5 Billion BCE), Fungi (c. 1.4 Billion BCE), Fish (c. 530 Billion BCE), Devonian Period (c. 417 Million BCE), Amazon Rainforest (c. 55 Million BCE), Photosynthesis (1845), Ecological Interactions (1859), Global Warming (1896), De-Extinction (2013).

A pair of clownfish nest among the protective tentacles of a sea anemone in a coral garden. The residing algae give the corals their beautiful colors.

Rice Cultivation

FEEDING ASIA. Rice is among the oldest and world's most important economic botanical food crop. It is the largest source of calories for the 3.3 billion people of Asia, providing 35–80 percent of their total caloric intake. But, while rice is nutritious, it is not sufficient to serve as the main food source. The worldwide popularity of rice as a food is attributed, in part, to its ability to be grown in areas as varied as flooded plains to deserts and in all continents, except Antarctica. China and India are the major rice-producing and consuming countries.

Some 12,000–16,000 years ago, rice grains were initially gathered and consumed by prehistoric people in the world's humid tropical and subtropical regions. Wild cultivated prototypes of rice, which descended from wild grasses, are members of the taxon family *Poaceae* (also called *Gramineae*). Based on genetic evidence, recent reports reveal that rice cultivation first occurred in China between 8,200 and 13,500 years ago. From China, cultivation spread to India, then to western Asia and Greece, brought by the armies of Alexander the Great in 300 BCE. The most popular cultivated rices are *Oryza satliva japonica* (Asian rice and, by far, the most common) and *Oryza glaberrima indica* (African rice), both of which were domesticated from a common origin.

The rice plants have an outer coating that protects the rice grain, the fruit of the plant. Seeds are milled to remove the chaff (outer husk) to produce brown rice. If milling is continued, and the rest of the husk and grain removed, white rice is left. Brown rice is more nutritious, containing proteins, minerals, and thiamine (vitamin B_1), while white rice mainly contains carbohydrates and is virtually devoid of thiamine. Beriberi results from a nutritional deficiency in thiamine, which has been historically endemic in Asian populations, who favor polished white rice because it has a longer shelf life and is not historically associated with poverty. Among cereals, rice is low in sodium and fat, and free of cholesterol, making it a healthy food choice.

SEE ALSO: Wheat: The Staff of Life (c. 11,000 BCE), Agriculture (c. 10,000 BCE), Botany (c. 320 BCE), Vitamins and Berberi (1912), Albumin from Rice (2011).

Rice is the world's most important food crop and provides the greatest proportion of calories to the people of Asia. Although this crop is typically grown on flooded plains, such as this one in Thailand, it can also be cultivated in deserts.

Mummification

Herodotus (484–425 BCE)

Mummies have long been of interest to scientists to achieve a better understanding of ancient cultures and learn the techniques of expert embalmers. This fascination also extends to nonscientists, who enjoy classic horror films—most notably *The Mummy* (1932), with Boris Karloff in the title role. Mummies are humans or animals whose bodies have been intentionally or naturally preserved long after their death. Under normal conditions, decomposition reduces the corpse to a skeleton within months. This process is hastened in hot and humid climates that favor bacterial breakdown. It can be retarded by chemicals that remove moisture from the body or in environments that are extremely cold, have very low humidity, or lack oxygen.

Many of these inhibiting conditions existed in ancient Egypt, where the practice of intentional mummification (to prepare the deceased for the afterlife) was so advanced that modern-day scientists are still seeking to learn their art and science. Evidence of the earliest intentional mummification dates back to approximately 2600 BCE, with the best-preserved specimens from the New Kingdom period of 1570–1075 BCE. The first surgical description appears in Herodotus's *The History* in which he describes the removal of the brain and all internal organs, except the heart, which was filled with spices, thought to be the center of a person's being and intelligence. Natron, a natural desert salt, was used to remove moisture from the body to hasten dehydration and prevent decomposition; the body was permitted to dry for seventy days and then wrapped in linen and canvas for protection. For religious reasons, some animals were also mummified, including sacred bulls, cats, birds, and crocodiles.

Examples of natural mummies have been found all over the world in the ice of glaciers, the arid desert, and oxygen-depleted peat bogs. Perhaps the most famous and best preserved of these is Ötzi the Iceman, who was found in 1991 in the Alps on the Italian-Austrian border; he was some 45 years of age when he died and lived about 3300 BCE. The most notable modern mummies are Vladimir Lenin and Eva Peron, who died in 1924 and 1952, respectively.

Ancient Egyptians perceived the preservation of the body after death, an integral practice of the Egyptian religion, as important to living well in the afterlife and as a status symbol for the wealthy. Conspicuous signs of wealth included more elaborate tombs and embalming procedures.

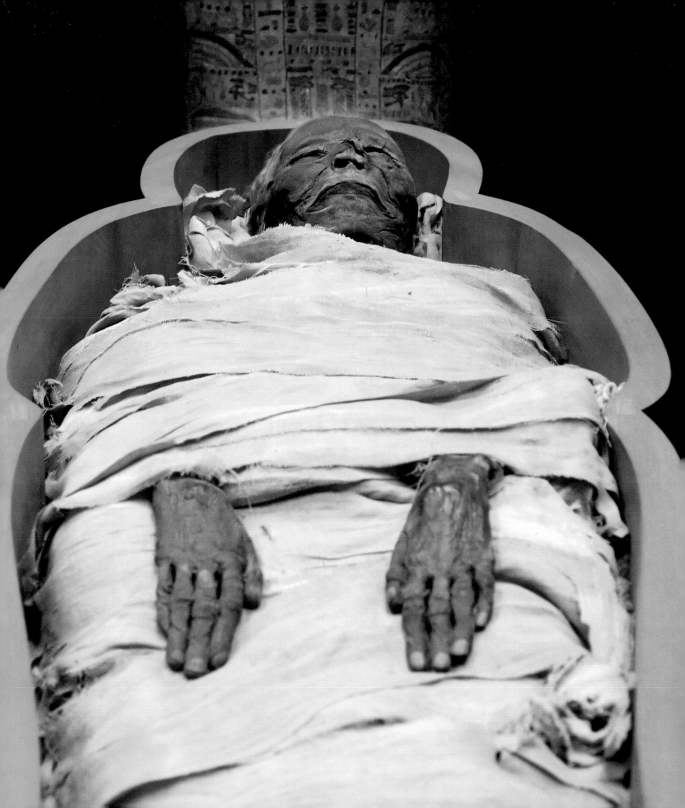

Animal Navigation

Paul Julius Reuter (1816–1899)

Humpback whales migrate 16,000 miles (25,000 kilometers) annually, feeding in polar waters during summers, and mating and calving in tropical and subtropical environs during the winter. Implicit in such long-distance migrations is an innate navigation mechanism, which has been most extensively studied in the domesticated homing pigeon, a bird derived from the Rock Pigeon (*Columba livia*).

The superb navigational skills of homing pigeons, also called messenger or carrier pigeons, have been known for milennia. Circa 2350 BCE, King Sargon of Akkad (present-day Iraq) ordered all messengers to carry homing pigeons, which would fly back to the king if they encountered any danger. In the eighth century BCE pigeons reported Olympic winners to the Athenians, and since then they have been important postal couriers during wartime, announcing the Duke of Wellington's victory at Waterloo in 1815 and receiving medals for service in World War II more than a century later. In the 1850s, Paul Julius Reuter, the founder of Reuters News Agency, used homing pigeons to deliver news and take advantage of stock prices faster than his competition.

When taken from their loft and after being released, homing pigeons have been shown to fly 1,100 miles (1,800 kilometers) away and find their way home. Far more intriguing is how they return home from sites they have never visited. Researchers have postulated theories accounting for their homing instinct based on map and compass models. The "compass" refers to an orienting mechanism, using the position of the sun.

The "map," a subject of intensive speculation, determines the bird's position relative to the location of the home loft. Visual clues, such as recognizable landmarks or unique ground features, are useful but only when birds (and some insects) are close to home. Birds are guided by the sun, moon, and stars, but can navigate under overcast conditions, albeit not as easily. Homing pigeons may navigate using low-frequency infrasound, less than 20 Hz, far below audible detection by humans. (Such infrasound is used for long-distance communication by elephants.) Greatest interest has been focused upon magnetoreception in which the homing pigeon detects and is guided by the earth's magnetic field by means of particles of magnetite (lodestone), an iron oxide, located in the area of their beaks or eyes.

SEE ALSO: Animal Migration (c. 330 BCE).

As adults, salmon spend one to five years in the open ocean and then return to the freshwater streams they were born in to spawn. It is believed that salmon use their keen sense of smell and an olfactory memory as navigation devices.

Four Humors

Hippocrates (c. 460–c. 370 BCE), **Galen** (c. 130–c. 200), **Avicenna** (980–1037)

For over two millennia, both Eastern and Western medical practitioners subscribed to the belief that a balance of four bodily humors ("fluids") affected our physical and mental well-being. The concept of the four humors originated in ancient Egypt and Mesopotamia, and was systematized and adapted to the practice of medicine—humorism—by Hippocrates during the fourth century BCE. Good health resulted when these fluids—blood, phlegm, black bile, and yellow bile—were in balance, while their excess or deficiency resulted in sickness.

This notion displaced the prevailing ancient belief that attributed diseases to supernatural evil spirits. Humorism gained ready acceptance over the centuries and was embraced as the basis for medical treatment by the Greeks and Romans, as promoted by Galen. He also proposed that humor excesses were responsible for such temperamental traits as being sanguine (blood), phlegmatic (phlegm), melancholy (black bile), and choleric (yellow bile). From the ancient world, it spread over the centuries, with local modifications, to the Muslims, to China and India, and Western European physicians. In *The Canon of Medicine* (1025), the famed Muslim physician Avicenna expanded upon the relationship between this imbalance to changes in temperament and disease and proposed its association with the master organs—the brain and heart. During the Elizabethan period, to maintain a balance in the fluids, which were believed to be produced by and circulating in the body, adjustments were made in the diet, exercise, clothing, and even bathing habits; bathing was perceived to be more harmful for men than women.

"Heroic medicine," as practiced by Western European-trained physicians, sought to restore this balance in humors by employing such drastic approaches as purging, inducing vomiting, and bleeding; George Washington's death in 1799 is generally believed to have been inadvertently hastened when his physicians bled him of ~125 ounces, or half, his total blood volume. Less dramatic treatment approaches involved the application of heat, cold, moisture, or dryness. Humorism continued to dominate thinking on well-being and the practice of medicine until the nineteenth century, when modern theories of medicine, based on advances in cellular pathology and biological and bacterial causes of disease, had gained acceptance.

SEE ALSO: Scientific Method (1620), Cell Theory (1838), Germ Theory of Disease (1890).

In ancient times, melancholy was attributed to an overabundance of the "black bile" humor, and the planets were believed to be able to correct the balance of humors. This stained glass, from the southern Netherlands (c. 1530), is entitled "The Planet Saturn Driving Out a Monk with a Pig's Head, or Time Banishing Melancholy."

Aristotle's *The History of Animals*

Aristotle (384–322 BCE), **Theophrastus** (372–287 BCE), **Carl Linnaeus** (1707–1778), **Richard Owen** (1804–1892)

Aristotle was among the most influential individuals who ever lived, and his contributions span the entire range of human experiences. Moreover, he originated entire new fields of study, and, for some two millennia, he was venerated as a quasi-religious authority, during which time his writings on living organisms were accepted as indisputable facts. He was born in Stagira in north Greece in 384 BCE, son of the court physician to the Macedonian royal family. After studying medicine, he was a pupil of Plato and tutored Alexander the Great. Aristotle established a school, the Lyceum, in Athens in 335 BCE, and remained its director until 323 BCE, when he was succeeded by his student, Theophrastus, the father of **botany**.

Among the fields of study Aristotle created was biology; up to one-third of his surviving writings are devoted to this subject. Many of his facts have been validated, but some were wrong—in particular, those dealing with the human body. He introduced the **scientific method**, conducting research based on observation and experimentation prior to formulating explanations. He maintained records on more than 500 species of animals he studied, including remarkably accurate descriptions of marine invertebrates. He examined fertilized eggs at different stages of development, and formulated the theory of epigenesis—that organs develop in a specific order. Aristotle drew a distinction between homologous and analogous body parts, predating Richard Owen by over twenty centuries.

In his monumental work, *The History of Animals*, Aristotle was first to classify animals into groups based on the similarities and differences of their physiology. Animals were grouped into those with blood (vertebrates) and without blood (invertebrates). He compared organs from different species and noted how they varied depending upon habitat. His Great Chain of Being classified living organisms into eleven levels, based on their scale of perfection at the time of their birth and nature of their souls; humans were at its apex, plants at the base, and a hierarchy of nature only improved upon in the eighteenth century by Linnaeus.

SEE ALSO: Botany (c. 320 BCE), Scientific Method (1620), Linnaean Classification of Species (1735), Theories of Germination (1759), Homology versus Analogy (1843), Embryonic Induction (1924), Phylogenetic Systematics (1950), Domains of Life (1990).

There was hardly an area of human inquiry that was not studied by Aristotle. This photo depicts Aristotle on a 5 drachmas coin issued in 1990. The drachma was the standard monetary unit of Greece until 2002, when it was replaced by the euro.

Animal Migration

Aristotle (384–322 BCE)

Among the earliest natural phenomena attracting the interest of the ancient Greeks, and for which references appear in the Bible, was the periodic movement of birds. Aristotle was one of the first to seek to explain these movements, noting the migration of pelicans, geese, swans, and doves to warmer climates in the winter. To account for the seasonal disappearance and reappearance of some birds, however, he postulated the transmutation of species—i.e., the seasonal change of one bird species to another. For example, he theorized that the redstarts and garden warblers seen in the summer transform themselves into robins and blackcaps, respectively, in winter.

We now know that members of all major animal species migrate to more favorable environments to take advantage of available food, water, and shelter that vary with seasons, or for reproductive purposes. The monarch butterfly travels 2,000 miles (3,219 kilometers) from southern Canada to central Mexico. Salmon lay eggs in fresh water streams, spend their adulthood in the ocean, and then return to the stream in which they were hatched to lay eggs before dying. The migration of locusts is legendary, as recounted in the Bible and Pearl Buck's *The Good Earth*, where they strip fields bare leading to famine. When the population of locusts becomes too large and their appetites voracious (eating their weight in plants daily), some locust members migrate to areas where food is more plentiful. The most common example of migratory behavior is seen in North America when birds leave their breeding range during the nonbreeding season and fly south with the approach of winter.

During their migration, animals use various navigational approaches to reach their destination. They may visualize environmental landmarks, and use the position of the sun, olfactory clues based on atmospheric odors, low-frequency infrasound, and magnetoreception (the detection of magnetic fields). A genetic component is thought to be associated with migratory behavior, but this is not well understood. Natural selection does favor those migratory birds that have more hollow bones (reducing their body weight), rounded wings, and changes in their heart rate and energy expenditures that permit them to make flights with greater energy efficiency.

SEE ALSO: Animal Navigation (c. 2350 BCE), Darwin's Theory of Natural Selection (1859), Animal Locomotion (1899).

Locusts are migratory members of the grasshopper family. Their migration follows a tremendous increase in insect population that cannot be sustained by the available resources. The insects have been known to travel thousands of miles before settling down to feed, consuming their weight in a day.

Botany

Aristotle (384–322 BCE), **Theophrastus** (372–287 BCE)

What Aristotle's writings were to zoology, Theophrastus's were to botany. Prior to Theophrastus, interest in plants was focused upon their use as foods and drugs. His two books on plants, written about 320 BCE, were the first scientific and systematic study of their nature and represented the primary sources of botanical knowledge during antiquity and into the Middle Ages. Under the direction of Pope Nicholas V, in about 1450, the classical works in the Vatican library were translated from Greek into Latin, with Theophrastus's published in 1483.

Born on the Greek island of Lesbos, he became a student and later friend of Aristotle at his Peripatetic School in Athens. When Aristotle was forced to leave Athens in 322 BCE, he bequeathed Theophrastus his writings. Theophrastus was also appointed his successor at the Lyceum, a position he very successfully held for thirty-five years, attracting, at one time, over 2,000 students.

At the Lyceum, Theophrastus had a garden with some 2,000 plants, which is thought to be the world's first botanical garden. He focused his attention on cultivated plants—some 500–550 in number—collected from the lands bordering the Atlantic to the Mediterranean. (Over 300,000 plants are now known to exist.) His personal observations and collections were supplemented by specimens and descriptions of plants gathered during the military travels of Alexander the Great in Asia, plants unknown in the Greek world; these included the cotton plant, pepper, cinnamon, and the banyan tree.

THE FATHER OF BOTANY. Theophrastus's *Enquiry into Plants* focused on the classification and description of plants and divided the plant kingdom into flowering (angiosperm) and nonflowering (gymnosperm) plants. *The Causes of Plants* examined the physiology, growth, and cultivation of plants, the latter aspect served as the foundation of horticulture. Collectively, these works included most aspects of botany: plant description and classification (trees, shrubs, undershrubs, herbs), and plant distribution, propagation, germination, and cultivation. In addition, he also repeatedly observed that different plants thrive in different locales based on environmental influences, a basic theme in ecology.

SEE ALSO: Wheat: The Staff of Life (c. 11,000 BCE), Agriculture (c. 10,000 BCE), Rice Cultivation (c. 7000 BCE), Aristotle's *The History of Animals* (c. 330 BCE).

Roughly 2,500 years ago, Theophrastus established the world's first botanical garden, in which grew approximately 2,000 plants.

Pliny's *Natural History*

Aristotle (384–322 BCE), **Theophrastus** (372–287 BCE), **Pliny the Elder** (23–79), **Georges-Louis Leclerc, Comte de Buffon** (1707–1788)

Courses with a "natural history" designation are generally absent from contemporary college catalogs, yet many larger cities throughout the world host museums of natural history, often tracing their founding to the nineteenth century. During this time, scientific study became more specialized and experimental, researchers were professionally trained and no longer talented amateur gentlemen, and the boundaries of natural history became more restricted than in Pliny the Elder's monumental work of the same name published in the year 77.

The influence of Pliny the Elder—a Roman lawyer, army and navy commander, and naturalist—was considered perhaps second only to Aristotle. The thirty-seven books that comprise Pliny's *Natural History* were intended to encompass all known information about the natural world and were based on the writings of the most distinguished authorities. Pliny assembled zoological data from Aristotle and plant-related material from Theophrastus. Books were also devoted to astronomy, geography, geology, mineralogy, and agriculture.

Pliny's work, with its broad scope of subject material assembled in a systematic manner, specific references to the hundreds of original sources and authors consulted, and an index of its contents, served as the prototype for subsequent encyclopedias. Although the work mixed fact with fiction, folklore, magic, and superstition, it remained the unchallenged source of natural history until the end of the fifteenth century. Far more accurate and limited in scope to the animal and mineral kingdoms was the thirty-six–volume *Histoire naturalle*, written from 1749 to 1788 by the French naturalist Georges-Louis Leclerc, Comte de Buffon.

Studies of nature were subdivided during the nineteenth century, and classified as either natural philosophy, which included physics and astronomy, and natural history, including biology (zoology and **botany**) and the geological sciences. Today, there is no universally accepted definition of natural history, but it generally refers to the study of plants and animals in their natural environment and emphasizes observation and description rather than experimentation.

SEE ALSO: Aristotle's *The History of Animals* (330 BCE), Botany (c. 320 BCE).

This statue of Pliny the Elder (Gaius Plinius Secundus) is on the façade of the Duomo cathedral of Como, the Italian city of his birth.

Skeletal System

Galen (c. 130–c. 200), **Andreas Vesalius** (1514–1564)

The earliest systematic description of the skeletal system appeared (c. 180) in the writings of Galen, which served as the unchallenged basis for the anatomical instruction until the sixteenth century. Galen commented on its protective and supportive role, noted that bone was composed of a hollow marrow, and, based on its color, believed that bone was made from sperm. Supported by his dissections of humans (and not Galen's apes), Andreas Vesalius, in his classic work, *De humani corporis fabrica* (1543) corrected Galen's many erroneous descriptions. By the eighteenth century, the human skeletal system was accurately described.

Animals have either exoskeletons or endoskeletons. Exoskeletons, commonly referred to as shells, protect their wearer's soft tissues against predators, provide support for these tissues, and serve as an attachment for muscles used for locomotion. In addition, exoskeletons may be sense organs, play a role in feeding and elimination, and, in terrestrial animals, provide a barrier against desiccation. Most shell-forming organisms made their appearance during the Cambrian period (542–488 million years ago), when there was a sudden change in the ocean's chemistry. The chemical composition of exoskeletons differs among species: In **insects**, spiders, and crustaceans, exoskeletons contain chitin, a glucose polymer similar to cellulose. In mollusks, calcium carbonate provides hardness and strength to the shells, while in microscopic diatoms, silica (silicon dioxide) influences the cell's float/sink equilibrium. When rigid exoskeletons are outgrown, a new one is grown below the old one, which is then shed (molted).

Endoskeletons are harder than exoskeletons and allow room for the growth of larger animals. They provide support—most notably in sponges and starfish, which would be shapeless in their absence—protection, and aid in the movement of their muscles of locomotion. The skeletal system of vertebrates consists of bone and cartilage. In mammals, the bones of the endoskeleton store calcium, and bone marrow is the site of the production of red and white **blood cells**. Chemically, bone is composed of calcium hydroxyapatite (providing rigidity) and collagen, an elastic protein. There are 206–208 bones in adult humans, while estimates in newborns range from 270–350.

SEE ALSO: Insects (c. 400 Million BCE), Leonardo's Human Anatomy (1489), Vesalius's *De humani corporis fabrica* (1543), Blood Cells (1658).

An 1857 photograph of a human and monkey skeleton with Reginald Southey (1835–1899), English physician and lifelong friend of the photographer, Charles Lutwidge Dodson, a.k.a. Lewis Carroll (1832–1898).

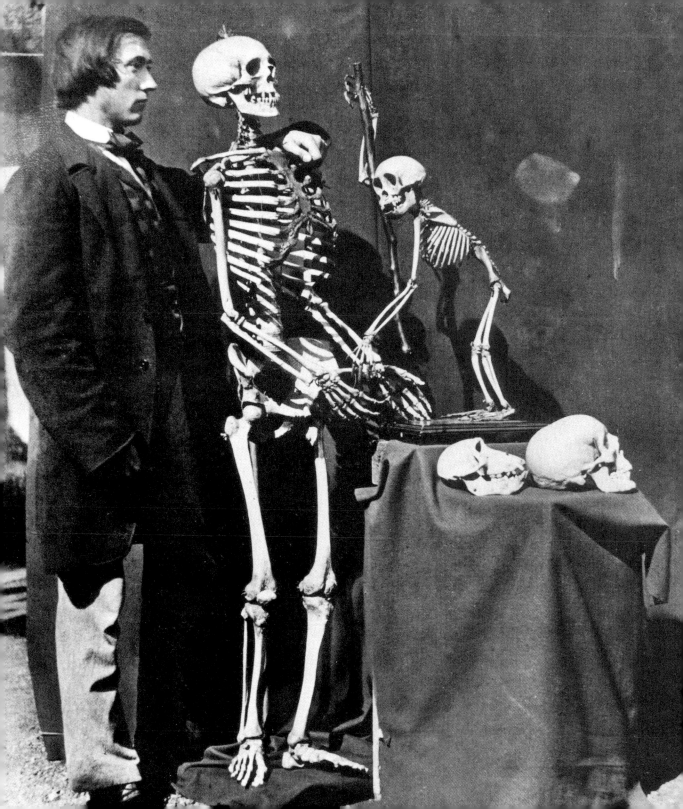

Pulmonary Circulation

Galen (c. 130–c. 200), **Ibn al-Nafis** (1213–1288), **Michael Servetus** (c. 1511–1553), **Realdo Colombo** (1516–1559), **William Harvey** (1578–1657), **Marcello Malpighi** (1628–1694)

The Greek physician Galen knew that blood was transported in blood vessels, that bright blood was carried in arteries and dark blood in veins, and that each had separate functions. His erroneous teaching that blood passed through the walls of the right ventricle (lower chamber) to the left side, where it picked up air and then traveled around the body, continued to be accepted for almost 1,000 years.

In 1242, the Arabian physician Ibn al-Nafis was the first to correctly describe the pulmonary circulation of blood in his book the *Commentary on the Anatomy of Canon of Avicenna*. He noted that there were no pores between the lower heart chambers or any direct pathways between them. Rather, he proposed that blood flowed from the pulmonary artery to the lungs, where it "mingled" with air and traveled through the pulmonary vein to the left side of the heart, from which it was distributed throughout the body. He also predicted the existence of pores between the pulmonary artery and vein, a prediction borne out four centuries later, when the Italian microscopist Marcello Malpighi first visualized capillaries.

The Spanish theologian and physician Michael Servetus was the first European to accurately describe pulmonary circulation in his *Restoration of Christianity* (1553), a theological work. As it was not a book of science or medicine, it was largely neglected; only three copies are extant. What was believed to be the last copy and its author were burned at the stake in Geneva, 1553, on the orders of John Calvin for Servetus's "heretical" writings and his denial of the Trinity and infant baptism.

Realdo (also Realdus) Colombo an Italian anatomist who worked with Michelangelo, made important anatomical findings, the most significant, in about the 1550s, his discovery of the pulmonary circuit. He proposed that venous (oxygen-deficient) blood traveled from the heart to the lungs, where it mixed with air, and then the blood returned to the heart. This discovery was of great value to William Harvey when he described blood circulation in his *De motu cordis* in 1628.

SEE ALSO: Vesalius's *De humani corporis fabrica* (1543), Harvey's *De motu cordis* (1628), Blood Cells (1658).

In 1553, the Spanish physician and theologian Michael Servetus was burned alive on a pyre of his own books. These included his "heretical" theological work, Restoration of Christianity, *which contained an early accurate description of pulmonary circulation.*

Leonardo's Human Anatomy

Galen (c. 130–c. 200), **Leonardo da Vinci** (1452–1519)

Leonardo da Vinci, without question a polymath and true genius, was also a pioneer human anatomist whose works, which continue to be studied for their accuracy, have been validated by the most modern imaging techniques. Studies of human anatomy had made few advances over the past millennium since Galen, who, unlike Leonardo, did not have access to human corpses. Prior to Leonardo, depictions of the human body focused on its external features with no details of its inner workings, which were verbally, but not pictorially, described.

Leonardo was provided access to corpses in Florence, the city of his birth. Starting in 1489, he dissected some twenty to thirty corpses—in good health, diseased, and deformed—over a twenty-year period. He prepared a human anatomy notebook in which he made accurate measurements of the proportions of all body parts, and these were depicted in drawings rendered in multiple views. For the hand and leg, for example, his drawings contained eight to ten layers, with the relationship between the different layers, and their arteries, muscles, ligaments, nerves, and bones portrayed.

He meticulously studied and drew different facial expressions to express an entire range of human emotions, and these appear in his most famous paintings. While much acclaimed is his drawing of a fetus *in utero*, correctly attached to its umbilical cord, his sketches of the female reproductive system contain a number of errors that are said to more accurately represent an animal than a human. Not content with drawings and sketches, he sought to better understand how the body worked. To this end, he prepared physical and mechanical models to simulate the function of body parts, such as how the valves of the heart opened and closed, and used such models in his drawings.

Leonardo envisioned that his work would be useful to medical practitioners and planned to publish his anatomical drawings as a treatise on human anatomy. But when he died in 1519, these papers were buried from view among his private possessions. After passing through many hands over the decades, they turned up in the British Royal Collection at the end of the seventeenth century, where they are now housed.

SEE ALSO: Vesalius's *De humani corporis fabrica* (1543), Placenta (1651).

The Vitruvian Man (c. 1490) was Leonardo's pen ink drawing based on the first-century BCE Roman architect Vitruvius, who believed that the ideal human body could be represented within the perfect geometric forms of a circle and a square.

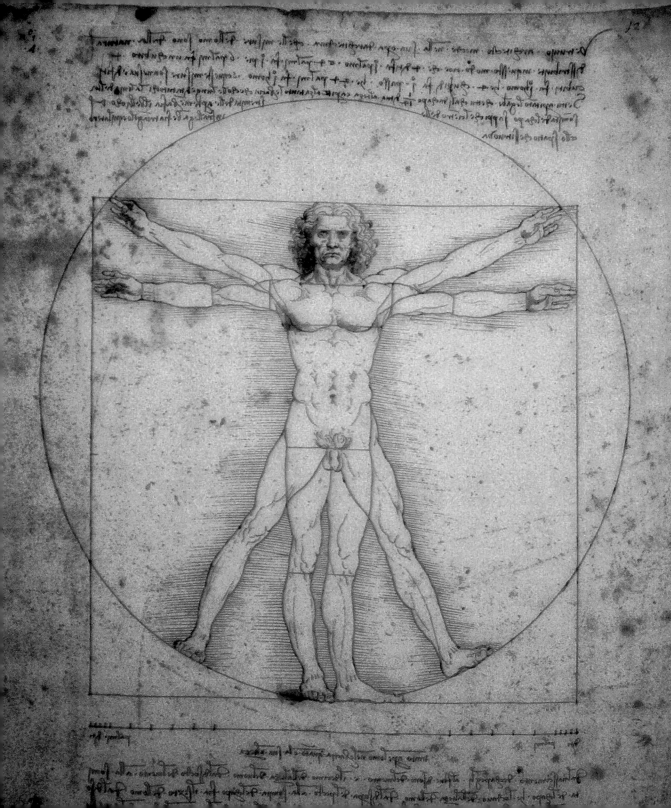

Sense of Hearing

Berengario Da Carpi (1460–1530), **Giulio Casserio** (1552–1616), **Heinrich Rinne** (1819–1868), **Hermann von Helmholtz** (1821–1894)

During the sixteenth century, a group of remarkable Italian anatomists studied and identified structures in the inner ear. Berengario Da Carpi described the ossicles (middle ear bones) in his book *Commentaria* (1521), and Giulio Casserio compared these bones in various animals. Heinrich Rinne described the conductive process between the tympanic membrane and the ossicles and used the tuning fork to differentiate causes of deafness (1855), and Hermann von Helmholtz postulated the perception of sounds and tones (1863).

Unlike the senses of smell, taste, and vision, hearing exclusively utilizes mechanical processes. An object produces sound when it vibrates, usually through air or water. Sounds transmitted as waves are characterized by their frequency (cycles/second, cps, or hertz, Hz), which is perceived as pitch, and by amplitude, the size of sound waves, sensed as volume.

The hearing process involves directing sound waves, sensing fluctuations in air pressure, and translating these fluctuations into signals interpreted by the brain. The outer ear collects sound waves and directs the sound to the tympanic membrane (eardrum), the boundary between the outer and middle ear. The sound enters the ear canal and causes the eardrum to vibrate. This air pressure is amplified by the ossicles, three tiny bones in the middle ear, which push on fluid in the inner ear through an opening in the cochlea.

The cochlea converts sound pressure waves into electrical impulses that are sent to the brain. The cochlea consists of three adjacent tubes that are shaped like a snail shell (hence, this structure's name), which are separated by membranes and lined by hair cells. When the hair cells are bent by the sound waves, they become excited and create impulses that are sent to the brain. The cochlea differentiates between the pitches and intensity of sounds based on vibrations in membranes along its length. Vibrations at the entrance of the cochlea respond most actively to high frequency sounds, while those at the opposite end reacts to sounds of low frequency. High amplitude (louder) sounds cause the membrane to vibrate more vigorously than low frequency (soft) sounds.

SEE ALSO: Sense of Taste (1974), Sense of Smell (1991).

The cochlea, a spiral-shaped cavity of the inner ear that contains nerve endings essential for hearing, takes its name from the Greek word kokhlias *("snail"). As such, it evokes the spiral shell of the garden snail (*Helix aspersa, *shown).*

Vesalius's *De humani corporis fabrica*

Galen (c. 130–c. 200), **Jan Stephen van Calcar** (1499–1546), **Andreas Vesalius** (1514–1564)

Knowledge of anatomy is a foundation course in medical education, viewed to be critical for the diagnoses and treatment of disease, as well as being essential for sculptors and painters. When the Flemish anatomist Andreas Vesalius assumed a professorship at the University of Padua, a major center of medical education in the sixteenth century, anatomy lectures were based on Galen's texts written almost fifteen hundred years earlier. These classical readings were followed by a dissection performed by a barber-surgeon as directed by the lecturer. Vesalius broke with tradition and performed the dissections on corpses himself, with his students surrounding the dissection table. But what Vesalius saw did not always correspond with Galen's time-honored descriptions.

Galen, the most accomplished of all medical scholars of antiquity and physician to the gladiators of Pergamon, had access to examining many human subjects. But because human dissections were banned in Ancient Rome, Galen prepared his anatomical drawings using Barbary apes, arguing that they were sufficiently similar to humans.

In 1543, at age 28, Vesalius published the first edition of *De humani corporis fabrica*, a work that included the complete structure of the human body, with the first detailed images of the internal organs. The book, containing two hundred woodcuts, was didactic and accurate, correcting Galen's errors. Always the perfectionist, Vesalius insisted that the artwork also be aesthetically pleasing. The final work was a true collaborative effort between the dissector and the illustrator, with the woodcuts historically attributed to Jan Stephen van Calcar, a student of the Italian Renaissance painter Titian.

Vesalius envisioned the book's readers to be not only physicians and anatomists but also artists. After some initial resistance to this challenge to Galen, the book led to Vesalius's fame and fortune and is today regarded as one of the most famous books in medicine and science. Of the approximate 500 copies originally printed, 130 are extant. In 1564, Vesalius drowned in a shipwreck in the Ionian Sea near the Greek Isle of Zakynthos (Zante), while returning from a pilgrimage to Jerusalem.

SEE ALSO: Skeletal System (c. 180), Leonardo's Human Anatomy (1489).

Frontispiece to Vesalius's De humani corporis fabrica, *the first complete, detailed, and accurate text on human anatomy.*

ANDREAE VESALII
BRVXELLENSIS, SCHOLAE
medicorum Patauinæ professoris, de
Humani corporis fabrica
Libri septem.

CVM CAESAREAE
Maiest. Galliarum Regis, ac Senatus Veneti gra
tia & priuilegio, ut in diplomatis eorundem continetur.

Tobacco

John Rolfe (1585–1622), James Bonsack (1859–1924)

Long before the first Europeans landed on the New World shores, Native Americans cultivated and smoked tobacco in religious ceremonies and used it to treat a wide variety of maladies. Cultivation sites in Mexico date back to 1600–1400 BCE. While the Spanish introduced tobacco to Europe in 1518, it was John Rolfe, an early English settler, who first successfully cultivated and exported tobacco as a lucrative cash crop in the Colony of Virginia in 1611. Prior to the twentieth century, tobacco was primarily chewed, snuffed, or smoked in pipes or cigars. In 1883, cigarettes were hand-rolled at four cigarettes per minute. That year, James Bonsack invented the automated cigarette-rolling machine that turned out two hundred per minute, causing the price to plummet. The coming decades saw the rise of the colossal American cigarette industry.

Tobacco is processed from the leaves of plants of the genus *Nicotiana*, a member of the *Solanaceae* (nightshade) family, whose cousins include potatoes, tomatoes, eggplant, peppers, and petunias. After being harvested, the leaves are dried, cured, aged, and blended with other varieties of N. *tobaccum* to produce distinctive flavors and tastes, and packaged. Cigarette US per capita consumption jumped from 54 in 1900 to 4,345 in 1963! In 1964, the US Surgeon General proclaimed that smoking was hazardous to the health.

Smoking is an equal-opportunity health hazard affecting almost every body organ. There is incontrovertible evidence that smoking increases the risk of diseases of the cardiovascular system (heart attack, stroke), lungs (emphysema, chronic bronchitis), and a wide range of cancers. The World Health Organization has identified smoking as the greatest cause of preventable deaths worldwide.

Most smokers know these dangers and yet continue to smoke. Why? They are addicted to nicotine, a naturally occurring, behaviorally active compound concentrated in the plant's leaves. In the 1970s, Brown & Williamson, a major cigarette manufacturer, developed Y–1, a crossbred mixture of N. *tobaccum* and N. *rustica* that doubled the nicotine content to 6.5 percent from 3.2–3.5 percent. Y–1 continued to be incorporated in their cigarettes from 1991 until 1999.

SEE ALSO: Plant-Derived Medicines (c. 60,000 BCE), Wheat: The Staff of Life (c. 11,000 BCE), Agriculture (c. 10,000 BCE), Rice Cultivation (c. 7000 BCE), Artificial Selection (Selective Breeding) (1760), Genetically Modified Crops (1982).

After being harvested, tobacco leaves are dried (cured) and aged to bring out their distinctive flavors and tastes. Tobacco can be cured by several methods, which take days to many weeks. The "flue cured" method for cigarette tobacco takes about a week and produces medium to high levels of nicotine.

Metabolism

Santorio Sanctorius (1561–1636), Hans Krebs (1900–1981)

For thirty years, Santorio Sanctorius, the Italian physiologist, physician, and inventor of the medical thermometer, meticulously weighed himself prior to and after engaging in all manner of life's activities—eating, drinking, fasting, eliminating, sleeping, and engaging in sex. These he published in 1614, in *Ars de statica medicina*, in which he described the first controlled experiment that introduced quantification to medical practice. Sanctorius noted that the weight of his feces and urine was less than the food he had ingested and attributed the difference to "insensible perspiration." So began the study of metabolism.

BUILDING UP AND BREAKING DOWN. One of the fundamental characteristics of all living beings is their use of energy to perform their activities. Metabolism, a word from the Greek meaning to "change" or "overthrow," refers to all chemical reactions in living beings that generate or utilize energy, and these are divided into anabolic and catabolic reactions. Anabolic reactions use energy for the biosynthesis (manufacture) of larger organic molecules as well as for the growth and differentiation of cells. Conversely, catabolic reactions involve the breakdown of these molecules to produce energy. Chemical reactions are organized into metabolic pathways in which one chemical is converted to another in a sequence, and these reactions are catalyzed by **enzymes**. The metabolic pathways involve carbohydrates, fats, proteins, and nucleic acids, and the nature of the chemicals in these pathways is very similar across many different species, ranging from microbes to humans.

Studies performed by Hans Krebs during the 1930s provided the basis for our fundamental understanding of metabolic pathways. Krebs was a German-born physician-biochemist who discovered the urea cycle, which described the pathway by which organisms remove ammonia formed in the body to the less toxic urea. After a Nazi order forbidding him as a Jew to practice medicine in Germany, he emigrated to England. There he made his most important discovery, the 1937 identification of the citric acid cycle (Krebs cycle), which described a series of chemical reactions used by *all* aerobic organisms to generate energy from carbohydrates, proteins, and fats. In recognition, he was awarded the 1953 Nobel Prize in Physiology or Medicine.

SEE ALSO: Plant Defenses against Herbivores (c. 400 Million BCE), Enzymes (1878), Inborn Errors of Metabolism (1923).

Adenosine triphosphate (ATP) is the "molecular unit of currency" for the transfer of energy within cells and represents the major source of energy for the metabolic reactions of most organisms. This image shows a three-dimensional representation of ATP.

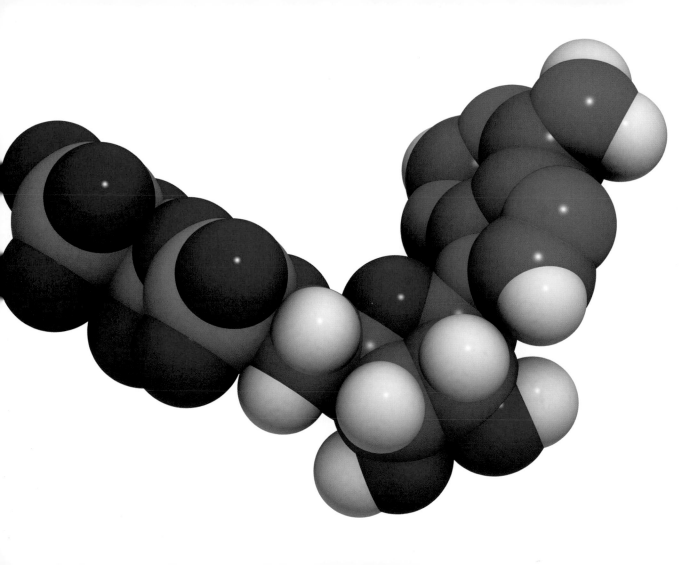

Scientific Method

Aristotle (384–322 BCE), **Francis Bacon** (1561–1626), **Galileo Galilei** (1564–1642), **Claude Bernard** (1813–1878), **Louis Pasteur** (1822–1895)

Formulation and fine-tuning of the scientific method has evolved over the ages and is based upon the contributions of many early distinguished scholars including Aristotle, who introduced logical deduction, a "top-down" approach, that is, starting with a theory or hypothesis and then testing that theory; Francis Bacon, the father of the modern scientific method, who in 1620 wrote *Novum Organum Scientiarum*, which proposed inductive reasoning as the foundation for scientific reasoning, a "bottom-up" approach in which specific observations led to the formulation of a general theory or hypothesis; and Galileo, who advocated experimentation rather than metaphysical explanations. In the mid-nineteenth century, Louis Pasteur elegantly utilized the scientific method when he designed experiments to disprove the theory of spontaneous generation.

In 1865, Claude Bernard, one of the greatest of all scientists, wrote *An Introduction to the Study of Experimental Medicine*, which he personalized by using his own thoughts and experiments. In this classic book, he examined the importance of the scientist bringing forth new knowledge to society, and he then proceeded to critically analyze what constituted a good scientific theory, the importance of observation rather than reliance on historical authorities and sources, inductive and deductive reasoning, and cause and effect.

When some nonscientists think of theories, such as the theory of evolution, not infrequently they use the term "theory" disparagingly and assume or imply that it imputes an unproven notion or a mere guess or speculation. Scientists, by contrast, use the term "theory" to refer to an explanation, model, or general principle that has been tested and confirmed and that explains or predicts a natural event. The scientific method follows a number of sequential steps and is an approach used to investigate phenomena or acquire new knowledge. Using a series of steps, it is based upon developing and testing a hypothesis that explains a given observation, objectively evaluating the test results obtained, and then accepting, rejecting, or modifying that hypothesis. A theory is broader and more general than a hypothesis and is supported by experimental evidence based on a number of hypotheses that can be tested independently.

SEE ALSO: Refuting Spontaneous Generation (1668), Darwin's Theory of Natural Selection (1859).

In his 1620 book, Novum Organum (New Method), *Francis Bacon proposed a scientific method of inquiry based on inductive reasoning, in which a generalization is built based on incremental data collection. This method was intended to improve upon Aristotle's deductive reasoning, in which specific facts are deduced from a generalization.*

Harvey's *De motu cordis*

Galen (c. 130–c. 200), **William Harvey** (1578–1657),
Marcello Malpighi (1628–1694)

In 1628, William Harvey published a heretical paper, *De motu cordis et sanguinis* (*On the motion of the heart and blood*), in which he proposed that blood was pumped by the heart in one direction through a closed system from arteries to veins and was then returned to the heart. Harvey based his theory not on speculation but on dissections and physiological experiments conducted on multiple species of living and dead animals and in human subjects. A critical piece of the puzzle was his observation that the valves present in the veins permitted blood to flow in only a single direction, toward the heart.

Why was Harvey reluctant to delay widely publicizing this concept, replete with convincing experimental evidence, years after first introducing it in his public 1615 Lumleian Lecture? His explanation challenged Galen's teachings of blood flow, promulgated 1,400 years earlier, which had been accepted as dogma by all authoritative scientists and physicians ever since. According to Galen, blood originated in the liver, after being formed from food. It then passed between the two lower chambers in the heart through invisible pores and was consumed by the organs of the body, serving as a nutrient. Blood was utilized at the same rate it was being produced. Based on his data and analysis, Harvey determined this to be a mathematical impossibility.

Harvey was the well-respected court physician to King James I and his son King Charles I, both of whom encouraged and supported his research. But by challenging the authority of Galen, the seventy-page *De motu cordis* generated controversy and animosity, particularly in Continental Europe, for some twenty years after it was published. An important missing link in Harvey's explanation was how blood flowed from the arteries to veins. He postulated the presence of capillaries, a fact validated by Marcello Malpighi in 1661.

De motu cordis is now considered to serve as the foundation for our understanding of the heart and cardiovascular system and among the most important publications in the history of biology and medicine. Called the father of modern physiology, Harvey was the first to supplement simple observation with experimental methodology and quantification.

SEE ALSO: Pulmonary Circulation (1242), Leonardo's Human Anatomy (1489), Scientific Method (1620), Placenta (1651).

During the course of his medical practice, William Harvey, shown in this engraving, examined four women accused of witchcraft. Notwithstanding the societal pressure upon him to discover suspicious body marks as evidence of their guilt, he provided testimony leading to their exoneration.

Mechanical Philosophy of Descartes

William Harvey (1578–1657), René Descartes (1596–1650)

Although René Descartes is best known for his philosophical writings and the development of mathematics, he also had a major impact on biological thinking. Called a pioneer of modern philosophy, he argued that the guarantors of truth are individuals and not the Church and was best known for "I think, therefore I am." Following the family tradition, he was educated to become a lawyer (which he never practiced), but even as a young student, his true passion was mathematics. He conceived the Cartesian coordinate system, by which a point in space could be expressed as a set of numbers, and he developed analytical geometry, linking algebra and geometry, which served as the basis for the formulation of calculus by Isaac Newton and Gottfried Leibniz in the 1660s.

In 1628, William Harvey used mechanical analogies to describe the circulation of blood, and this is said to have inspired Descartes to formulate his mechanical philosophy. It was a mathematical and mechanistic approach that influenced Descartes's perception of biology and that dominated much physiological research later during the nineteenth and twentieth centuries. In his 1637 *Discourse on the Method*, Descartes sought to explain everything in nature, with the exception of the human mind, in terms of principles of mechanics, mathematics, matter, and motion. The only realities were those that could be measured, such as size, shape, position, duration, and length; everything else, including the senses, was subjective and only existed outside the mind of the individual and had no physical reality. Just as the universe was a machine, so too was the organism's body that contained the parts, arrangements, and movements needed to walk, eat, breath, and carry out all other functions.

Although Descartes recognized the distinction between animate and inanimate beings, he viewed animals to be no more than machines because, unlike humans, they lacked a soul that was responsible for intellect, volition, and conscious experiences; animals were incapable of using language, and they could not reason. He identified the pineal gland, located in the center of the brain, to be the "seat of the soul," that acted through nerves to control the body.

SEE ALSO: Harvey's *De motu cordis* (1628), Circadian Rhythms (1729), Hypothalamic-Pituitary Axis (1968).

With the exception of the human mind, Descartes sought to explain everything in nature in mechanical terms, including animals, which he compared to robots or complex machines.

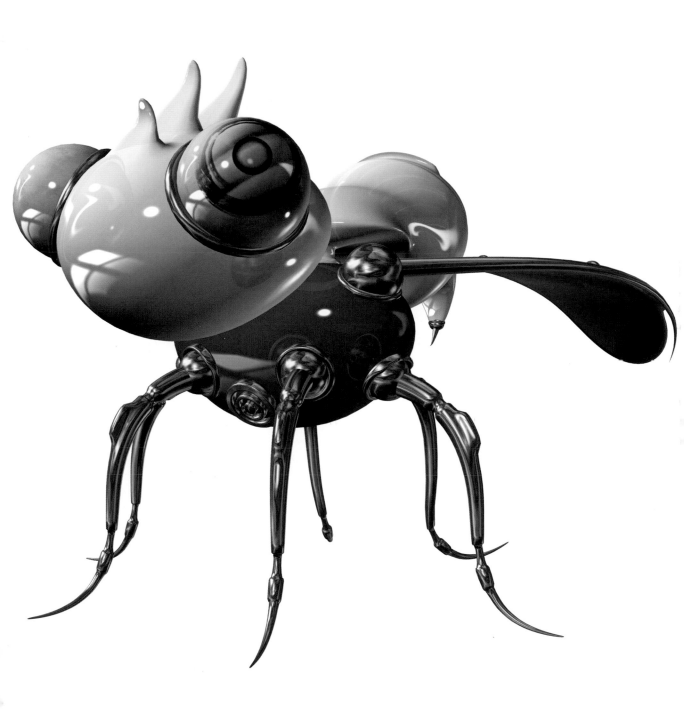

Placenta

Aristotle (384–322 BCE), **Galen** (c. 130–c. 200), **Leonardo da Vinci** (1452–1519), **William Harvey** (1578–1657)

Interest in the mysteries, importance, and functions of the placenta goes back to ancient sources and scholars and continues to the present day. In Egypt, a sculpture depicts the royal placenta, with an attached umbilical cord, and refers to it as the Pharaoh's "soul" or "secret helper." The kingdom's success was thought to be dependent upon the sovereign's health and preservation of his soul. The Hebrew Bible refers to the placenta as being the "Bundle of Life" and "External Soul." The placenta (from the Greek "flat cake") aroused the interest of Aristotle and Galen, the greatest scholars of ancient times. In about 340 BCE, Aristotle initiated examination and naming of the membranes surrounding the fetus, but because of differences among species and his studies using animal subjects, some of his incorrect conclusions were perpetuated for over a millennium.

In about 1510, Leonardo da Vinci focused his genius on depictions of human anatomy, including the fetus. These drawings included the uterus with its blood vessels, fetal membranes, and umbilical cord. He also stated that fetal blood vessels were not continuous with that of the mother, an ongoing subject of question and conjecture dating from ancient times to the eighteenth century. In his 1628 medical classic *De motu cordis*, William Harvey provided the foundation for our modern understanding of the physiology of the circulatory system and heart.

In 1651, Harvey extended these studies to consider, in detail, fetal circulation and its relationship to the mother. He also raised a very basic question: How does the fetus survive and breathe in the mother's womb for many months and yet, perishes within a very short time after delivery if unable to breathe? Since mother and fetus have two separate circulatory systems, he postulated that fetal nourishment and air are provided from the fluid within the amniotic sac surrounding the fetus. We now know that from the fourth week of development until birth, placental circulation carries nutrients, respiratory gases, and waste materials between the embryo, fetus (after the ninth week), and the mother.

SEE ALSO: Leonardo's Human Anatomy (1489), Harvey's *De motu cordis* (1628), Ovaries and Female Reproduction (1900), Progesterone (1929).

This engraving from the Usual Medicine Dictionary *(1885) by Dr. Paul Labarthe shows a fetus about to be born.*

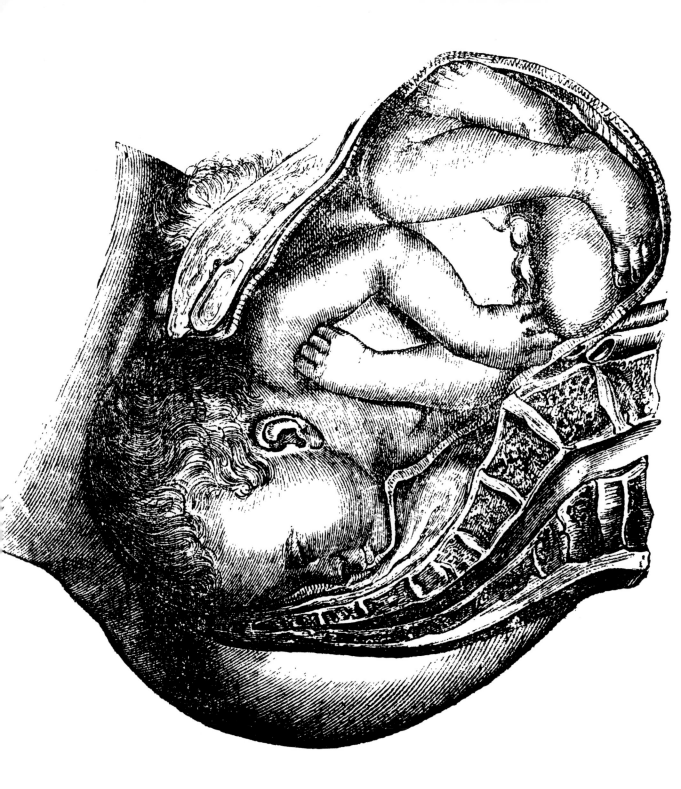

Lymphatic System

Thomas Bartholin (1616–1680), Olauf Rudbeck the Elder (1630–1702)

A SCANDINAVIAN DISCOVERY. Whereas credit for the discovery of the lymphatic system is controversial, it is clear that both claimants were Scandinavian and members of distinguished academic families. The father and the son of Thomas Bartholin were anatomists, as was he, at the University of Copenhagen. After his brother informed him that the thoracic duct had been found in dogs, Bartholin searched for one in the cadavers of two criminals, whose bodies were kindly provided by the king for this purpose. In 1652, he publicly announced that he had discovered the lymphatic system in humans and described it as distinct and independent.

Bartholin's claim for precedence was challenged by Olauf (variously spelled Olof and Olaus) Rudbeck the Elder, who enjoyed a distinguished career as a scientist and physician. Rubeck presented his discovery of the lymphatic system at the court of Queen Christina of Sweden in 1652, but failed to produce a written report until the following year and, without doubt, after Bartholin. (Cinema buffs will recall that the role of Queen Christina was portrayed by Greta Garbo in a 1933 eponymous film.)

Rudbeck was also a historical linguist with what some would describe as considerable imagination. From 1679 to the year of his death in 1702, he prepared a 3,000-page, four-volume work entitled *Atlantica*. In it he sought to establish that Sweden was Atlantis—the legendary island described by Plato in 300 BCE—and that Swedish was the language of Adam and from which Latin and Hebrew originated. His theories were subject to some criticism and even ridicule in his native Scandinavia, at a time when Sweden was one of the great European powers.

The lymphatic system is a network of organs and lymph nodes, ducts, and vessels that produce and remove lymph from tissues and send it to the bloodstream. The thoracic duct, the major lymphatic vessel in the body, collects and channels lymph from the lower portion of the body. Lymph is a milky fluid that contains lymphocytes, a major component of the immune system, and chyle, which consists of lymph and fats. The lymphatic system defends the body against infections and the spread of tumors, and also collects and removes interstitial fluids that surround cells.

SEE ALSO: Innate Immunity (1882), Adaptive Immunity (1897).

A 1661 oil painting of Queen Christina of Sweden (1626–1689) by Dutch-born artist Abraham Wuchters (1610–1682). A highly complex and storied woman, Christina became queen in 1633, befriended philosopher René Descartes, refused to marry, abdicated the throne in 1654, converted to the Catholic faith, and spent most of her remaining years in Rome.

Blood Cells

Antonie van Leeuwenhoek (1632–1723), **Jan Swammerdam** (1637–1680),
Gabriel Andral (1797–1876), **Alfred Donné** (1801–1878),
Paul Ehrlich (1854–1915)

Blood has played a central role in the lives of ancient people, appearing in their religious beliefs, myths, health, and as a symbol of courage and sacrifice. Across many cultures, and surviving to this day, is the belief that blood is a symbol of family relationships, as well as a tribal and natural bond. For the ancient Greeks, blood was the essential nutrient of life, the center of life itself, the soul, and death was, by contrast, final and irrevocable, when blood was lacking. The exceptions were the immortal gods and demons, who were without blood but not dead. The Greeks did not commonly engage in blood sacrifices, unlike some other cultures, such as the Anglo-Saxons and Norsemen, who deemed that blood transferred the power of its source.

The importance of blood extends to other cultures and periods. The Judaic and Islamic scriptures forbid the consumption of blood, while some Christian churches view wine as the symbol of the blood of Jesus. When a man has a nosebleed in some East Asian cultures, it is a sign of sexual desire, and the Japanese have classified personality traits on the basis of **blood types**. The Gothic novelist Bram Stoker might have been inspired by the New World vampire bat, which feeds exclusively on blood, when he created his eponymous character in the 1897 novel *Dracula*.

Scientists have studied how blood transports nutrients and oxygen to cells and carries waste materials from the cells for ultimate disposal from the body. Jan Swammerdam, a Dutch biologist, was the first to see red blood cells under a microscope in 1658. Antonie van Leeuwenhoek described their size and shape and illustrated them in 1695. Around 1840, the French professor of medicine Gabriel Andral described white blood cells and was a pioneer in blood chemistry and scientific hematology, which integrated clinical and analytical medicine. Several years later, the French physician Alfred Donné observed the first blood platelet. Lastly, Paul Ehrlich's wide-ranging scientific accomplishments include an 1879 staining technique for differential counting of white blood cells.

SEE ALSO: Leeuwenhoek's Microscopic World (1674), Hemoglobin and Hemocyanin (1866), Blood Types (1901), Blood Clotting (1905).

Bram Stoker's Dracula *(1897)—which was partially inspired by blood-sucking vampire bats—was set in the Transylvania region of Romania, where this souvenir was found in 2007.*

Refuting Spontaneous Generation

Aristotle (384–322 BCE), **Francesco Redi** (1626–1697), **Lazzaro Spallanzani** (1729–1799), **Louis Pasteur** (1822–1895)

In his book, *The History of Animals*, written over 2,000 years ago, Aristotle proclaimed that while some organisms arise from similar organisms, others, such as insects, arise spontaneously from putrefying earth or vegetable matter. Each spring, the ancients observed that the Nile overflowed its banks leaving behind muddy soil and frogs that were not present in dry times. In Shakespeare's *Antony and Cleopatra*, we learn that crocodiles and snakes are formed in the mud of the Nile. The concept that some living beings could arise from nonliving inanimate matter, which Aristotle called *spontaneous generation*, remained essentially unchallenged until the seventeenth century. After all, it was commonly observed that maggots appeared to arise from decaying flesh.

In 1668, the Italian physician and poet Francesco Redi devised an experiment that questioned the validity of spontaneous generation and the origin of maggots from rotting meat. Redi placed meat in three widemouthed jars, which he set aside for several days. In the one that was open, flies reached the meat and laid their eggs. In another jar that was sealed, no flies or maggots were found. The mouth of the third was covered with gauze, preventing flies from entering the jar containing meat but upon which they laid their eggs that hatched into maggots.

One century later, Lazzaro Spallanzani, an Italian priest and biologist, boiled broth in a sealed container and permitted the air to escape. While he did not see the growth of any living organism, the question persisted as to whether air was an essential factor required for spontaneous generation to occur.

In 1859 the French Academy of Sciences sponsored a contest for the best experiment that would conclusively prove or disprove the validity of spontaneous generation. In his winning entry, Louis Pasteur placed boiled meat broth in a swan-necked flask, with the neck curved downward. This allowed the free flow of air into the flask, while preventing the entrance of airborne microbes. The broth-containing flask remained free of growth, and the concept of spontaneous generation was believed to be relegated to history.

SEE ALSO: Origin of Life (c. 4 Billion BCE), Aristotle's *The History of Animals* (c. 330 BCE), Cell Theory (1838), Germ Theory of Disease (1890), Miller-Urey Experiment (1953).

Louis Pasteur, a French microbiologist and chemist, made significant discoveries relating to the germ causes of disease, vaccinations, fermentation, and pasteurization.

Phosphorus Cycle

Hennig Brand (c. 1630–c. 1692), **Carl Wilhelm Scheele** (1742–1786), **Johan Gottlieb Gahn** (1745–1818)

Phosphorus was the first element discovered that was not known in the ancient world. In 1669, the German alchemist Hennig Brand was searching for the philosopher's stone, which could transform base metals, such as lead, into gold or silver. He boiled down urine, yielding solid phosphorus, which emitted a pale-green glow. One hundred years later, Johan Gottlieb Gahn, a Swedish chemist-metallurgist, extracted phosphorus from calcium phosphate in bone, which became its major source until the 1840s. At about the same time, another Swede, the pharmacist Carl Wilhelm Scheele, discovered a method for mass-producing phosphorus, which enabled Sweden to become one of the world's principal manufacturers of matches.

Phosphorus is essential for living organisms. It is a component of **deoxyribonucleic acid (DNA)** and ribonucleic acid (RNA), as well as adenosine triphosphate (ATP), involved in energy transfer. In combination with lipids to form phospholipids, it is the core of cell membranes. Calcium phosphate provides strength to bones and teeth.

Of all the **biosphere**'s recycled elements, phosphorus is the scarcest. Much of the earth's phosphorus is found as a phosphate (phosphorus + oxygen) in rock and sedimentary deposits from which it is released into the sea by weathering and mining. Whereas inadequate phosphorus slows or stunts the growth of **algae**, excess phosphorus causes an uncontrolled overgrowth.

Humans added phosphates to household detergents and fertilizers in the mid-twentieth century, which had a major negative impact that disrupted the natural balance of the phosphate cycle. Phosphate runoffs into lakes and streams can cause algal blooms—rapidly growing dense populations of algae. When algae die, they are consumed by bacteria, a process that depletes vast amounts of oxygen from the water and leads to the death of fish and other aquatic organisms from oxygen starvation. Outflows from municipal sewage treatment plants also contribute phosphates to water. US states began to ban this household use of phosphates in the 1970s.

SEE ALSO: Prokaryotes (c. 3.9 Billion BCE), Algae (c. 2.5 Billion BCE), Land Plants (c. 450 Million BCE), Coral Reefs (c. 8000 BCE), Deoxyribonucleic Acid (DNA) (1869), Biosphere (1875), Energy Balance (1960), *Silent Spring* (1962).

"The hand that rocks the cradle can also rock the Axis." This 1942 photograph shows a woman supporting the wartime effort by making dies for bombs. During World Wars I and II, incendiary weapons containing white phosphorus were used.

Ergotism and Witchcraft

Louis-René Tulasne (1815–1885)

On January 20, 1692, three preteen girls were brought to trial in colonial Salem, Massachusetts, accused of being witches. These charges were based on their blasphemous outcries, convulsive seizures, and trance-like states, which the local physician diagnosed as signs of bewitchment. These children were convicted; two were hanged and one spared the noose by dying in prison. By the end of the year, twenty individuals were accused and executed as witches. Witch trials were not unique to Salem, and it has been reported that, from 1450 to 1750, some 40,000 to 60,000 people were tried and condemned to death as witches in Europe; the last reported European witch burning occurred in Poland in 1793. During this period, in Europe and North America, supernatural influences pervaded all aspects of life and were blamed for illness and misfortune.

Many scholars have attributed the symptoms exhibited by the three girls, and many of those accused, to ergotism, a disease of rye and other cereals caused by the fungus *Claviceps purpurea*. The ergot kernel, the sclerotium, develops when spores of *Claviceps* infect cereal plants. This fungal infection resembles pollen grain growing into an ovary during plant fertilization. During the cool and damp periods in early spring, the rye and grain-like ergot are harvested and milled together; rye was a staple crop of the poor. Ergot poisoning occurs in humans and other mammals, in particular, grazing cattle.

Ergotism was common in France, where the climatic conditions were propitious for the growth of ergot; it caused the death of 40,000 in southern France in 944. The primary symptoms are convulsions and gangrenous ergotism, in which constriction of blood flow to the extremities can cause extreme burning pain, gangrene, and the loss of the extremities. In 1670, a French doctor named Thuillier determined that ergotism was not an infectious disease but rather resulted from the consumption of ergot-contaminated rye. The life cycle of ergot was established by the French mycologist Louis Tulasne in 1853. Alkaloids extracted from ergot have been used medically to treat migraine headaches and to induce uterine contractions and to control bleeding after childbirth.

SEE ALSO: Fungi (c. 1.4 Billion BCE), Plant-Derived Medicines (c. 60,000 BCE), Agriculture (c. 10,000 BCE).

The painting Witches Sabbath *(1798) by Spanish artist Francisco Goya (1746–1828) represents the devil as a goat.*

Leeuwenhoek's Microscopic World

Antonie van Leeuwenhoek (1632–1723), **Robert Hooke** (1635–1703)

In 1674, Dutch scientist Antonie van Leeuwenhoek discovered a world with more than a trillion trillion inhabitants that was previously unknown to humans; in that year he saw single-celled organisms he called *animalcules* and *beasties*. Although one of the most famous biological scientists, the founder of microbiology, he had only a modest formal education, wrote in only a single language—his native Dutch, and never wrote a book or a scientific paper.

Leeuwenhoek was born and spent virtually his entire life in Delft, Holland, and was a contemporary of the famous Dutch painter Johannes Vermeer. He established a business as a fabric merchant but his true love was his hobby: grinding lenses. He was said to have been inspired to take up microscopy after reading English polymath Robert Hooke's 1665 book *Micrographie*, in which he popularized the microscope and its use. Hooke was the first to have observed microscopic sections of cork, which he termed "cells."

Starting in 1673, at the age of forty, and for the next fifty years until the day of his death, Leeuwenhoek corresponded with the Royal Society in London, writing hundreds of letters in informal Dutch describing his microscopic observations. These included protists (1674), bacteria (1676), as well as blood capillaries, muscle fibers, plant tissues, and sperm cells from a variety of species. These highly detailed observations were made possible because of his skill grinding lenses, achieving magnifications of up to 275 times the actual specimen size, with clear and bright images; by contrast, earlier microscopes could only attain twenty to thirty times magnification. Over his lifetime, Leeuwenhoek handmade 400–500 lenses and some twenty-five microscopes, always maintaining secrecy regarding his expert techniques.

The use of lenses for the purposes of magnification dates back to the ancient Assyrians and Romans. The first compound microscope, in which more than one lens is used, was invented in about 1590, and these were essential tools used by Hooke and subsequent biologists into the twentieth century. Modern light microscopes can effectively magnify 1,000–2,000 times, while **electron microscopes** used in biology magnify up to two million times!

SEE ALSO: Spermatozoa (1677), Cell Nucleus (1831), Cell Theory (1838), Electron Microscope (1931).

A 1982 postage stamp from Transkei, South Africa, bears an illustration of Antonie van Leeuwenhoek, the first scientist to observe and describe single-celled organisms.

Antonie van Leeuwenhoek 1632-1723 father of the microscope

20c TRANSKEI

C2.4 1982

Spermatozoa

Antonie van Leeuwenhoek (1632–1723), Lazzaro Spallanzani (1729–1799), Oscar Hertwig (1849–1922)

During the seventeenth and eighteenth centuries, one of the great scientific questions of philosophical and religious interest was that of procreation, in particular, that of humans. Some argued that the egg was the seed that gave rise to the animal, while others believed it was semen. The nature of semen when impregnating the egg was perceived to be ethereal and variously described as a spirit, a vapor, or an odor, but not physical. In 1677, the Dutch microscopist Antonie van Leeuwenhoek examined semen from various animal species, as well as his own—acquired not by sinfully "abusing" himself, he asserted, but from conjugal coitus—and found many spermatozoa, but at that time did not associate them with impregnation. However, in 1683, he concluded that "man comes not from an egg but from an animalcule in the masculine seed" and that parts of the egg were transferred to the spermatozoon.

The Italian priest-biologist Lazzaro Spallanzani accepted the preformation theory, which asserted that all living beings were created by God at the beginning and were encapsulated within the first female of their species. The new individual within the egg was preformed and, under the influence of semen, expanded. In 1768, Spallanzani was first to describe that both the solid fraction of semen and the egg were essential for reproduction. However, he failed to recognize the role of spermatozoa in the reproductive process.

In the 1870s, there were two views regarding the process of fertilization: the spermatozoa made contact with the egg and, by transmitting a mechanical vibration, stimulated its development; alternatively, the spermatozoa physically penetrated the egg and mixed its chemical components with the egg yolk. In 1876, Oscar Hertwig, a German embryologist, sought to study this problem using the sea urchin, because it was transparent, had a finely divided yolk, and lacked a membrane. He was able to microscopically visualize spermatozoa as they entered the egg and fused with its nucleus. Moreover, Hertwig observed that only a single spermatozoon was required to fertilize the egg, and that once the spermatozoon (sperm) entered the egg, a membrane barrier was formed precluding penetration by additional spermatozoa.

SEE ALSO: Leeuwenhoek's Microscopic World (1674), Theories of Germination (1759), Germ-Layer Theory of Development (1828), Meiosis (1876).

In the mid-eighteenth century, Spallanzani proposed that all humans were encapsulated within the first female (Eve) and expanded under the influence of semen. This 1913 statue, Eve and the Serpent, *by the Belgian sculptor Albert Desenfans (1845–1938) is located in Josaphat Park in Brussels.*

Miasma Theory

Giovanni Maria Lancisi (1654–1720), **William Farr** (1807–1883), **John Snow** (1813–1858), **Florence Nightingale** (1820–1910), **Robert Koch** (1843–1910)

Although references appear in ancient Greek writings, the miasma (or miasmatic) theory of infectious diseases originated during the Middle Ages and continued to enjoy favor in Europe, India, and China until the late nineteenth century. According to this theory, vapor, mist, or noxious air originating from decomposing organic matter (miasmata), entered the body and caused such diseases as cholera, the Black Death (bubonic plague), typhoid, tuberculosis, and malaria ("bad air"). During the black plagues in Europe, plague doctors would visit their patients garbed in goggles, protective clothing, and a hood with a long beak filled with perfumes to combat the odors of decayed flesh. During this time, sewage was removed and discarded far from the city and swamps drained to eliminate foul odors.

In his 1717 work, *On the Noxious Effluvia of Marshes*, the Italian epidemiologist and physician to popes, Giovanni Lancisi, described a correlation between the presence of mosquitoes and prevalence of malaria and offered what is arguably the most articulate description of the miasma theory. During the early 1850s, a cholera pandemic beset London that was concentrated in undrained, filthy, odoriferous areas near the banks of the River Thames that were inhabited by the poor. The medically trained William Farr, serving as the assistant commissioner for the 1851 London census, attributed the transmittal of cholera to noxious air. He was supported by Florence Nightingale, social reformer and founder of modern nursing, who advanced the cause of sanitary and fresh-smelling hospitals. By contrast, physician and epidemiologist John Snow, while not knowing the cause of cholera (the germ theory had yet to be formulated), rejected the miasma theory. By finding clusters of cases, Snow convincingly traced the source of the 1854 cholera outbreak in Soho, central London, to a contaminated water source.

Support for the miasma theory of disease dissipated after German physician Robert Koch's rediscovery of the cholera bacterium (1882) and formulation of the **germ theory of disease** (1890). Although the miasma theory was discarded, it led to greater emphasis on public health and the construction of sanitation facilities, and swamps and marshes were drained for malaria control.

SEE ALSO: Germ Theory of Disease (1890), Endotoxins (1892), Malaria-Causing Protozoan Parasite (1898).

This image of a plague doctor appears in a 1721 work by Jean-Jacques Manget (1652–1742), a physician and writer from Geneva.

Habit des Medecins, et autres personnes qui visitent les Pestiferés, Il est de marroquin de levant, le masque a les yeux de cristal, et un long néz remply de parfum

Circadian Rhythms

Jean-Jacques d'Ortous de Mairan (1678–1771), **Jürgen Aschoff** (1913–1998), **Colin Pittendrigh** (1918–1996)

In 1729, the French scientist Jean-Jacques d'Ortous de Mairan noted that mimosa plants folded and unfolded their leaves following a regular schedule, over a 24-hour period, even when kept in total darkness. This was the first scientific description of circadian rhythms (CR). These daily light-dark cycles appear at approximate 24-hour intervals and are named *circadian* ("around a day"). CR govern rhythmic biological changes and are self-sustaining even in the absence of environmental cues. Synchronization of an organism to both the external and internal environment is vital to that organism's well-being and, indeed, its survival.

CR are not unique to plants. It was studied in the 1950s in *Drosophila* (fruit flies) by Colin Pittendrigh and in humans by Jürgen Aschoff and was also observed in fungi, animals, and cyanobacteria (blue-green algae). CR are important for the sleep-wake cycle and the feeding patterns of all animals and are accompanied by more subtle patterns involving changes in gene activity, brain wave activity, hormone production and release, and cell **regeneration**. Disruption of CR has adverse effects on health, as is observed in those with "jet lag" who experience fatigue, disorientation, and insomnia.

In mammals, the circadian master clock is found in the suprachiasmatic nucleus (SCN), which is located in the hypothalamus. Information about incoming light is processed by the SCN and transmitted from the retina in the eye to the pineal gland. The pineal controls the production and secretion of the hormone melatonin, which peaks at night and ebbs during the daytime. Working in concert with the biological clock and serving as the molecular basis for CR is an internal clock, which consists of genes that oscillate in their activity and, hence, their protein products, and these regulate biological functions in the body.

In a manner similar to that seen in animals, in the absence of environmental signals, such as changes in light, temperature, or humidity, plants exhibit circadian fluctuations. These have been seen in their photosynthetic activity as well as leaf movements, opening of flowers, germination, growth, and enzyme activity. Once again, the underlying basis for CR appears to revolve around the activity of genes.

SEE ALSO: Regeneration (1744), Phototropism (1880), Secretin: The First Hormone (1902), Thyroid Gland and Metamorphosis (1912), Insulin (1921), REM Sleep (1953), Hypothalamic-Pituitary Axis (1968).

The first recorded observation of the circadian clock occurred in 1729 when a French scientist noticed how mimosa plants opened and folded their leaves over 24-hour cycles.

Blood Pressure

William Harvey (1578–1657), Stephen Hales (1677–1761)

Stephen Hales was clergyman who made his first scientific discoveries after age 50 and proceeded to become a leading English scientist of his era. He considered his most important work to be the design of a ventilation system for use on ships and in prisons and made seminal discoveries in plant physiology.

During the course of studying the flow of sap in a vine, it was necessary for Hales to stem this flow to prevent plant damage. He tied a piece of bladder over the cut and, unexpectedly, noted that the force of flowing sap caused bladder expansion; Hales applied the same approach to measuring blood pressure and built upon the observations of William Harvey. During pioneering studies on the heart in the early decades of the seventeenth century, Harvey reported that blood pulsated as it flowed from a severed artery, as if influenced by rhythmic pressure.

FROM PLANT TO HORSE. Hales's first experimental subjects were horses. He tied a live horse lying on its back to a barn door, and placed a brass tube into its femoral artery, with the flexible windpipe of a goose used to connect the pipe to a nine-foot-long glass tube. When he untied the ligature around the artery, the blood rose to a height of over eight feet. He then studied variables that maintained and influenced blood pressure, such as the volume of blood pumped from the heart (cardiac output) and the capacity of blood to flow through the smallest blood vessels (peripheral resistance). Hales estimated cardiac output by determining the product of the heart rate and the volume of the heart, which he determined using a wax mold. Variations in peripheral resistance were evaluated by injecting various substances—as brandy and saline solutions— into an isolated heart and measuring the rate of the output; these differences were attributed to the diameter of the capillaries. He described these results in his 1733 work *Haemastaticks*.

Plants were Hales's primary interest, and he applied what he observed in animals to his study of plants. More significantly, he observed analogies between the plant and animal worlds, as the role of sap in plants and blood in animals.

SEE ALSO: Harvey's *De motu cordis* (1628).

Human cardiovascular system, depicting the heart, arteries, and veins.

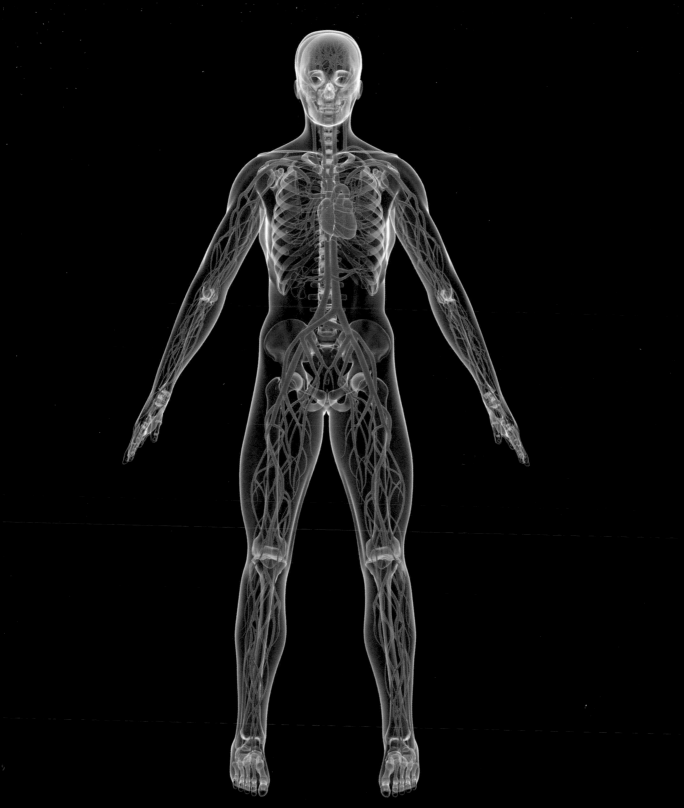

Linnaean Classification of Species

Aristotle (384–322 BCE), **Theophrastus** (372–287 BCE), **Carl Linnaeus** (1707–1778), **Charles Darwin** (1809–1882)

What do mountain lions, pumas, panthers, and catamonts have in common? These are but four of more than a dozen names given in the United States to the same animal, *Felis concolor*. When out and about in nature, we generally refer to plants and birds by their common names, but such names can be misleading. Crayfish, starfish, silverfish, and jellyfish are not related to each other and are not fish.

Classification goes back to ancient times. Aristotle grouped animals based on their mode of reproduction, while Theophrastus classified plants by their uses and methods of cultivation. In his first edition of *Systema Naturae* (1735), the Swedish botanist and physician Carl Linnaeus introduced a new approach to taxonomy (the science of naming and classifying plants and animals). First, he assigned latinized names to plants and animals, based on a binomial nomenclature (genus and species) that uniquely designated each living organism—a system that is still used. For example, the genus *Canis* includes the closely related dogs, wolves, coyotes, and jackals, with each unique member assigned a species name. Moreover, Linnaeus developed a multi-level hierarchical classification in which higher "ranks" would incorporate successive groups at lower levels. Related genera would be grouped into families—*Canis* and *Vulpes* (foxes) are grouped together in *Canidae*. Based on the Linnaean classification, the most inclusive rank was the kingdom, of which he counted two: animal and plant.

The Linnaean classification assigned organisms to different categories based on their physical characteristics and presumed natural relationships, and this was predicated on the then-prevailing Biblical interpretation that plants and animals were originally created in the form that they now exist. One century later, Darwin presented convincing evidence that two extant animals or plants might have had a common ancestor or that extinct organisms may have been the ancestors of those extant. Contemporary classifications are based on **phylogenetic systematics**, which incorporates relationships that include both extant and extinct organisms.

SEE ALSO: Aristotle's *The History of Animals* (c. 330 BCE), Botany (c. 320 BCE), Darwin's Theory of Natural Selection (1859), Ontogeny Recapitulates Phylogeny (1866), Phylogenetic Systematics (1950), Domains of Life (1990), Protist Taxonomy (2005).

A signboard of Methodus plantarum sexualis *(1736), a work by Georg Dionysius Ehret (1708–1770), a German botanist best known for his botanical illustrations. This image depicts the twenty-four classes of plant sexual systems devised by Linnaeus.*

Clariss: LINNÆI. M.D.
METHODUS plantarum SEXUALIS
in SISTEMATE NATURÆ
descripta

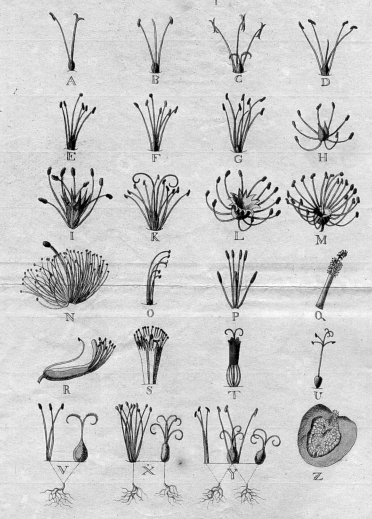

Monandria.

Diandria.

Triandria.

Tetrandria.

Pentandria.

Hexandria.

Heptandria.

Octandria.

Enneandria.

Decandria
Dodecandria
Jcofandria

Polyandria

Didynamia

Tetradinamia

Monadelphia.

Diadelphia.

Polyadelphia.

Syngenefia.

Gynandria.

Monoecia.

Dioecia.

Polygamia.

Cryptogamia

G.D. EHRET. Palat-heidelb:
fecit & edidit

Lugd. bat: 1736

Cerebrospinal Fluid

Hippocrates (c. 460–c. 370 BCE), **Galen** (c. 130–c. 200), **Emanuel Swendenborg** (1688–1772), **Domenico Cotugno** (1736–1822)

The presence of "water" surrounding the brain was first recognized by Hippocrates and referred to by Galen as an "excremental liquid" in the ventricles (central cavity). For the next sixteen centuries, the scientific world was silent about or oblivious to cerebrospinal fluid (CSF).

Interest in CSF was renewed during the mid-1700s by Emanuel Swendenborg, a Swedish scientist, metallurgist, theologian, and mystic. After completing his studies and European travels, he returned to Sweden in 1715 and spent the next two decades working on scientific and engineering projects, such as a description of a flying machine. By his own admission, he was not an experimental scientist but rather preferred contemplating "facts already discovered, and eliciting their causes." These included ideas about the nervous system—in particular, the brain. In a manuscript he prepared from 1741–1744, Swendenborg referred to CSF as "spirituous lymph" and a "highly gifted juice" that is dispensed from the roof of the fourth ventricle to the medulla oblongata and spinal cord; this paper was finally published in translation in 1887. At age 53, he experienced a spiritual awakening and devoted the remainder of his life to theological issues. His best-known work is a book on the afterlife, *Heaven and Hell* (1758).

Domenico Cotugno, an Italian physician and anatomy professor at the University of Naples, first described the circulation of the CSF by decapitating cadavers, standing them on their feet, and observing its flow; CSF has been referred to as Liquor Cotunni in his honor. CSF is formed in a central portion of the brain called the *choroid plexus*, and from there circulates and provides the brain and spinal cord with nutrients and carries away metabolic waste materials. Another major function involves serving as a shock absorber to provide protection to the head from sudden jerking or blows, thereby preventing the brain and skull from coming into contact; this cushion may be insufficient to guard against brain damage caused by automobile accidents or sports injuries. The CSF also provides buoyancy and supports the weight of the brain in the skull.

A major function of the cerebrospinal fluid is to serve as a shock absorber protecting the brain from injury that results from sudden jerking movements and blows to the head.

Regeneration

Aristotle (384–322 BCE), **Antonie van Leeuwenhoek** (1632–1723), **Abraham Trembley** (1710–1784), **Charles Bonnet** (1720–1793)

Stories of regeneration can be traced to Greek mythology. As punishment for the theft of fire, Zeus sentences Prometheus to be bound to a rock, where each day, an eagle feeds on his liver; the liver grows back, and the daily feedings continue. Another, the second of the Twelve Labors of Hercules, involves slaying the Lernaean hydra—each time a head is decapitated, it is replaced with two more. Less dramatic reports of regenerating lizard tails appear in the writings of ancient Greek scientists, including Aristotle.

Until the eighteenth century, biologists were largely content with observing and cataloguing the natural world. The Swiss naturalist Abraham Trembley was likely the first experimental biologist to manipulate living organisms and observe the consequences. While serving as a tutor for the children of a distinguished Dutch family, Trembley discovered polyps (*Chlorohydra viridissima*) in a fresh water pond. He found them intriguing when observing that the number of arms on individual organisms varied in number. When he cut the polyp in half, it regenerated into two fully complete individuals, and when cut into multiple pieces, multiple individuals materialized. After creating a seven-headed polyp, he called it a *hydra*, after the Greek mythological creature. In other experiments, he grafted two individuals together and a fused, single individual resulted. He detailed these experiments and his findings in a 1744 book. Trembley initially believed that polyps were plants, but seeing their movement caused him to revise this assessment. When he first found these strange creatures, Trembley was unaware of their earlier description by Antonie van Leeuwenhoek in 1702–1703, as one of his "animalculum."

Although Trembley's findings were lauded by much of the scientific community, they were not universally embraced. The ability of the dissected hydra to be regenerated into a complete replica of the original organism was contrary to the prevailing precept of preformation, that is, the embryo develops from pre-existing parts. Among the early doubters was Trembley's cousin Charles Bonnet, he, too, a Swiss naturalist. But, Bonnet became convinced in 1745, when he observed similar regeneration in worms.

SEE ALSO: Leeuwenhoek's Microscopic World (1674), Cell Theory (1838), Embryonic Induction (1924).

The hydra was a mythological animal with legendary powers of regeneration. Hercules and the Hydra is a c. 1475 painting by Antonio del Pollaiolo (c. 1429–1498), an Italian painter and sculptor.

Theories of Germination

Aristotle (384–322 BCE), **William Harvey** (1578–1657),
Casper Friedrich Wolff (1733–1794)

The development of the embryo, or germination, remained the subject of debate for almost two millennia from the time of Aristotle until the eighteenth century. Aristotle proposed two possible alternatives: preformation and epigenesis, with each having formidable advocates.

Preformation was based on a Biblical interpretation of creation. At the time of conception, the embryo had a complete set of organs that were too small to be seen and that were located in either the mother's egg or the father's semen. It was believed that during the process of development, each body part increased in size. In the seventeenth century, it was further proposed that the preformed germs of all plants and animals originated within the original parents of each species (i.e., Adam and Eve) and, hence, no new living beings were being created *de novo*. Preformation was the prevailing view from about 1675 until end of the eighteenth century.

Aristotle favored the theory of epigenesis: each individual began entirely new as an undifferentiated mass in the egg and gradually differentiated and grew, with male semen providing the form or soul that guided this developmental process. Although supported by William Harvey, epigenesis gained little traction during the seventeenth century.

The German physiologist and embryologist Casper Friedrich Wolff revitalized the theory of epigenesis and became its leading advocate. Microscopically studying the chick embryo, he saw no evidence supporting the notion that a preformed miniature becomes enlarged. Instead, he saw continuous growth and gradual development of the chick. In his 1759 doctoral dissertation *Theoria Generationis*, Wolff described that organs of the body did not exist at the beginning of the generation process but formed from undifferentiated matter through a series of incremental steps. To bolster his arguments, he also showed that a plant root, despite its differentiated tissues, is capable of regenerating a new plant even after the stem and roots are removed. Wolff's vigorous support of epigenesis and rejection of preformation only generated controversy about his theory in the scientific community, and damaged his personal career. His findings were later validated and served as the foundation for the **germ-layer theory** in 1828.

SEE ALSO: Spermatozoa (1677), Germ-Layer Theory of Development (1828), Embryonic Induction (1924).

A nine-week (seven weeks post ovulation) human embryo obtained from an ectopic pregnancy. In obstetrical practice, pregnancy is dated from the first day of the last menstrual period, which is about two weeks prior to the ovulation.

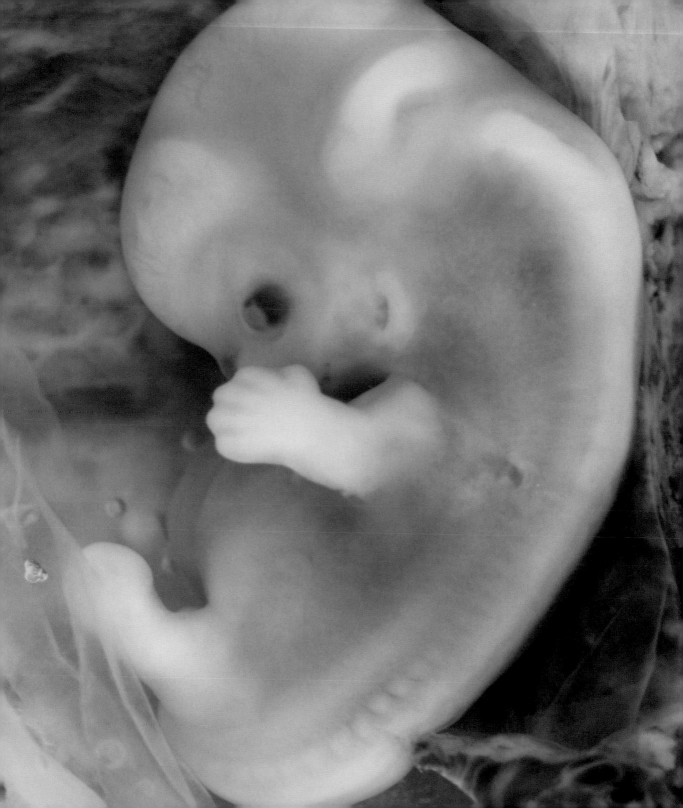

Artificial Selection (Selective Breeding)

Abu Rayhan Biruni (973–1048), **Robert Bakewell** (1725–1795),
Charles Darwin (1809–1882)

A fundamental building block Charles Darwin used in conceptualizing his **theory of natural selection** was selective breeding, and he specifically cited the pioneering work of Robert Bakewell in this field. Darwin noted that many domesticated animals and plants were developed by intentionally breeding individuals with special prized traits.

Selective breeding, a term coined by Darwin, was practiced by the Romans 2,000 years ago and was described by the Persian polymath Abu Rayhan Biruni during the eleventh century. However, it was Bakewell, a leading figure during the British Agricultural Revolution, who introduced it on a scientific basis. Bakewell was born into a family of English tenant farmers and spent his early years traveling on the Continent, learning farming methods. Upon his father's death in 1760, he took control of the farm and transformed its grasslands for cattle grazing by using his innovative breeding techniques, irrigation, flooding, and fertilizing pasturelands. He then turned his attention to livestock and, through selective breeding, produced the New Leichester sheep lineage. Characterized as large and fine boned, this breed's long lustrous wool was extensively exported to North America and Australia. Today, Bakewell's legacy is not his breeds but his breeding methods.

Desirable traits are specific to the species being bred, and individual members are crossbred to obtain a hybridized product with these characteristics. Plants are commonly bred for high crop yields, a fast growth rate, and resistance to disease and negative climatic conditions. For chickens, breeding objectives might include the quality and size of the eggs, the meat, and the production of young birds likely to successfully reproduce. Aquaculture, involving fish and shellfish, has yet to achieve its full potential. Breeding objectives include an increase in growth and survival rates, meat quality, and resistance to disease, and for shellfish, also shell size and color.

SEE ALSO: Wheat: The Staff of Life (c. 11,000 BCE), Agriculture (c. 10,000 BCE), Domestication of Animals (c. 10,000 BCE), Rice Cultivation (c. 7000 BCE), Darwin's Theory of Natural Selection (1859), Mendelian Inheritance (1866), Genetically Modified Crops (1982).

A champion bull is being led in the ring at an agricultural show in Scotland, perhaps anticipating another blue ribbon to add to his collection.

Animal Electricity

Luigi Galvani (1737–1798), **Alessandro Volta** (1745–1827),
Giovanni Aldini (1762–1834)

One can imagine the shock that Luigi Galvani experienced in 1786 when seeing a dead frog's leg violently contracting. Galvani had seen such contractions numerous times before, since, for the past decade, he had been conducting experiments using electricity to study physiological responses of the frog. But, on this day, his assistant used a metal scalpel to touch the exposed nerve of a *dissected* leg, which was placed on a table after being used in an earlier experiment. In a subsequent experiment, Galvani observed that when two different metals were placed in contact with a nerve or a muscle, an electric current was generated and the muscle contracted.

Galvani, an Italian physician, anatomist, and physiologist, was a graduate and now respected faculty member at the University of Bologna's medical school. On the basis of his experiments, he concluded that nerves and muscles contained an electrical fluid, which was analogous to electricity, and dubbed it "animal electricity." He postulated that an electrical field was generated in the brain that passed in the blood from the nerves to muscles and was responsible for the contractions. Italian physics professor Alessandro Volta's initial enthusiasm for Galvani's animal electricity turned to skepticism and then outright opposition. Volta, who taught at the University of Pavia, accepted the results but (correctly) rejected the notion of animal electricity. He alternatively explained that when two different metals come in contact, a current is generated. He called this metallic electricity *Galvanism*.

Both men influenced the future course of science and perhaps literature. Galvani's experiments were among the first in electrophysiology (bioelectricity), the electrical properties of living cells, and Volta's studies led to a voltaic pile, an early battery. The terms *galvanize* and the electrical unit *volt* are named in their honor. Among Galvani's staunch supporters was his nephew Giovanni Aldini, a physics professor. Aldini continued his uncle's experiments and in 1803 performed a widely publicized public demonstration of electrical stimulation of a deceased criminal's limb. Although Mary Shelley left no record of this connection, it has been suggested that her *Frankenstein* (1818) was inspired by Aldini's attempts at human reanimation.

SEE ALSO: Nervous System Communication (1791), Action Potential (1939).

An illustration of Dr. Frankenstein's (unnamed) monster, from an 1831 edition of the work. The monster was animated by a powerful electrical current.

T. Holst, del.   W. Chevalier, sculp.

FRANKENSTEIN.

*"By the glimmer of the half-extinguished
light, I saw the dull, yellow eye of the
creature open; it breathed hard, and a
convulsive motion agitated its limbs.
*** I rushed out of the room."*

Page 43.

Gas Exchange

Antoine-Laurent Lavoisier (1743–1794)

In 1789, the French nobleman and chemist Antoine Lavoisier demonstrated the importance of oxygen and carbon dioxide in breathing, accomplishments that failed to spare this pioneer of modern chemistry from the guillotine's blade in 1794. Details describing the role of these gases in the metabolism of energy-rich compounds, such as carbohydrates and fats, and the chemical reactions in cellular respiration, would emerge in the twentieth century.

In all living organisms—from unicellular bacteria to mammals—the process of respiration involves an exchange of gases, in opposite directions across a respiratory surface, between the external environment and the interior of the organism. Respiration or breathing requires oxygen input and removal of carbon dioxide, the end product of **metabolism**. This gas exchange across a respiratory surface occurs by diffusion, whereby gases move downhill from regions of higher concentration to lower concentration. The nature of the process is similar but the respiratory surface differs among species.

In single-celled organisms, such as bacteria, gases readily cross their cell membrane, while in earthworms and **amphibians**, the exchange is across the skin. On the body surface of insects, spiracles open into breathing tubes called *trachea*. **Fish** must extract oxygen that is dissolved in water. As fish swim, water enters the mouth and across gills, which have very extensive surface areas for gas diffusion and are rich in capillaries that are closely spaced. As oxygen passes over the gills in one direction, carbon dioxide-containing blood flows in the opposite direction and is removed from the body. In mammals, both oxygen and carbon dioxide are transported around the body in blood, into and out of adjacent capillaries, with gas exchange across the alveoli, the tennis court-sized respiratory surface.

By contrast, in plants during **photosynthesis** (in light), there is an intake of carbon dioxide and release of oxygen. In respiration (in darkness), oxygen enters the plant and carbon dioxide is released. Gases diffuse through the stomata, pores on the underside of leaves, and across the mesophyll, the respiratory surface located on the inside of the leaf.

SEE ALSO: Fish (c. 530 Million BCE), Amphibians (c. 360 Million BCE), Metabolism (1614), Photosynthesis (1845), Enzymes (1878), Mitochondria and Cellular Respiration (1925), Energy Balance (1960).

A goldfish actively engaged in oxygen–carbon dioxide gas exchange across its gills.

Nervous System Communication

Luigi Galvani (1737–1798), **Julius Bernstein** (1839–1917)

The earliest animals spent their lives on the bottom of the sea filtering water for particles of food—not unlike present-day sponges. With that lifestyle, they neither sensed nor responded to changes in their environment and, therefore, even a rudimentary nervous system would have been superfluous. In time, diffusely distributed nerve nets evolved in jellyfish-like animals, and they could sense touch and detect chemicals but responded with their entire bodies without more specific spatial discrimination.

Then, some 550 million years ago, the hypothetical urbilaterian was said to have appeared. It exhibited bilateralism, with both sensory structures and nervous tissue concentrated at its anterior (front) end and connected to a nerve trunk that was capable of communicating with distant parts of the body. It is believed that vertebrates, worms, and insects all descended from this common ancestor, the urbilaterian, whose fossil remains have not been discovered. Over the next several hundred million years, the nervous system of descendent animals evolved to be able to coordinate body functions and act upon changes in their external environment and within their bodies.

Ancient Greeks knew that the brain influenced muscle, and they believed that information from nerves was conducted by animal spirits. By the mid-eighteenth century, scholarly attention was directed to **animal electricity**. Based on studies conducted in frogs, in 1791 the Italian physician-physicist Luigi Galvani provided definitive proof that an electrical current in nerves was responsible for muscle contractions. The German physiologist Julius Bernstein, in 1902, proposed that this flow of current within nerve cells (neurons) was based on a voltage difference inside and outside the cells resulting from an uneven distribution of charged particles.

In accordance with the **neuron doctrine**, each neuron is a discrete unit and is separated from adjacent neurons and muscles by synapses, which are physical gaps. Whereas electrical impulses are responsible for communication over long distances within the neuron, chemicals (**neurotransmitters**) carry messages over the short distances across the synapses. The release of a neurotransmitter occurs in response to electrical impulses and transmits the message to that nerve or muscle.

SEE ALSO: Medulla: The Vital Brain (c. 530 Million BCE), Animal Electricity (1786), Neuron Doctrine (1891), Neurotransmitters (1920), Action Potential (1939).

This illustration depicts the transmission of messages across a synapse (gap) between two neurons and mediated by chemicals called neurotransmitters.

Paleontology

Xenophanes (c. 570–c. 478 BCE), **Shen Kuo** (1031–1095), **Georges Cuvier** (1769–1832), **Charles Lyell** (1797–1875), **Charles Darwin** (1809–1882)

Paleontology is the study of fossils—the remains of ancient life forms whose impressions were preserved in rock. Fossils have been used to support the existence of dragons, Noah's Flood and other catastrophic events, and Darwin's theory of evolution. Based on his study of fossil seashells, the ancient Greek philosopher Xenophanes argued that the dry land on which they were found was once under sea. Shen Kuo, an eleventh-century Chinese naturalist, postulated a theory of climate change based on finding petrified bamboo.

The 1700s was a period in which fossils were first actively collected and classified. Toward the end of that century, the French naturalist and zoologist Georges Cuvier recognized that fossils were remnants of living forms from the past, and the fossil record was based on the sequence in which fossils built up in strata (sedimentary rock layers). Cuvier noted that the fossils in the older strata were more dissimilar to current life forms and while some species disappeared and became extinct, new ones appeared. In 1796, in one of the earliest papers in paleontology, Cuvier reported, based on living and fossil skeletal remains, that African and Indian elephants were different species and that the mastodon was even different from the others. He also concluded that large **reptiles** lived prior to **mammals**, based on the strata in which the fossils were found.

Cuvier actively opposed pre-Darwinian theories of evolution, and consistent with Biblical teachings and the story of the Great Flood, he argued that a single catastrophic event destroyed life and was replaced by existing life forms. In his extremely successful and influential three-volume *Principles of Geology* (first published in 1830–1833), Charles Lyell, a Scottish lawyer-turned geologist, challenged this view and gained public acceptance of uniformitarianism, namely, that changes in the Earth occurred by natural, imperceptibly gradual processes. This concept had a profound impact on Charles Darwin, who collected scores of fossils during his five-year voyage on the *Beagle* and, after reading Lydell's works, came to view evolution as being biological uniformitarianism, which occurred over generations but too slowly to be perceived.

SEE ALSO: Reptiles (c. 320 Million BCE), Mammals (c. 200 Million BCE), Darwin and the Voyages of the *Beagle* (1831), Fossil Record and Evolution (1836), Darwin's Theory of Natural Selection (1859).

This fossilized tooth of a mako shark was washed from Miocene cliffs and found along the Potomac River in Westmoreland State Park in Virginia. The Miocene epoch dates from approximately 23 million to 5 million years ago.

Population Growth and Food Supply

William Godwin (1756–1836), **Thomas Malthus** (1766–1834), **Charles Darwin** (1809–1882), **Alfred Russel Wallace** (1823–1913)

During the latter years of the eighteenth century, English reformers such as William Godwin and his fellow Utopians foresaw a virtually limitless improvement of societal life, where population growth would produce more workers, leading to greater national wealth, prosperity, and a higher quality of life for all. However, one dissonant voice foresaw dire consequences arising from an unbridled expansion of the population. An essay appeared in 1798 predicting that the expanding population, especially among the lower socioeconomic classes, would exceed the available supply of food by the middle of the nineteenth century.

The author of "An Essay on the Principle of Population" was the thirty-two-year-old English political economist and demographer Reverend Thomas Malthus, who was fascinated (one might say obsessed) with all aspects of populations—including births, deaths, and the ages at which marriage and childbirth occurred. While the food supply was increasing arithmetically (1, 2, 3, 4, 5 . . .), Malthus projected that the human population was expanding at a geometric rate (2, 4, 8, 16, 32 . . .), which, if not brought under control, would, in short order, lead to poverty and starvation. He advocated "preventive checks" to reduce the birth rate including marrying at a later age, birth control, and abstaining from procreation. Failing these, "positive checks," such as disease, war, disasters, and starvation, would augment the death rate. His essay's popularity was exemplified by the appearance of a final sixth edition in 1826, but happily his predictions never materialized, as they failed to anticipate the coming of the agricultural revolution.

Malthus's writing was, however, the spark that independently influenced both Charles Darwin and Alfred Russel Wallace in their respective evolving theories of natural selection that were presented some twenty years later. Darwin, as he acknowledged in his autobiography, transferred Malthus's concepts from humans to the natural world. He recognized that all living species routinely over-reproduce and that, in a challenging environment, only some species and individuals possess a trait that enables them to have a selective advantage to survive, reproduce, and transmit this trait to their offspring.

SEE ALSO: Darwin's Theory of Natural Selection (1859), Factors Affecting Population Growth (1935), Green Revolution (1945).

This illustration from the time of the Great Famine of 1876–78 depicts "The Famine in India: Natives Waiting for Relief at Bangalore," the original caption from the Illustrated London News *(1877).*

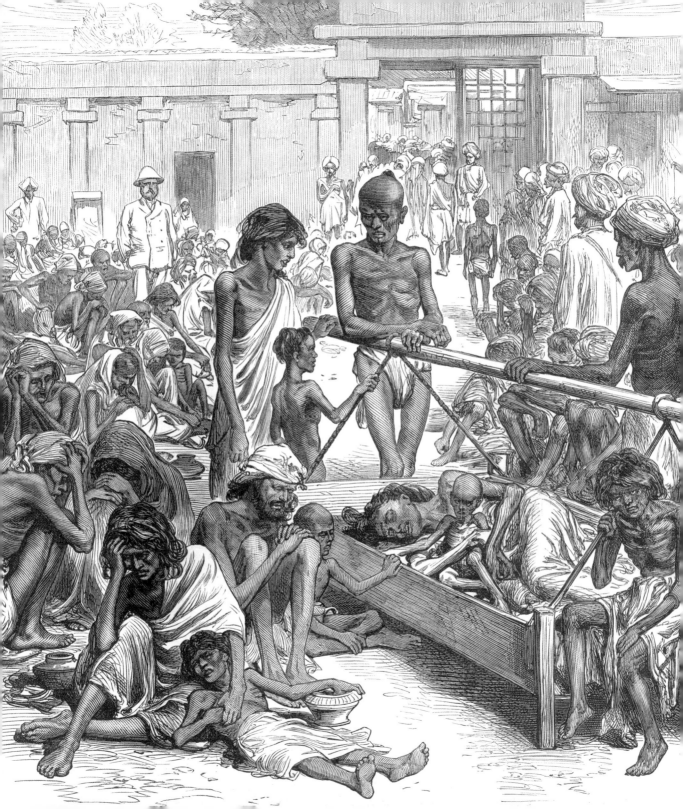

Lamarckian Inheritance

Erasmus Darwin (1731–1802), Jean-Baptiste Lamarck (1744–1829)

From the time of ancient Greeks to the early Christian era, concepts of evolution were openly discussed. This dialog ended during the Middle Ages and was replaced by the dogma of Biblical scriptures, including the fixed and unchanging nature of living organisms since creation. With the discovery and mounting accumulation of fossils in the 1700s, a number of prominent naturalists began questioning whether life-forms were fixed since creation or had evolved.

Although the theory of the inheritance of acquired characteristics is inextricably linked to Lamarck, it was first explored by the ancient Greeks and expanded by a leading eighteenth-century intellectual, Erasmus Darwin, grandfather of Charles, in his two-volume *Zoonomia* (1794–1796). Here, the Earth's age was described in millions of years, in contrast to Irish Bishop Ussher's 1654 calculation that creation was in 4004 BCE.

Jean-Baptiste Lamarck was a soldier in the French army, an acclaimed botanist, and, in his time, the foremost expert on invertebrates (a term he coined). In his most famous work, *Philosophie zoologique* (1809), he argued that rather than living beings evolving from a series of successive catastrophies and recreations, there were gradual alterations. He theorized that as the environment changed, organisms needed to change to survive. If a body part was used more than previously, that part would increase in size or strength during its lifetime with the enhancement transmitted to the offspring. So, for example, if a giraffe stretched its neck to reach higher leaves, its neck would increase in length. Its progeny would inherit the longer neck and, with continued stretching, would make it still longer over generations. Similarly, he contended that wading birds would have evolved long legs by stretching them to remove their bodies from water. By contrast, disused body parts would shrink and disappear, thus explaining how snakes lost their legs.

Long before his death, Lamarck's theory was challenged and rejected by religious and scientific communities; he died blind, in poverty, and seemingly unremembered. More recently, Lamarckism has been reexamined in terms of **epigenetics**, where traits are inherited by mechanisms not involving genes.

SEE ALSO: Paleontology (1796), Fossil Record and Evolution (1836), Darwin's Theory of Natural Selection (1859), Germ Plasm Theory of Heredity (1883), Epigenetics (2012).

Lamarck used the giraffe as a prime example of acquired physical characteristics being transmitted from one generation to the next. He believed that the giraffe's long neck resulted from its ancestors increasingly stretching their necks while trying to reach the uppermost leaves of a tree.

Germ-Layer Theory of Development

Karl Ernst von Baer (1792–1876), **Christian Heinrich Pander** (1794–1865),
Robert Remak (1815–1865), **Hans Spemann** (1869–1941)

Casper Friedrich Wolff provided evidence supporting the epigenetic theory of generation—namely that, after conception, each individual begins as an undifferentiated mass in the egg and gradually differentiates and grows. Wolff's theory (1759) was largely disregarded by the scientific community; however, during the following century, it was revisited and served as the foundation for the germ-layer theory.

In 1815, the Estonian-born Karl Ernst von Baer attended the University of Würzburg, where he was introduced to the new field of embryology. His anatomy professor encouraged him to pursue research on chick embryo development but, unable to pay for the eggs or hiring an attendant to watch the incubators, he turned the project over to his more-affluent friend Christian Heinrich Pander, who identified three distinct regions in the chick embryo.

Von Baer extended Pander's findings in 1828 to show that in all vertebrate embryos, there are three concentric germ layers. In 1842, the Polish-German embryologist Robert Remak provided microscopic evidence for the existence of these layers and designated them by names still in use. The ectoderm or outermost layer develops into the skin and nerves; from the endoderm, the innermost layer, comes the digestive system and lungs; and between these layers, the mesoderm, is derived blood, heart, kidneys, gonads, bones, and connective tissues. It was subsequently determined that while all vertebrates exhibit bilateral symmetry and have three germ layers, animals that display radial symmetry (hydra and sea anemone) have two layers, while only the sponge has a single germ layer.

Von Baer proposed other principles in embryology: General features of a large group of animals appear earlier than the specialized features seen in a smaller group. All vertebrates begin development with skin that differentiates to scales in fish and reptiles, feathers in birds, and hair and fur in mammals. In 1924, Hans Spemann's discovery of **embryonic induction** explained how groups of cells form particular tissues and organs.

SEE ALSO: Theories of Germination (1759), Embryonic Induction (1924), Induced Pluripotent Stem Cells (2006).

Eggs, such as this one-week-old chicken egg, are candled to observe the development of a chick or duck embryo and veins. Candling is performed in a darkened room with the egg perched on a light.

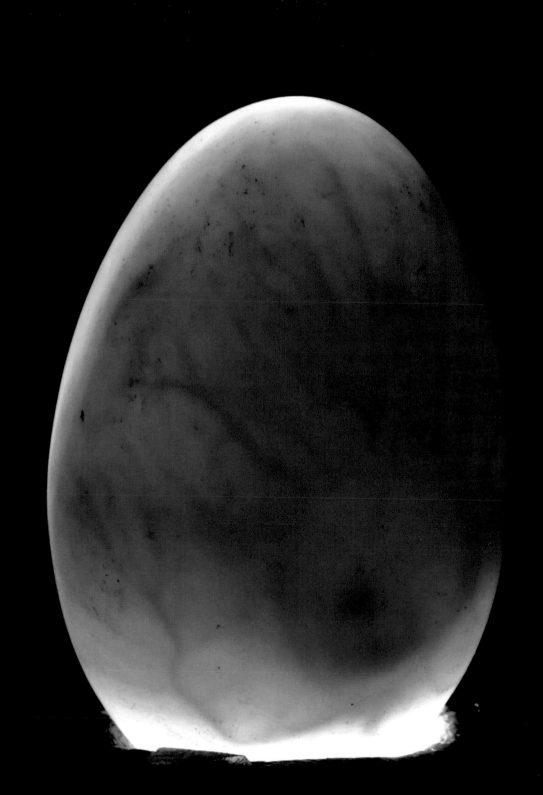

Cell Nucleus

Antonie van Leeuwenhoek (1632–1723), **Franz Bauer** (1758–1840), **Robert Brown** (1773–1858), **Matthias Schleiden** (1804–1881), **Oscar Hertwig** (1849–1922), **Albert Einstein** (1879–1955)

During the 1670s, the Dutch microscopist Antonie van Leeuwenhoek was the first to see a world previously unknown, which included the fibers of a muscle, bacteria, sperm cells, and the nucleus in the red **blood cell** of salmon. The next reported sighting of a cell nucleus was in 1802 by Franz Bauer, an Austrian microscopist and botanical artist. However, credit for its discovery is generally assigned to the Scottish botanist Robert Brown. When studying the epidermis (outer layer) of an orchid, he saw an opaque spot that was also present during an early stage of pollen formation; he called this spot a *nucleus*. Brown first described its appearance to his colleagues at a meeting of the Linnean Society of London in 1831 and published his findings two years later. Both Brown and Bauer thought that the nucleus was a cell structure that was unique to monocots, a plant group that includes orchids. In 1838, the German botanist Matthias Schleiden, the co-discoverer of the **cell theory**, first recognized the connection between the nucleus and cell division, and in 1877, Oscar Hertwig demonstrated its role in fertilization of the egg.

CARRIER OF GENETIC MATERIAL. The nucleus, the largest organelle within the cell, contains chromosomes and **deoxyribonucleic acid (DNA)**, and regulates cell **metabolism**, cell division, gene expression, and protein synthesis. The nuclear envelope—a double membrane surrounding the nucleus and separating it from the rest of the cell—is in continuity with the rough endoplasmic reticulum, the site of protein synthesis.

At the time of his 1831 discovery, Brown was an established botanist. Earlier in his career, from 1801–1805, he collected 3,400 species of plants while in Australia and described and published reports of 1,200 of these. In 1827, he reported on microscopic pollen grains (and later other particles) moving continuously and randomly through a liquid or gas medium colliding with one another. An explanation of this Brownian motion came in 1905, when Albert Einstein explained that it resulted from molecules of water that were not visible hitting visible pollen grain molecules.

SEE ALSO: Metabolism (1614), Blood Cells (1658), Leeuwenhoek's Microscopic World (1674), Spermatozoa (1677), Cell Theory (1838), Deoxyribonucleic Acid (DNA) (1869), Genes on Chromosomes (1910), DNA as Carrier of Genetic Information (1944), Ribosomes (1955), Cell Cycle Checkpoints (1970).

Interior three-dimensional image of an animal cell, with the nucleus as the large, round organelle in the center.

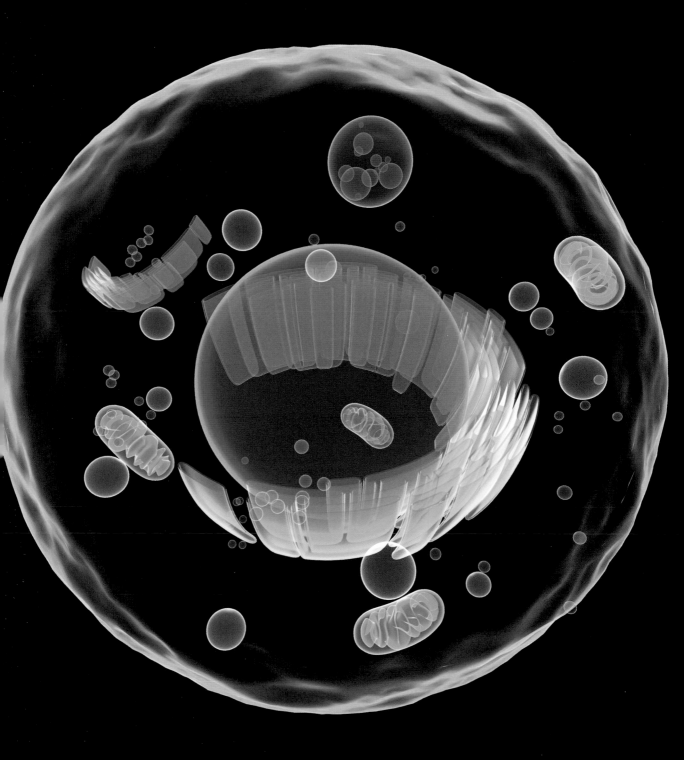

Darwin and the Voyages of the *Beagle*

Charles Darwin (1809–1882)

There is little to suggest that prior to 1859, Charles Darwin would rank among the most important biologists, and that his *Origin of Species* (1859) perhaps would be the most significant book written on science. His father was a financially and socially successful physician, and his mother was the daughter of Josiah Wedgwood, founder of the pottery company bearing his family name. Charles's grandfather was Erasmus Darwin, a distinguished eighteenth-century intellectual. Neither his year of medical studies nor his bachelor's studies at Cambridge were marked with distinction. His time was spent exploring nature and hunting.

Captain Robert FitzRoy was looking for a "gentleman passenger" who could serve as a recorder and collector of biological samples on a five-year voyage of the *HMS Beagle* that was intended to circumnavigate the globe, with emphasis on charting the South American coastline. The twenty-two-year-old Darwin was selected for this unpaid position because of his keen interest in the natural sciences but, as important, he could serve as a socially equal companion to the captain who was but four years his senior. When Darwin set sail in 1831, he shared the belief of most Europeans in the divine creation of the world and the unchanging nature of its inhabitants.

When not seasick, Darwin was diligently observing and collecting animals, marine invertebrates, insects, and fossils of extinct animals. He also experienced an earthquake in Chile. The most memorable segment of his voyage was the five weeks he spent on the Galápagos Islands, ten volcanic islands some 600 miles (1,000 kilometers) west of Ecuador. Among his many collectables were four mockingbirds caught on four islands; he noted that each was different. He also brought back to England fourteen finches whose beaks differed in size and shape. When Darwin returned to England in 1835, he was a well-recognized naturalist, a reputation enhanced by his presentations, papers, and a popular work entitled *Journal of Researches* (renamed *The Voyage of the Beagle*).

SEE ALSO: Fossil Record and Evolution (1836), Darwin's Theory of Natural Selection (1859), Continental Drift (1912).

Topographical and bathymetric map of the Galápagos Islands, located west of Ecuador, where Darwin found fourteen finches whose beaks were different in size and shape—an observation that proved to be a major building block in his theory of natural selection (1859).

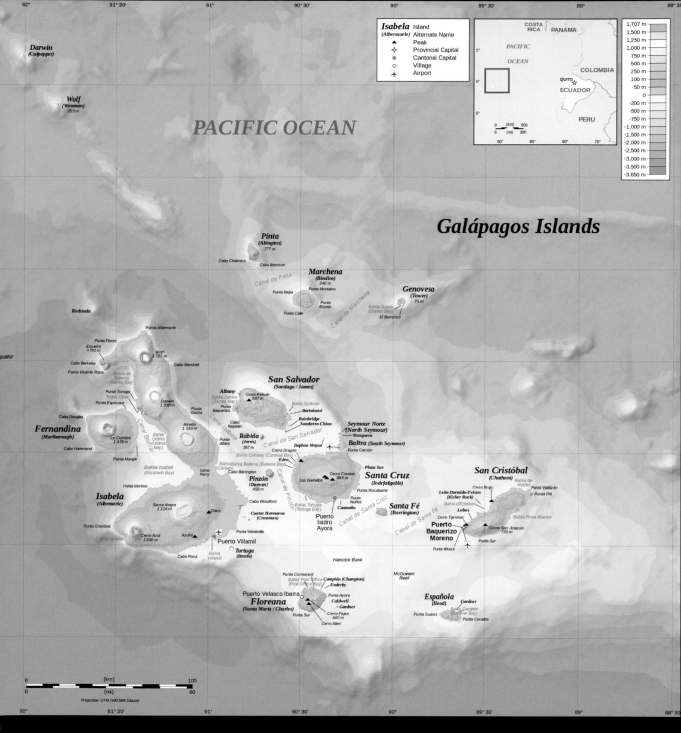

Anatomy Act of 1832

Robert Knox (1791–1862), William Burke (1792–1829)

William Burke and William Hare were vilified in 1828, and yet they were instrumental in advancing instruction in anatomy for generations of studies. Until 1832, schools of medicine and anatomy in England and Scotland were deficient in their major instructional tool: cadavers for dissection and anatomical instructional demonstrations. Notwithstanding Church protestations that an intact body was requisite to entering heaven, the Parliament enacted the Murder Act of 1751. Medical schools were allocated one body per year for purposes of dissection, with that body restricted to murderers who were condemned to execution and dissection. The supply remained grossly inadequate.

In 1810, an Anatomy Society was established, with member anatomists, surgeons, and physiologists, virtually all of whom were of the upper social strata and wealthy. They proposed being given access to unclaimed cadavers of paupers from workhouses. The outcry from the poor prevailed, and the proposal was rejected. To meet the still-unfulfilled need, anatomists resorted to purchasing bodies from "resurrectionists," without questioning their source. Body snatching became so common that family members routinely stood watch over the graves of loved ones after interment, with a watchful eye for gravediggers.

Burke and Hare devised a simpler and more direct method of obtaining their cadavers. Operating in Edinburgh, they murdered sixteen individuals and sold their bodies to Dr. Robert Knox, a highly respected surgeon and anatomist, who was reported to be the most popular private lecturer of anatomy in that city. After being apprehended, Hare was given an opportunity "to turn King's evidence" against his partner in murder and was granted immunity from prosecution. Burke was sentenced to hanging and dissection, and his skeleton is currently on display at the University of Edinburgh's Anatomy Museum. In his confession, Burke swore that Knox was unaware of the source of the bodies, and, although never prosecuted, Knox's prosperous career in Edinburgh unceremoniously ended. The notoriety associated with the murders led to the passage of the Anatomy Act of 1832, which enabled doctors, anatomy teachers, and medical students to obtain and study cadavers that were legally donated or authorized by government authorities.

SEE ALSO: Leonardo's Human Anatomy (1489), Vesalius's *De humani corporis fabrica* (1543).

This painting of body snatchers hangs on the wall of the Old Crown Inn in Penicuik, Scotland—an inn that Burke and Hare allegedly frequented.

Human Digestion

William Beaumont (1785–1853), **Alexis St. Martin** (1802–1880)

In 1822, Alexis St. Martin, a nineteen-year-old French Canadian employed by the American Fur Company to transport furs by canoe, was accidentally shot in his abdomen at close range by a musket loaded with birdshot. He was treated by William Beaumont, a United States Army surgeon posted at Fort Mackinac, on Mackinac Island, the site of several battles to control the Great Lakes during the War of 1812.

Beaumont treated his patient's wounds but frankly expressed little hope that he would survive for more than thirty-six hours. St. Martin did survive but was left with a fistula (hole) in his stomach the size of a man's finger, an opening which left to its own could never heal. Beaumont, who was licensed to practice medicine after completing a two-year apprenticeship (which was not unusual at the time), saw in St. Martin's fistula a unique opportunity to observe and study human digestion in the stomach for the first time.

He began his digestive studies several years after the accident. In a typical experiment, pieces of food attached to a string were placed in the hole in the stomach, and its digestion was observed. Samples of gastric (stomach) acid were removed, and this led Beaumont to conclude that chemical processes (and not mechanical factors) were responsible for breaking down food into nutrients. In 1833, Beaumont published the results of some 240 experiments he conducted on St. Martin in a 280-page book, *Experiments and Observations on the Gastric Juice and the Physiology of Digestion*.

Doctor and patient parted company for the last time in 1833. Beaumont died in 1853 at the age of sixty-seven after he slipped on ice and sustained a severe head injury while leaving a patient's home. Contrary to expectations, St. Martin lived fifty-eight years after his accident and outlived his physician by over a decade, notwithstanding heavy drinking in his later years. Over the years, an unanswered question was whether Beaumont chose not to perform the simple operative procedure of closing the fistula so that he could continue his experiments on digestion.

SEE ALSO: Enzymes (1878), Associative Learning (1897), Secretin: The First Hormone (1902)

Anatomy of the human stomach, schematically represented with gears and cogs working to digest food with gastric juices.

Fossil Record and Evolution

Georges Cuvier (1769–1832), **Richard Owen** (1804–1892),
Charles Darwin (1809–1882)

Prior to the nineteenth century, uncovered fossilized skeletal remains appeared to differ rather abruptly and dramatically in form and without apparent intermediate transitions. This was widely interpreted as support of creationism and the view that no animal species had ever become extinct. When Cuvier studied fossilized mammalian skeletons in 1796, he rejected the concept of evolution. By contrast, analogous fossilized skeletons were one of the major linchpins Darwin used when formulating his theory of evolution.

Georges Cuvier, the great French naturalist-zoologist, combined his knowledge of **paleontology** with his expertise in comparative anatomy when comparing the fossilized remains of mammals with their living counterparts. In 1796, Cuvier presented two papers; one comparing living elephants with extinct mammoths and, in the other, the giant sloth and the extinct Megatherium found in Paraguay. His findings and many of the geological features of the earth, he believed, could best be explained by several catastrophic events, causing the extinction of many animal species and followed by successive creations. He was a major proponent of catastrophism and highly critical of evolution.

Charles Darwin's voyage on the *Beagle* in the early 1830s took him to Patagonia, where he found the fossilized remains of mastodons, Megatheria, horses, and the large armadillo-like Glyptodons. Upon returning to England in 1836, Darwin took the fossils and his detailed notes to anatomist Richard Owen. Owen determined that these remains were more closely related to living mammals in South America than anywhere else. (He later rejected **Darwin's theory of natural selection**.) In his *Origin of Species* (1859), Darwin noted the importance of these fossils and acknowledged that while "missing links" or transitional forms between the fossilized and living forms might never be found, and represented the greatest objection to his conclusions, nevertheless, the evidence strongly supported his theory of evolution. In 2012, a collection of 314 fossil slides collected by Darwin and his peers were rediscovered in a corner of the British Geological Survey, after being lost for more than 150 years.

SEE ALSO: Devonian Period (c. 417 Million BCE), Paleontology (1796), Darwin and the Voyages of the *Beagle* (1831), Darwin's Theory of Natural Selection (1859), Radiometric Dating (1907), Continental Drift (1912), De-Extinction (2013).

The first discoveries of fossil remains of extinct mammals in the 1790s challenged support for the concept that living organisms were unchanged since the time of creation. This image is of an ammonite, an extinct marine invertebrate classified as a mollusk, whose name was inspired by tightly coiled rams' horns.

Nitrogen Cycle and Plant Chemistry

Jean-Baptiste Boussingault (1802–1887), **Hermann Hellriegel** (1831–1895), **Martinus Beijerinck** (1851–1931)

Discovered in 1772, nitrogen constitutes some 78 percent of the earth's atmosphere—four times that of oxygen—and is an essential component of amino acids, proteins, and nucleic acids. Through a series of mutually beneficial interrelationships, nitrogen in decomposing plant and animal material is made available as a soluble plant nutrient and then converted to a gaseous form and returned to the atmosphere.

That nitrogen must be reduced (fixed before use) by plants or animals was determined by the French agricultural chemist Jean-Baptiste Boussingault. From 1834 to 1876, at his farm in Alsace, France, he established the world's first agricultural research station, applying chemical experimental methods to the fields. Boussingault also determined the nature of nitrogen's movement between plants, animals, and the physical environment, and studied such related problems as soil fertilization, crop rotation, plant and soil fixation of nitrogen, ammonia in rainwater, and nitrification.

In 1837, Boussingault disproved the general belief that plants absorbed nitrogen directly from the atmosphere and showed that they did so from the soil as nitrates. The following year, he discovered that nitrogen was essential for both plants and animals, and that both herbivores and carnivores obtain their nitrogen from plants. His chemical findings laid the foundation for our current understanding of the nitrogen cycle.

In 1888, the German agricultural chemist Hermann Hellriegel and the Dutch botanist and microbiologist Martinus Beijerinck independently discovered the mechanism by which leguminous plants utilize atmospheric nitrogen (N_2) and soil microbes convert it to ammonia (NH_3), nitrates (NO_3), and nitrites (NO_2). Symbiotic (mutualistic) nitrogen-fixing bacteria, such as *Rhizobium*, acting in plants of the legume family—including soybeans, alfalfa, kudzu, peas, beans, and peanuts—enter the root hairs of the root system of the plant, multiply, and stimulate the formation of root nodules. Within the nodules, the bacteria convert nitrogen to nitrates, which are utilized for growth by the legumes. When the plant dies, the fixed nitrogen is released, making it available for use by other plants, and thereby fertilizing the soil.

SEE ALSO: Prokaryotes (c. 3.9 Billion BCE), Land Plants (c. 450 Million BCE), Agriculture (c, 10,000 BCE), Plant Nutrition (1840), Ecological Interactions (1859).

This World War II poster promotes the harvesting of legumes, which provide a food source and utilize atmospheric nitrogen to fertilize the soil.

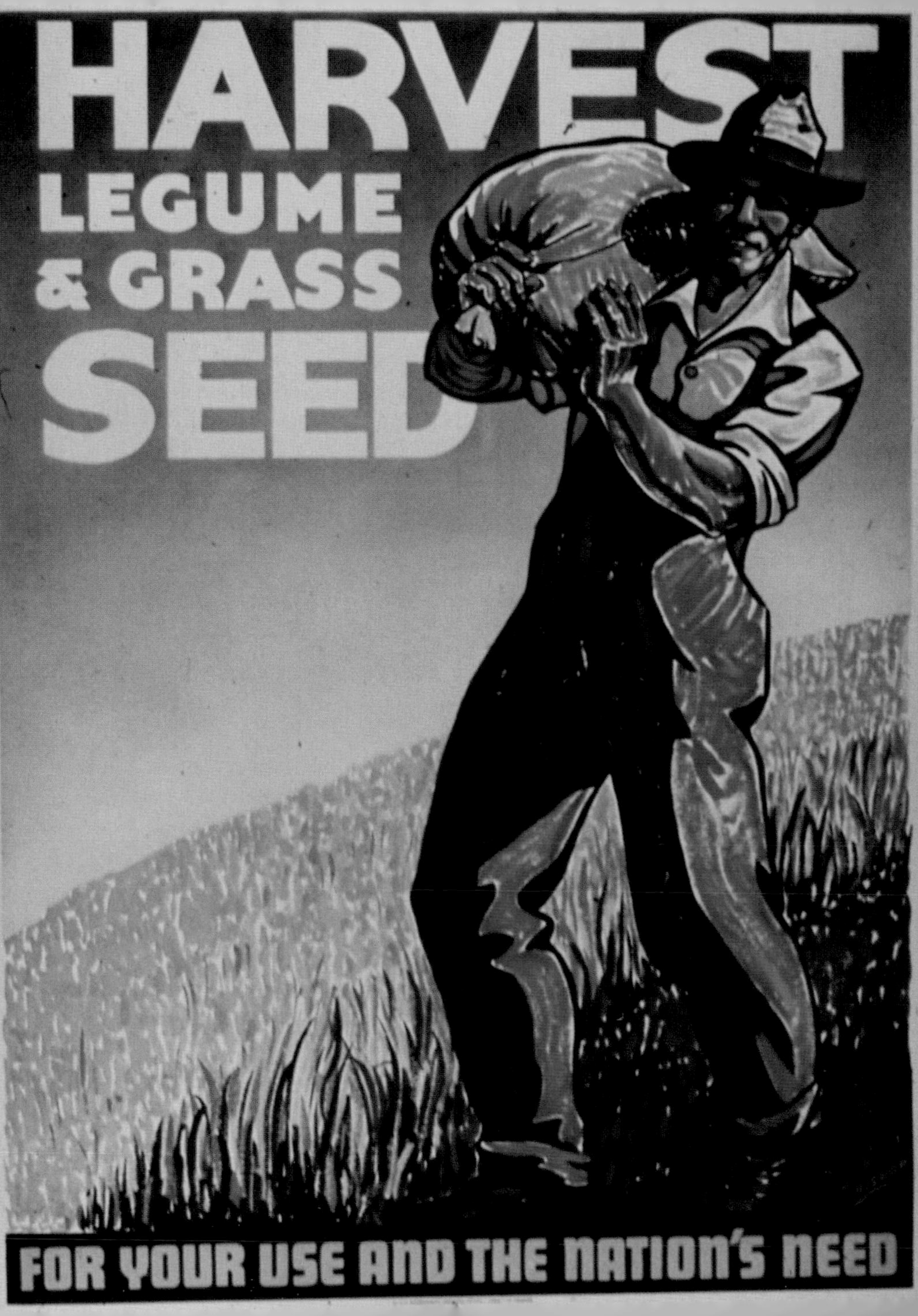
HARVEST
LEGUME
& GRASS
SEED
FOR YOUR USE AND THE NATION'S NEED

Cell Theory

Antonie van Leeuwenhoek (1632–1723), **Robert Hooke** (1635–1703), **Matthias Schleiden** (1804–1881), **Theodor Schwann** (1810–1882), **Rudolf Virchow** (1821–1902)

The importance of the cell theory in biology has been likened to the atomic theory in chemistry and physics. Just as the atom is the basic unit of matter, the cell is the basic unit of life. Both are the most fundamental principles central to their respective sciences. The groundwork for formulation of the cell theory dates back to 1665, when Robert Hooke discovered the cell in a slice of cork. A decade later, Antonie van Leeuwenhoek viewed living single-cell organisms under a microscope using his handmade lenses capable of magnifying some 275 times. Over 160 years passed before the two German friends, Schleiden and Schwann, were enjoying after-dinner coffee and sharing notes regarding their cell research.

In 1838, the botanist Matthias Schleiden proposed that every structural component of plants is composed of cells, and the following year, Theodor Schwann, a zoologist, reached a similar conclusion for animals. Their original cell theory consisted of three fundamental precepts: all living organisms are composed of cells; the cell is the basic unit of the structure and function in all living organisms; and all cells arise from other pre-existing cells, added in 1855 by Rudolf Virchow.

While the first three precepts remain, the following refine and expand the theory: Cells contain hereditary information (DNA), which is transmitted from cell to cell during division. The chemical composition of all cells in a given species is basically the same; and energy flow (**metabolism** and biochemistry) occurs within cells.

Unlike Schleiden, Schwann and Virchow continued advancing the frontiers of science and medicine. Schwann discovered the sheath surrounding nerve fibers (Schwann cells), isolated pepsin, a stomach **enzyme** that breaks down proteins; and coined the term *metabolism* for chemical changes in living tissues. Virchow was a leader in modern pathology. He promoted the use of the microscope and standardizing autopsy procedures, and he also founded the field of social medicine, which seeks to understand how social and economic factors influence health and disease.

SEE ALSO: Metabolism (1614), Leeuwenhoek's Microscopic World (1674), Refuting Spontaneous Generation (1668), Meiosis (1876), Enzymes (1878), Mitochondria and Cellular Respiration (1925), DNA as Carrier of Genetic Information (1944).

The ability to inspect single-celled organisms under a microscope—such as the Gram-positive bacterium Bacillus *shown in this illustration—allowed scientists to uncover the structure of the cell, the fundamental unit of life.*

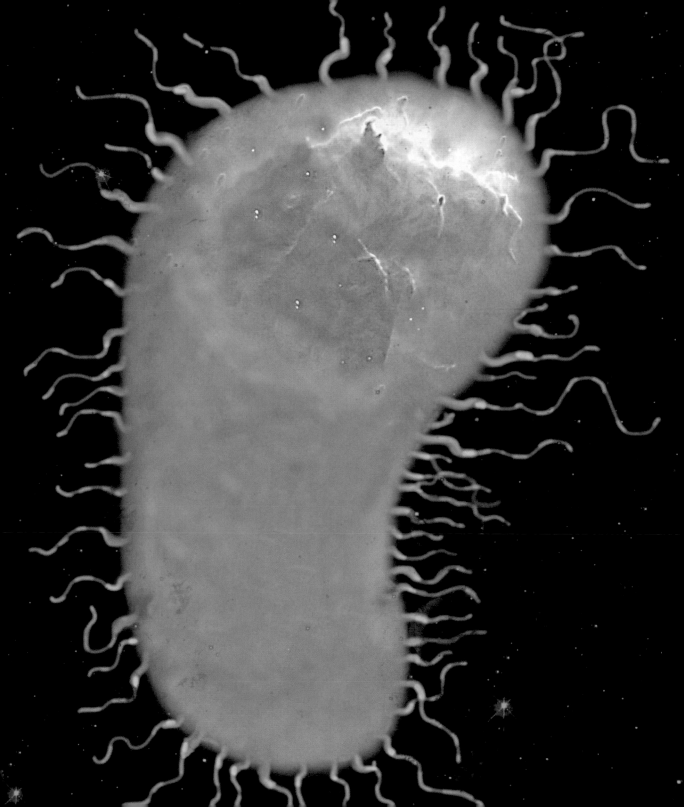

Plant Nutrition

John Woodward (1665–1728), **Nicolas-Théodore de Saussure** (1767–1845),
Justus von Liebig (1803–1873), **Julius von Sachs** (1832–1897)

In 1699, John Woodward, an English naturalist, showed that spearmint, grown in water of various states of purity, thrived best in park wastewater containing garden mold. Woodward concluded that plants were not formed from water (as Aristotle believed) but of "particles" contained in the soil. In 1804, the Swiss chemist and plant physiologist Nicolas-Théodore de Saussure, seeking to determine the nature of Woodward's "particles," observed that the increase in plant mass could not solely be attributed to its uptake of water, but also to the absorption of carbon dioxide. This discovery was one of the fundamental pieces leading to an understanding of **photosynthesis** several decades later.

The German chemist Justus von Liebig compared plant growth after the addition of various minerals to the soil. In 1840, he formulated the Law of the Minimum (often simply called "Liebig's Law"), which states that plant growth is not determined by increasing the total amount of plentiful nutrients, but rather is dependent upon the presence of the scarcest resource, which is the limiting factor to its growth. He found that plants required carbon, hydrogen, and oxygen to grow—these supplied by the air and water—as well as phosphorus, potassium, and nitrogen, which are obtained from soil minerals.

Arguably, the most important plant physiologist of the second half of the nineteenth century was the German botanist Julius von Sachs who, in about 1860, determined that six mineral elements—nitrogen, phosphorus, potassium, calcium, magnesium, and sulfur—are macronutrients needed by the plant in relatively large amounts and are used as building blocks in the plant's structure. In 1923, an additional eight essential elements were identified as micronutrients that are only required in trace quantities.

It is now known that plants use inorganic elements as nutrients, which are formed from the weathering of rock minerals and the decaying of organic matter, animals, and microbes. Some of these mineral elements are essential; that is, the plant cannot complete its life cycle in their absence and other elements cannot substitute for them when they are not present. Still other mineral elements are considered beneficial, playing a role in nonessential plant functions.

SEE ALSO: Phosphorus Cycle (1669), Nitrogen Cycle and Plant Chemistry (1837), Photosynthesis (1845), Phototropism (1880).

This 1849 painting Heinkehr vom Feld (Return from the Field) *was the work of German artist Friedrich Eduard Meyerheim (1808–1879).*

Urine Formation

William Bowman (1816–1892), Carl Ludwig (1816–1895)

A critical body function in living organisms involves balancing water intake and with its loss. This balance is determined, to a large measure, by the volume and composition of the urine, the composition of which varies and mirrors the water requirements of the organism. Freshwater animals excrete very dilute urine, while marine animals, seeking to conserve water, secrete highly concentrated urine. Depending upon their habitat, terrestrial animals generally retain water and secrete concentrated urine.

The kidney is responsible for filtering blood. In most mammals, blood plasma is filtered by the nephrons in the kidneys, with most of the water and useful materials returned to the blood stream and conserved by the body. The remaining excess water and the waste products of **metabolism**—including urea (from amino acid metabolism)—remain in the urine and are eliminated. Amphibians and fish do not retain great amounts of water and, therefore, excrete large volumes of dilute urine containing the water-soluble urea. By contrast, in most birds, reptiles, and terrestrial insects, the end product of amino acid metabolism is the water-insoluble uric acid. The urine of birds and reptiles is a white suspension of uric acid that is mixed with fecal material prior to elimination.

Based upon microscopic examination, William Bowman, English physician and histologist, studied the structure of the kidneys. In 1842, he identified the glomerular capsule (now called "Bowman's capsule") as the beginning of the nephron, the function unit of the kidney. The capsule is the keystone in Bowman's filtration theory of urine formation and the basis for our current understanding of kidney function. In 1844, Carl Ludwig proposed that **blood pressure** forced fluids out of the kidney capillaries into the nephrons. This fluid contained all the components of plasma but proteins, and water was returned to the bloodstream to concentrate the urine. Ludwig, one of the greatest physiologists, taught that the functions of living organisms were dictated by chemical and physical laws and not by special biological laws and divine influences. More specifically, he argued that urine was formed in the kidneys from a filtration process and not by vital forces, as suggested by Bowman.

SEE ALSO: Metabolism (1614), Blood Pressure (1733), Homeostasis (1854), Osmoregulation in Freshwater and Marine Fish (1930).

The nephron (shown), the basic structural and functional unit of the kidney, filters the blood and returns to the blood what is needed, excreting the remainder as urine.

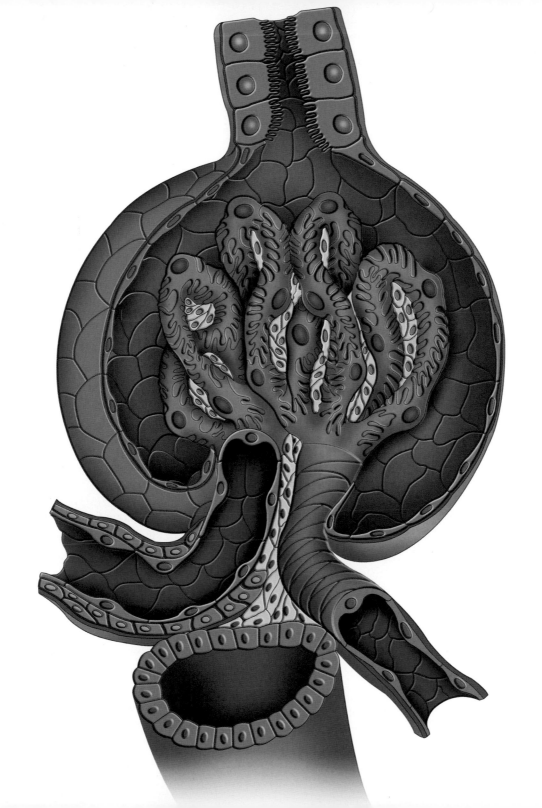

Apoptosis (Programmed Cell Death)

Carl Vogt (1817–1895), **Walther Flemming** (1843–1905), **Sydney Brenner** (b. 1927), **John Foxton Ross Kerr** (b. 1934), **John E. Sulston** (b. 1942), **H. Robert Horvitz** (b. 1947)

There is "a time to be born and a time to die . . ." Each day, the body's cells—particularly, the skin and **blood cells**, are being produced anew. Since their total number must be kept constant, a mechanism must be in place to maintain a balance and remove any redundant cells. This mechanism, programmed cell death (PCD), is an orderly, highly regulated process that functions to keep normal cell division (**mitosis**) in check. There are other conditions in which removal of cells—such as those that are old, diseased, or damaged from exposure to toxic materials or radiation—benefits the organism. During menstruation, the body sheds the lining of the uterus. By contrast, inadequate PCD can lead to the spread of cancer or, for instance, babies born with fingers still joined.

Responding to signals arising from outside and within the cell to initiate PCD, the cell undergoes a reduction in size as its components break down and condense. These cell fragments (apoptotic bodies) become enclosed in a membrane, walling them off and preventing them from damaging nearby cells. Phagocytic cells then engulf and destroy the fragments.

Carl Vogt, a German biologist working in Switzerland, was first to describe the concept of PCD in 1842 while studying the development of tadpoles. In 1885, Walther Flemming, another biologist, described this phenomenon in more precise detail; Flemming's fame rests on his discovery of mitosis and chromosomes, one of the most significant discoveries in cell biology and in all of science. Interest in PCD was revived in 1965, when the Australian pathologist John Foxton Ross Kerr first described the ultramicroscopic characteristics of PCD and how it was a normal process that differed from necrosis caused by tissue injury. It was Kerr who was first to refer to PCD as apoptosis, from the Greek "falling off," as petals or leaves. During the 1970s, John E. Sulston, H. Robert Horvitz, and Sydney Brenner, working at Cambridge University and studying the genetic sequence of roundworms, gained an understanding of apoptosis at a molecular level; the three were co-recipients of the 2002 Nobel Prize.

SEE ALSO: Blood Cells (1658), Cell Theory (1838), Homeostasis (1854), Mitosis (1882), Cell Cycle Checkpoints (1970).

Apoptosis is a normal body process intended to remove redundant cells. This three-dimensional illustration shows an apoptotic cell that will be discarded from the body because it is old, diseased, or damaged.

Venoms

Cleopatra (69–30 BCE), **Charles Lucien Bonaparte** (1803–1857)

Animals are armed with an array of weapons for use when hunting for prey or defending themselves or their brood when under attack. In addition to keen vision, claws, teeth, horns, tough exteriors, webs, and fleetness of foot or fin, some vertebrates and invertebrates have chemical weapons that they deploy offensively or defensively to combat potential predators. These chemicals are venoms, which are toxic substances directly introduced into the victim's bloodstream (envenoming) by bite, sting, or insertion of a spine or other sharp part. Among vertebrates, the best known and studied venomous animals are snakes. The first genes for venom may have evolved from lizards, the closest relative of snakes.

Of the 3,000 snake species, about 600 are venomous. Snakes use their venom to protect themselves or to directly kill or immobilize their prey. A poison gland containing venom is located in the back of the head; it is connected by a duct to a hollow fang. In addition to toxic chemicals, the venom also contains saliva, the digestive juice common to most land vertebrates. Venoms may contain twenty or more ingredients, the principal ones being neurotoxins and hemotoxins, with some venoms containing a mixture. The protein nature of venoms was first discovered in 1843 by Charles Lucien Bonaparte, nephew of Napoleon.

Neurotoxins, contained in the venom of cobras and coral snakes, affect nerves and muscles and cause paralysis at the nerve-muscle junction; death results from heart or respiratory failure. Although legend has it that Cleopatra committed suicide after sustaining the bite of an asp (Egyptian cobra), recent scholarship points to death after ingestion of a poisonous mixture. Hemotoxins, used effectively by rattlesnakes and other pit vipers, prevent blood clot formation or precipitate breakdown of existing clots, causing extensive blood loss and ensuing shock that disables the victim and its escape. Other hemotoxins cause almost immediate clotting, resulting in stroke and heart attack.

Chemicals extracted from snake venoms have been used medically to treat hypertension, stroke, and heart attack, and are being evaluated for relief of severe pain and the treatment of melanoma, diabetes, Alzheimer's, and Parkinson's.

SEE ALSO: Medulla: The Vital Brain (c. 530 Million BCE), Nervous System Communication (1791), Blood Clotting (1905).

The Eastern diamondback rattlesnake (Crotalus adamanteus) is the most dangerous venomous snake in North America, with human mortality rates after being bitten ranging from 10 to 30 percent. Its venom has proteolytic and hemotoxic properties—i.e., it damages tissues and destroys red blood cells—while promoting uncontrolled bleeding.

Homology versus Analogy

Richard Owen (1804–1892), Charles Darwin (1809–1882)

Do the human's arm, cat's limb, bat's wing, and seal's flipper have anything in common? There are no similarities with respect to their functions—lifting, walking, flying, and swimming—but careful analysis reveals commonality in their fundamental construction. Each of these mammalian forelimbs (or pentadactyls) consists of a long bone, connected to two smaller bones, linked to a number of even smaller bones, attached to approximately five digits. What about the wings of insects and **birds**? They all have the same function—flying—but they bear no structural resemblance to one another.

In 1843, the famed but highly controversial English biologist and comparative anatomist Richard Owen sought to explain why similar structures could have dissimilar functions and why similar functions could have dissimilar structures. He referred to the mammalian forelimbs as homologies—that is, the same organ has the same basic structures but different functions in various species. By contrast, the wings of insects, bats, and birds have a similar function but evolved separately by different pathways. Owen's definitions were modified by Charles Darwin to incorporate an evolutionary explanation. In homologies, the basic structure has evolved from a common ancestor to serve dissimilar functions in an adaptation to different environmental situations. In contrast to homologies, analogous structures have similar characteristics, but have evolved independently from dissimilar ancestors, to meet environmental challenges, a process referred to as *convergent evolution*.

Vestigial structures are difficult to explain in the absence of a common ancestor. Eye bulbs in blind, cave-dwelling salamanders, the human appendix, and the pelvic girdle of whales have no function in their extant species but are homologous to a functioning structure in an ancestral species.

Homologies exist at both a structural and a molecular level. The genetic code—the sequence of nucleotides in DNA and RNA that determines the order of amino acids in the biosynthesis of proteins—is nearly identical in all organisms from bacteria to humans. Similarly, common genes exist across living organisms. The universal nature of the genetic code and genes provide additional support for evolution from a common ancestor.

SEE ALSO: Birds (c. 150 Million BCE), Aristotle's *The History of Animals* (c. 330 BCE), Fossil Record and Evolution (1836), Darwin's Theory of Natural Selection (1859), Phylogenetic Systematics (1950).

The wings of birds, bats, and insects are analogous structures, which have similar functions but dissimilar frameworks that evolved independently from very different ancestors to meet a common environmental need.

Photosynthesis

Jan Ingenhousz (1730–1799), **Joseph Priestley** (1733–1804),
Julius Robert Mayer (1814–1878)

Photosynthesis is of critical importance for the survival of living organisms because it captures the energy of the sun and converts it into the chemical energy required to carry out biological processes. In its absence there would be little food or organic matter, and most organisms would cease to exist in an atmosphere devoid of oxygen. The chemical equation that summarizes the process of photosynthesis is:

$$6\ CO_2 + 12\ H_2O + Light \rightarrow C_6H_{12}O_6 + 6\ O_2$$

In the process, carbon dioxide (CO_2) from the environment enters the stomata (tiny pores) on the underside of leaves, where it is joined by water that has traveled from the roots of plants and is transported up to the leaves through vascular bundles (veins). Sunlight is absorbed by chlorophyll, a green pigment located in chloroplasts, cell structures that are the locus of photosynthesis. Photosynthesis occurs in two stages: light reactions and dark reactions. In the light reaction, sunlight is converted to chemical energy and stored in the form of adenosine triphosphate (ATP) and NADPH, a high-energy electron-carrying molecule. In the dark reaction stage, carbon dioxide, ATP and NADPH are converted to the sugar glucose ($C_6H_{12}O_6$), which is stored in plant leaves, and oxygen is released through the stomata into the environment.

Discovery of the process of photosynthesis began in 1771 with the studies by the English clergyman-scientist Joseph Priestley, who burned a candle in a closed container until the air (later found to be oxygen) within the container could no longer support combustion. Priestley then placed a sprig of mint in the container and, after several days, the candle could, once again, burn. In 1779, the Dutch physician Jan Ingenhousz repeated Priestley's experiment and showed that light and tissues from a green plant were required to restore the oxygen. The German physician-physicist Julius Robert Mayer in 1845 formulated the concept that solar energy is stored as chemical energy in organic products formed during photosynthesis. (Mayer was also the earliest to state the first law of thermodynamics dealing with the conservation of matter.)

SEE ALSO: Algae (c. 2.5 Billion BCE), Land Plants (c. 450 Million BCE), Metabolism (1614), Gas Exchange (1789), Plant Nutrition (1840), Phototropism (1880), Mitochondria and Cellular Respiration (1925).

The survival of living organisms depends upon the process of photosynthesis, which provides organic food from inorganic molecules in the presence of sunlight and oxygen. Chlorophyll, a green pigment, gives leaves their color and is critical in this process.

Optical Isomers

Jöns Jakob Berzelius (1779–1848), **Friedrich Wöhler** (1800–1882),
Louis Pasteur (1822–1895)

An accepted truth among nineteenth-century chemists was that the properties of chemical compounds could only change if they had different elements. This dogma was radically altered in 1828 when the German chemist Friedrich Wöhler synthesized silver cyanide, which had the same elements as silver fulminate, but different properties. Two years later, the Swedish chemist Jöns Jakob Berzelius found that urea and ammonium cyanide had the same chemical composition but different properties; he named this phenomenon *isomerism*.

MIRROR IMAGES. In 1848, Louis Pasteur, French chemist and microbiologist, observed that rotated polarized light passed through a solution containing tartaric acid naturally present in yeast deposits in wine but not tartaric acid synthesized in the laboratory. Using a pair of tweezers and a microscope, Pasteur discovered two sets of laboratory-prepared tartaric acid crystals and found that while both sets rotated polarized light to the same degree in solution, light was rotated in opposite directions. One set of isomers rotated polarized light counterclockwise to the left [designated levo-, L-, or (-)], while the other rotated light clockwise to the right [dextro-, D-, or (+)]. These two optical isomers (called *enantiomers*) are analogous to our left and right hands—mirror images that cannot be superimposed upon one another. Enantiomers have the same elements but these elements are differently arranged in space around a central carbon atom. When equal amounts of D- and L- enantiomers are present in a solution (a *racemic mixture*), each cancels the other and polarized light is not rotated. With this discovery, Pasteur first gained scientific recognition.

Enantiomers assume great importance in biological systems and the properties of some drugs. Amino acids, the building blocks of proteins and **enzymes**, are all L-enantiomers, with D- rare in nature. Living organisms can only incorporate L-amino acids into proteins, and only L- is biologically active. By contrast, sugars, which make up carbohydrates, are dextro. Different enantiomers of drugs can exhibit different activity or toxicity. L-DOPA is effective for the treatment of Parkinson's disease, while the D-enantiomer provides no benefit and contributes to DOPA's toxicity. Methamphetamine occurs as two enantiomers, with D- ten times more active in its brain-stimulating effects.

SEE ALSO: Enzymes (1878), Amino Acid Sequence of Insulin (1952).

This image depicts microcrystals of tartaric acid as seen in polarized light. The ability of two isomers to rotate polarized light in different directions is a function of differences in their three-dimensional structures. In nature, sugars in carbohydrates rotate polarized light to the left, while amino acids rotate polarized light to the right.

Testosterone

Arnold Adolph Berthold (1803–1861), Charles-Édouard Brown-Séquard (1817–1894)

When the German physiologist Arnold Adolph Berthold castrated roosters in 1849, the results were likely not surprising. He was undoubtedly aware that as early as 2000 BCE male farm animals were castrated to make them more amenable to carrying out their chores. Moreover, Roman emperors who feared assassination, like Constantine in the fourth century, were reported to surround themselves with unaggressive eunuchs.

Berthold, working at the University of Gottingen in Germany, castrated prepubescent male chickens that, upon maturity, failed to exhibit the characteristic physical and behavioral signs associated with roosters. He also castrated adult roosters and observed that they stopped fighting among themselves, had a loss of sex drive, and stopped crowing. Then, after he placed the testicles into a rooster's body cavity, its normal behavior was restored. With these experiments, Berthold established himself as a pioneer in the discipline of endocrinology, demonstrating the role of the gonads in the development of secondary sex characteristics.

Four decades passed before Charles-Édouard Brown-Séquard picked up the trail where Berthold left off. Brown-Séquard was a highly distinguished Mauritian-born physiologist and neurologist who taught in London, Paris, and Cambridge, MA (Harvard). He had conducted research on the physiology of the spinal cord and postulated that substances secreted into the bloodstream have effects on distant organs. (These hypothetical substances, the hormones, were to be discovered several decades later.) In 1889, Brown-Séquard authored a paper that appeared in *Lancet*, one of the world's leading medical journals. In it, he reported that after injecting himself with a liquid extract prepared from the testicles of dogs and guinea pigs, he experienced mental and physical rejuvenation, feeling many years younger than his age of seventy-two. Regrettably, Brown-Séquard had experienced a classic placebo response. Notwithstanding promotional claims to the contrary, carefully controlled studies in recent years have failed to produce any such rejuvenating effects in aging men. Brown-Séquard, Robert Louis Stevenson's neighbor in London, was said to be the inspiration for *Dr. Jekyll and Mr. Hyde*.

SEE ALSO: Ovaries and Female Reproduction (1900), Secretin: The First Hormone (1902), Progesterone (1929).

The British artist Francis Smith (1722–1822) painted Kisler Aga, Chief of the Black Eunuchs and First Keeper of the Serraglio *between 1763 and 1779. The word* eunuch *typically refers to a castrated man who accordingly has little testosterone. In ancient times, eunuchism was often practiced to render slaves less aggressive and more servile.*

Trichromatic Color Vision

Johannes Kepler (1571–1630), **Thomas Young** (1773–1829), **Hermann von Helmholtz** (1821–1894), **Max Schultze** (1825–1874)

Light detection in animals varies from a simple photosensitive organ in flatworms, which only provides a measure of the direction and intensity of light, to birds of prey that can detect rabbits from altitudes of 6 to 9 miles (10 to 15 kilometers). In the visual system of vertebrates, the lens focuses on an object and activates photoreceptive cells on the retina. These cells convert patterns of light into neuronal signals, which are transmitted along the optic nerve to the visual cortex in the back of the brain, and then to higher cerebral centers for information processing.

Prior to the seventeenth century, the gross structure of the eye had been established and, thereafter, attention was directed to function. In 1604, the physicist-astronomer Johannes Kepler determined that the retina—and not the cornea, as was previously believed—was responsible for the detection of light. Almost two centuries later, the English polymath Thomas Young focused his attention on the eye. Young, a Renaissance man, was only a physician and physicist, but he also made significant contributions in language and music, and was among the first to decipher some inscriptions on the Rosetta Stone. In 1793, he described the ability of the eye to focus on near and distant objects depending upon muscles that changed the shape of the lens. First hypothesized by Young in 1802, the famed German physicist Hermann von Helmholtz in 1850 developed the theory of trichromatic color vision, in which there are three sets of color-perceiving elements in the retina: red, green, and blue. The Young-Helmholtz theory serves as the basis for primate color vision.

During the 1830s, the retina was found to contain two types of cells called *rods* and *cones* because of their shape when seen under a microscope. After studying and comparing the eyes of nocturnal and daytime birds, in 1866, the microscopic anatomist Max Schultze discovered that the cones detected color, while the rods were highly sensitive to light. As might be anticipated, different animals have different numbers and relative proportions of each cell type. In 1991, a third type of photoreceptor was discovered that governs the body's circadian clock.

SEE ALSO: Circadian Rhythms (1729), Localization of Cerebral Function (1861).

The great horned owl (Bubo virginianus) *is the most widely distributed owl in the Americas. Having eyes almost the size of human eyes, their retinas contain many rod cells for excellent night vision. Owl eyes don't move in their sockets, but these birds of prey can swivel their heads 270 degrees, enabling them to look in any direction.*

Homeostasis

Claude Bernard (1813–1878), Walter B. Cannon (1871–1945)

Claude Bernard is acknowledged to be one of the greatest of all biologists and is the father of modern experimental physiology. He was the first scientist given a state funeral in France. Among his many accomplishments were studies on the role of the liver in carbohydrate **metabolism**, pancreatic secretions in digestion, the influence of the involuntary nervous system on regulating **blood pressure**, and the nature of the toxicity of carbon monoxide and curare. In his classic work, *Introduction to the Study of Experimental Medicine* (1865), he described the nature of scientific research and the scientist. His greatest contribution, however, was his 1854 formulation of the *milieu intérieur*, now referred to as homeostasis (from the Greek "standing still"), which is considered to be one of the unifying principles of modern biology.

Bernard noted that animals reside in two environments: an exterior environment and an interior environment (*milieu intérieur*). Primitive life forms evolved in the sea, which provided a relatively stable external environment. But with evolution, these life forms moved to unstable terrestrial soundings with respect to the ambient temperature, salt and water composition, and pH. Their survival required adaption mechanisms for keeping their internal environment stable in the face of such changes. Homeostasis is the ability to maintain a constant internal environment in response to an external environmental change. Those life forms and their progeny that successfully achieved homeostasis survived; those that didn't succumbed.

Bernard's concept of *milieu intérieur* largely languished until the early years of the twentieth century when it was renamed *homeostasis* and popularized by W. B. Cannon, of "fight-or-flight fame," in his 1932 book, *The Wisdom of the Body*. In it, Cannon described the effort of multiple organs working cooperatively to maintain homeostasis. We now know that the nervous and hormonal systems play a major role in maintaining the homeostatic balance or steady states. To maintain homeostasis, including body temperature, blood sugar levels, and pH of the blood and body fluids, the body employs **negative feedback** systems in which it responds in an opposite direction to a change.

SEE ALSO: Medulla: The Vital Brain (c. 530 Million BCE), Metabolism (1614), Blood Pressure (1733), The Liver and Glucose Metabolism (1856), Thermoreception (c. 1882), Negative Feedback (1885).

As these wild geese fight, significant changes occur in their cardiovascular systems and carbohydrate metabolism. Their endocrine and nervous systems play major roles in restoring and maintaining the homeostatic balance in their bodies after the conflict concludes.

The Liver and Glucose Metabolism

Claude Bernard (1813–1878)

In 1843, Claude Bernard—who was to become one of the greatest physiologists—noted that after the digestion of sugar cane or starch, glucose was formed, and that it was readily absorbed. He put aside this observation until 1848, when he found glucose in animal blood samples even after the animals were placed on carbohydrate-free diets or had been fasted for several days. This led him to ponder whether the body could produce glucose. High levels were found in blood in the hepatic vein, which exits the liver; similarly, assay of livers from mammals, birds, reptiles, and fish also revealed the presence of glucose, though not in other organs. Bernard concluded that the liver was the source of blood glucose.

One morning in 1849, Bernard found an animal liver that had not been discarded after an experiment performed the previous day. By chance, he analyzed the liver and discovered that the glucose content was higher than in the earlier study. This was Bernard's first indication that the liver was producing and not simply storing glucose. But these findings challenged two other prevailing biological beliefs: that organs had one *and only one* biological function, and that the liver synthesized bile. In addition, it was well accepted at the time that plants, but not animals, could manufacture nutrients.

Bernard hypothesized that glucose was stored in an unknown glucose-starting molecule, which he called *glycogen*, but he was unable to isolate it and proceeded to tackle a broad expanse of other scientific challenges. These included studying the mechanisms underlying carbon monoxide poisoning and curare's paralysis of voluntary muscle, alcoholic fermentation, and spontaneous generation. In 1856, he returned to his search for the elusive glycogenic factor when he observed a white, starch-like substance in liver. This substance—glycogen—was built up from glucose. When needed, glycogen was broken down to glucose, keeping blood sugar levels constant, thereby completing the glucose **metabolism** cycle. Thus, the digestive system could not only break down complex molecules to simple ones but could also build up simple molecules to complex ones.

SEE ALSO: Metabolism (1614), Human Digestion (1833), Homeostasis (1854), Insulin (1921).

An undated photograph of the great French physiologist Claude Bernard.

Microbial Fermentation

Louis Pasteur (1822–1895), Eduard Buchner (1860–1917)

The fermentation process to produce alcoholic beverages goes back about 12,000 years. Since wine, beer, and bread were basic staples of the European diet, and these are made with yeast, scientific attention was focused in its direction. Yeast was long known to be an integral component of the fermentation process, but the nature of that role was alternatively thought to be either the result of causing chemical instability in the substrate (as grapes) or a physical process.

In 1837 and 1838, three scientists independently came to the same conclusion: yeast was a living organism. Starting in 1857, and over the next twenty years, Louis Pasteur conducted a series of studies establishing that fermentation involved living organisms, namely bacteria and yeast—studies intended to solve practical problems. The first series of experiments involved lactic acid, the simplest type of fermentation. Pasteur observed that when the sugar *lactose* was fermented in the presence of the bacterium *lactobacilli*, lactic acid was formed, and that lactic acid was responsible for causing old milk to have a sour taste, the same phenomenon that gives yogurt its sour taste.

During the 1860s, Emperor Napoleon III called upon Pasteur to investigate a major French crisis, the souring of wines. Pasteur's solution was to heat fermented wine to 140°F (60°C), a temperature that kills the microbes responsible for spoiling the wine but not sufficiently high to alter its taste. This same process, called *pasteurization*, was later successfully used for beer and vinegar. (Pasteurization of milk was first to appear in the US in 1893.) Pasteur's interest in fermentation and microbes led to his contributions to the development of the **germ theory of disease**.

Pasteur was unsuccessful in his efforts to extract the principle from yeast that was responsible for fermentation. In 1897, the German chemist Eduard Buchner showed that living yeast cells were not required for fermentation to occur but rather the "press juice," from a cell-free extract, was sufficient—a discovery for which he received the 1907 Nobel Prize. This extract was an *enzyme*, a word meaning "in yeast."

SEE ALSO: Leeuwenhoek's Microscopic World (1674), Enzymes (1878), Germ Theory of Disease (1890).

The first practitioners of fermentation were hunter-gathers seeking to produce wine and beer, and the role of microorganisms in the fermentation process was discovered by Louis Pasteur in the mid-nineteenth century. This image depicts modern winery steel tanks.

Darwin's Theory of Natural Selection

Charles Lyell (1797–1875), **Thomas Malthus** (1766–1834),
Charles Darwin (1809–1882), **Alfred Russel Wallace** (1823–1913)

On the Origin of Species by Means of Natural Selection was over twenty years in its formulation and was based on a number of disparate sources and observations that Charles Darwin had the genius to integrate. While sailing on the HMS *Beagle* (1831–1835), he read *Principles of Geology*, wherein Charles Lyell proposed that the fossils embedded in rock were imprints of living beings millions of years old that no longer inhabited the earth nor resembled extant beings. In 1838, Darwin read Thomas Malthus's *An Essay on the Principle of Population*, in which Malthus postulated that the rate of growth of the population was far exceeding the food supply and that, if unchecked, would have catastrophic consequences. Darwin also considered the practice of farmers who selected their best animal stock for breeding (**artificial selection**). The fourteen finches he found on the Galápagos Islands were similar in all respects, with the exception of the size and shape of their beaks, which were adapted to the available supply of food on their island.

Darwin was not the first to conceive of evolution, but the others lacked a coherent theory to explain its occurrence. His theory was based on natural selection. In nature, there is competition within the species for the limited resources. Those living beings that have the most favorable traits that are best adapted to their environments are most likely to survive and reproduce and pass their favorable traits on to offspring. Thus, over many generations, the species that has arisen from a common ancestor "descends with modifications."

In the 1840s, Darwin sketched the outline of his natural selection theory in an essay. Anticipating a storm of protest to greet his anti-Creation theory, he hesitated to go public but, over the next decade, continued to gather additional evidence to bolster it. In 1858, Darwin learned that a fellow naturalist, Alfred Russel Wallace, had independently developed a theory of natural selection that was strikingly similar to his own. Darwin rapidly completed his book, *Origin of Species*, which appeared in 1859 and proved to be a runaway best seller and a classic in scientific literature.

SEE ALSO: Artificial Selection (Selective Breeding) (1760), Population Growth and Food Supply (1798), Darwin and the Voyages of the *Beagle* (1831), Fossil Record and Evolution (1836), Mendelian Inheritance (1866), Evolutionary Genetics (1937).

An 1869 photograph of Charles Darwin taken by Julia Margaret Cameron (1815–1879), who was known for her photographs of British celebrities.

Ecological Interactions

Charles Darwin (1809–1882)

Ecology examines the relations between living organisms and their environment, and it is not surprising that the relationship between or among species sharing the same ecosystem affects one another. At one end, the nature of this interaction can benefit one species at the expense of the other, to the other extreme each can benefit from the interaction. In his *Origin of Species* (1859), Darwin noted that the struggle for survival was greatest among members of the same species because they possess similar phenotypes and niche requirements.

What's in a Relationship? Predation and parasitism are situations in which only one species profits from the interaction, while another species pays the price. Predation represents the ultimate extreme of an ecological interaction, in which one species captures and feeds on another, as an owl kills a field mouse or the carnivorous pitcher plant catches insects. In a somewhat less extreme instance—parasitism—one species (the parasite) benefits at the expense of the other (host), which derives no benefit from the interaction, as when tapeworms inhabit the intestines of a vertebrate host. Intracellular parasites, such as protozoa or bacteria, often rely upon a vector to transport the parasite to its host; the *anopheles* mosquito conveys the **malaria-carrying protozoan parasite** to its human host, for example.

In commensalism, one species receives benefit from another, which does not suffer adversely from the interplay. The remora, a tropical open-ocean-dwelling fish, lives symbiotically with sharks and eats the shark's leftover food. The fierasfer is a small, slender fish that lives inside the cloacal cavity (the lower end of the alimentary canal) of the sea cucumber to protect itself from predators.

The most equitable of all interactions is mutualism, in which each species provides resources or services to the other resulting in mutual benefit. Lichen is a plant that results when a green alga lives symbiotically with a fungus, where the fungus gains oxygen and carbohydrate from the alga, which reciprocally obtains water, carbon dioxide, and mineral salts from the fungus.

SEE ALSO: Algae (c. 2.5 Billion BCE), Fungi (c. 1.4 Billion BCE), Nitrogen Cycle and Plant Chemistry (1837), Darwin's Theory of Natural Selection (1859), Malaria-Causing Protozoan Parasite (1898).

In this example of mutualism, a cleaner shrimp is cleaning parasites from the mouth of a moray eel. The fish benefits by having the parasites removed, and the shrimp gains the nutritional value of the parasites.

Invasive Species

In 1859, to satisfy his weekend hunting obsession, an Australian settler imported twenty-four rabbits from England and released them on his property. The hybrid offspring resulting from inbreeding were hardy and vigorous, and the climatic conditions and habitat were so agreeable for breeding that Australia's rabbit population reached ten billion by the 1920s—thus validating sayings about their reproductive prowess. This population explosion has had a devastating impact on the local ecology; the rabbits eat native plants that normally feed livestock and cause erosion of topsoil. The Aussies have resorted to hunting, trapping, and poisoning to eliminate them, and in 1907 constructed a 2,000 mile (3,200 kilometer) rabbit-proof fence in Western Australia to contain them, without success. Biological approaches have included the introduction of bacteria and the far more successful *myxoma* virus, which eliminated their population by more than 95 percent in some areas, before the emergence of resistance. The current rabbit population is approximately 200 million.

UNWELCOME GUESTS. Rabbits are but one example of an invasive species—plants, animals, or microbes—that are introduced into a new ecosystem where they are not native, and they outcompete the native species causing their decline or elimination. Invasive species, which may be introduced accidentally or intentionally, are very competitive, highly adaptive to their new environs, and extremely successful at reproducing. If the new environment lacks predators, their spread will go unchecked.

The cannibal snail (*Euglandina rosea*, rosy wolfsnail), a native of the southeastern United States, was intentionally introduced to Hawaii in 1955 to eliminate another invasive species, the giant African land snail. This effort was not successful but the native O'ahu tree snail proved to be collateral damage and was almost hunted to extinction. The predatory cannibal snail is now considered the greatest threat to indigenous snails in Hawaii.

Zebra mussels originated in the Balkans and Poland, and first appeared in North America in Lake St. Clair in 1988, after their accidental discharge in Canadian waters. These highly effective filter feeders deplete algae and small animals that are the food of native species, interfere with the feeding of native mollusks, and tenaciously attach to hard surfaces, including water intake pipes.

SEE ALSO: Ecological Interactions (1859).

Carpobrotus edulis—*also called the highway ice plant, pigface, and sour fig—is native to the Cape region of South Africa. However, the succulent plant has become an invasive species in the Mediterranean and parts of California and Australia.*

Localization of Cerebral Function

René Descartes (1596–1650), **Franz Joseph Gall** (1758–1828), **Paul Broca** (1824–1880), **Eduard Hitzig** (1838–1907), **Gustav Fritsch** (1838–1927), **David Ferrier** (1843–1928)

The notion that the brain influenced thoughts and emotions can be traced back to the ancient Greeks. The French philosopher René Descartes believed that the soul was located in the pineal gland in the brain's center, but the Church's traditional teachings prevailed: the mind was created by God and did not have a physical location. In the late 1790s, the German neuroanatomist Franz Gall broke with tradition and formally proposed that the brain was not a homogeneous mass but rather that different intellectual activities originated in different parts of the brain. Gall's views were vigorously denounced by both the Church as being antireligious and by scientists for lack of proof. Gall is now best remembered for teaching that the skull's shape was a reliable indicator of personality and the development of mental and moral faculties—more specifically, twenty-seven faculties. Gall's cranioscopy devolved into phrenology, a lucrative opportunity for quacks pursuing this pseudoscience during the early decades of the nineteenth century.

The French physician and anatomist Paul Broca first demonstrated that a physiological function could be ascribed to a specific anatomical cerebral localization. In 1861, he performed an autopsy on "Tan," a patient who had experienced a progressive loss of speech and paralysis but not comprehension or mental function. (He was called "Tan" because no matter what the question, "Tan" was his only response.) The autopsy revealed a specific lesion in the frontal lobe of his cerebral cortex, an area important for speech production. More convincing evidence was to follow.

While working in the Prussian army in the 1860s, neurologist Eduard Hitzig noted that application of an electric current to the skull of wounded soldiers elicited involuntary eye movements. In 1870, he explored this phenomenon further with Gustav Fritsch, an anatomist. Administering an electrical current to the cerebral hemisphere (more specifically, the motor cortex) caused involuntary muscle contractions of specific parts of a dog's body. Finally, in 1873, the Scottish neurologist David Ferrier, using electrical stimulation and lesions, was able to construct a map demonstrating which localized areas of the cortex controlled motor functions.

SEE ALSO: Mechanical Philosophy of Descartes (1637), Brain Lateralization (1964).

In Vienna, c. 1812, Franz Gall argued that that the shape of an individual's skull was a reliable indicator of personality and the development of some twenty-seven mental and moral faculties.

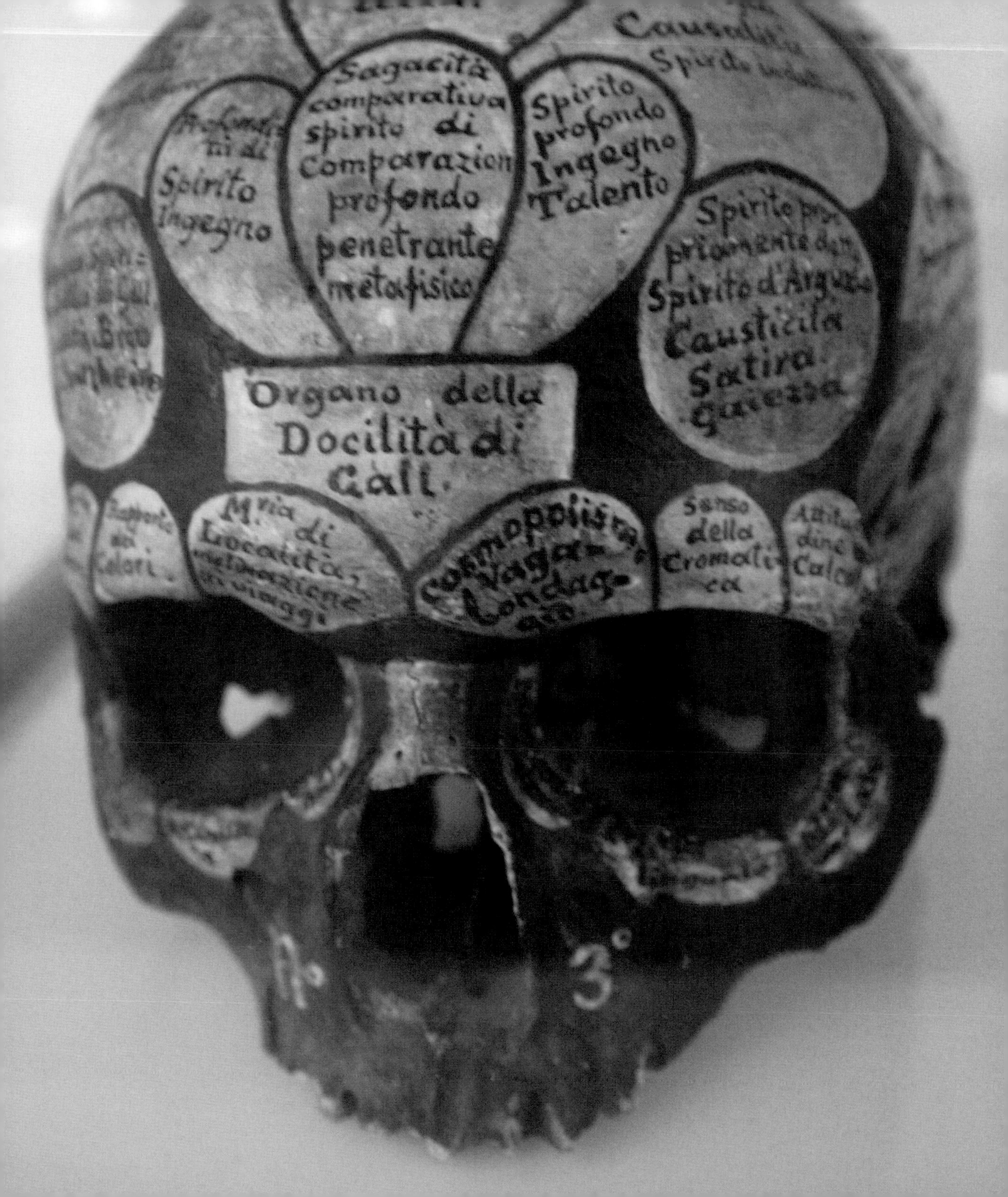
Sagacità comparativa spirito di Comparazion profondo penetrante metafisico
Spirito profondo Ingegno Talento
Spirito Ingegno
Organo della Docilità di Gall.
Causticita
Satira
Senso della Cromatica
N° 3°

Biological Mimicry

Fritz Müller (1821–1897), Henry Walter Bates (1825–1892)

TRICKERY IN NATURE. In 1862, the English explorer-naturalist Henry Bates returned from a decade-long exploration of the Brazilian **Amazon rainforest** and, after examining almost one hundred species of butterflies, he reported his unusual findings. Of particular interest were two distant families bearing a close resemblance: one family, Heliconidae, is brightly colored and unpalatable to birds; the other family, Pieridae, is also colorful but palatable to predators. Bates surmised that the coloration of the unpalatable species proclaimed a warning to potential predators of their poor taste based on the birds' previous experience. He also noted that some palatable butterflies bearing a close resemblance to unpalatable ones were avoided by predators. This has been called *Batesian mimicry*.

Other favorable evolutionary adaptations involving mimicry have been observed in nature. The mimic gains an advantage from its resemblance to another organism (the "model") and a third party mistakes the mimic for the model, such as the harmless colubrid snake that mimics the characteristic "hood" of the Indian cobra's threat display. The third party may be a potential predator or a potential prey of the mimic. Following observations by Bates, other examples have been seen in plants and animals, and in some cases, plants mimic animals and vice versa. The most common examples of mimicry are based on appearance, but sound, smell, and behavior are also imitated.

The German zoologist Fritz Müller noted in 1878 that two unrelated and unpalatable species of butterflies had similar color patterns, and each had adequate defense mechanisms—an apparent exception to Batesian mimicry. But once a predator had learned to avoid a butterfly with one color pattern, it would avoid all other species with a similar pattern (*Müllerian mimicry*). Animals exhibiting *aposematism* transmit a warning signal (pronounced color, sound, odor, or taste) to potential predators that they possess a secondary and more potent defense mechanism—as do the brightly colored poison dart frog or skunk. Aggressive mimicry can be used to escape the detection of potential prey; similarly, in an example of inter-sexual mimicry, the male cuttlefish camouflages itself as a female in order to escape the detection of other males and get closer to females. And some plants, including orchids, mimic female bees and wasps to attract males, which results in pollination of the plant.

SEE ALSO: Amazon Rainforest (c. 55 Million BCE), Animal Coloration (1890).

This illustration shows four forms of Heliconius numata *(top), along with two forms of the poisonous H. melpoméne (bottom, right) and the two corresponding forms of its poisonous mimic, H. erato (bottom, left). The correspondence in warning colors between* H. melpomene *and* H. erato *is a demonstration of Müllerian mimicry.*

Mendelian Inheritance

Charles Darwin (1809–1882), **Gregor Mendel** (1822–1884)

Charles Darwin's 1859 revolutionary work, *Origin of Species*, proposed a theory of evolution based on mutation and natural selection. But neither Darwin nor his contemporaries could explain how favorable traits were inherited. Gregor Mendel, an obscure Augustinian monk who taught gymnasium (high school) science, working on the grounds of a monastery in Brno (now in the Czech Republic), provided the answer and the foundation for the science of genetics.

Mendel sought to trace hereditary characteristics in successive generations of hybrids. He used common garden peas (*Pisum sativum*) because they were inexpensive, easily cultivated in large numbers, and their pollination could be controlled. Moreover, there were many distinct contrasting varieties, with respect to such traits as their color, shape of seed and pod, and plant height. Mendel found that the offspring of bred plants inherited alternative forms of the trait (e.g., tall or short) from each parent. When the inherited traits were different, one trait was dominant and expressed in outward appearance, while the other was recessive and hidden. (Decades later, it was determined that traits are passed from parents to their offspring by gene transmission.) Mendel later studied peas that differed with respect to two traits and observed that each trait was independently transmitted to their offspring, not affecting the transmission of any other trait. These findings appeared in his 1866 paper, "Experiments on Plant Hybrids."

Mendel was very familiar with Darwin's *Origin*, and his personal copy of the German translation was heavily annotated. Less clear is whether Darwin ever read or was acquainted with Mendel's paper, although he, too, was interested in variation and breeding in peas. Mendel's paper was highly mathematical in presentation, which would not have attracted Darwin's interest. Moreover, although Darwin could read German, it was arduous for him to do so. Had Darwin read Mendel, it would have undoubtedly provided him with insights as to how the benefits of natural selection were transmitted to subsequent generations. Sadly, Darwin and the rest of the scientific world remained oblivious to Mendel's paper until 1900, when it was rediscovered.

SEE ALSO: Lamarckian Inheritance (1809), Darwin's Theory of Natural Selection (1859), Meiosis (1876), Genetics Rediscovered (1900), Hardy-Weinberg Equilibrium (1908), Genes on Chromosomes (1910), Evolutionary Genetics (1937), The Double Helix (1953).

Developed in 1905 by the English mathematician-geneticist Reginald C. Punnett (1875–1967), Punnett squares predict the appearance of offspring after mating plants (here, Mendel's flowers) or animals, and they are used by biologists to determine the probability that an offspring will have a particular trait.

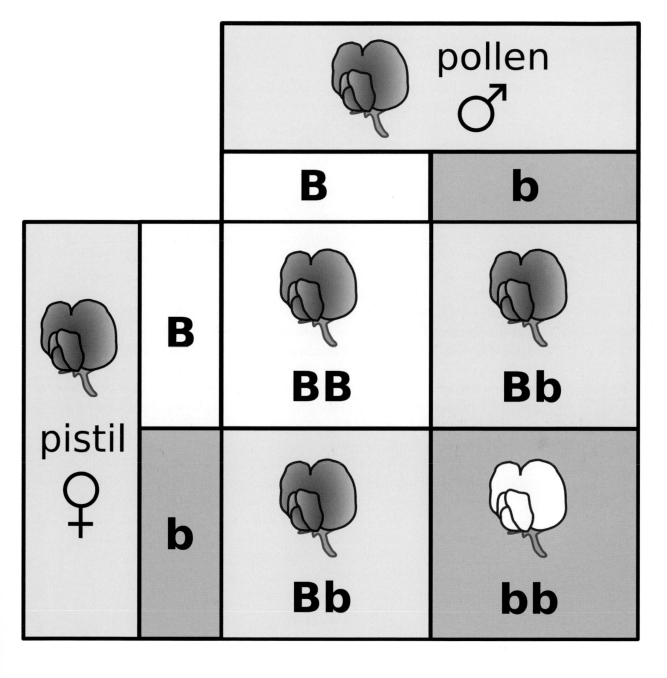

Ontogeny Recapitulates Phylogeny

Antoine Étienne Serres (1786–1868), **Charles Darwin** (1809–1882),
Ernst Haeckel (1834–1919)

The name of Ernst Haeckel is inextricably linked to the theory that ontogeny recapitulates phylogeny (ORP). Haeckel was a German biologist, naturalist, and accomplished illustrator, who served as professor of comparative anatomy at the University of Jena for forty-seven years. During that time he discovered, described, and named thousands of plants and animals and was a renowned illustrator of invertebrates and a pioneer in developmental biology, studying the processes by which organisms grow and develop from single-cell zygotes to adulthood.

Haeckel's theory of ORP was based on concepts originated by the French embryologist Étienne Serres about four decades earlier. In part, Serres proposed that higher animals trace the embryological stages that are analogous to adult stages seen in lower animals. In his *Origin of Species*, Charles Darwin recognized the importance of embryonic development in understanding evolution, a theory actively supported by Haeckel.

After studying the embryos of a number of species, in particular chicks and humans, in 1866 Haeckel proposed ORP—also called the recapitulation theory and the biogenic law—stating that the development of embryos of every species (ontogeny) fully repeats the evolutionary development of that species (phylogeny). He drew direct comparisons between slits and arches seen in the neck of developing human and chick embryos and the gill slits and gill arches in adult fish, leading to his conclusion that all three have a common ancestry. Similarly, in advanced stages of human embryonic development, a tail is present that is lost prior to birth.

To support his theory, Haeckel created drawings of embryos of different species, depicting their progression from the earlier to the later stages of development and the evolvement from their similarity to diversity. These drawings, emphasizing the similarities between early stages of different species, were criticized as being oversimplified, grossly exaggerated, and inaccurate. While elements of his theory are true, ORP as proposed has been discredited and is now rejected by modern biologists. Nevertheless, many of us may have used biology textbooks that cite ORP or, more commonly, used reproductions of Haeckel's drawings of embryos as evidence to support the theory of evolution.

SEE ALSO: Darwin's Theory of Natural Selection (1859).

Ernst Haeckel used drawings such as this, tracing the development of Asteroidea *(common starfish), in his 1904 book,* Art Forms of Nature, *to support his theory that ontogeny recapitulates phylogeny.*

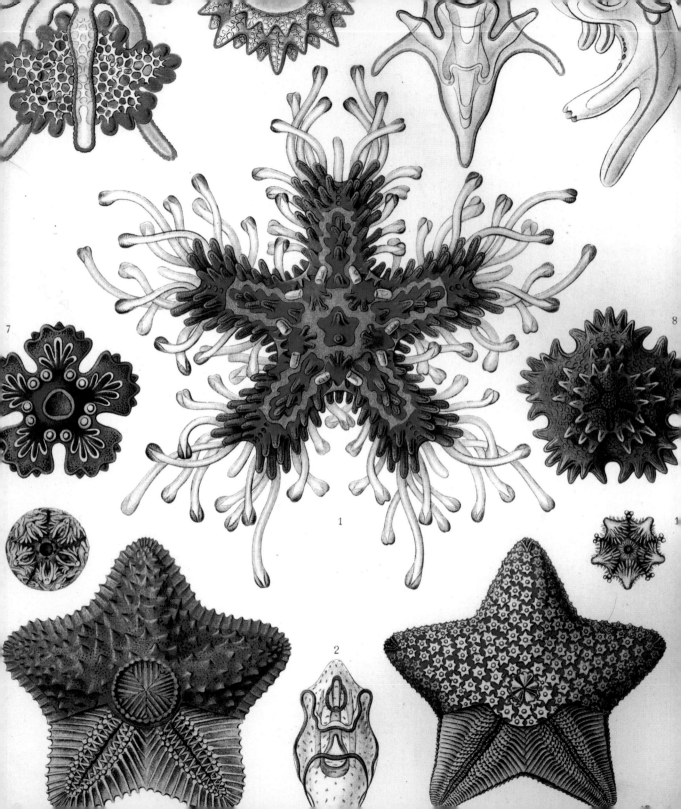

Hemoglobin and Hemocyanin

Friedrich Ludwig Hünefeld (1799-1882), **Felix Hoppe-Seyler** (1825–1895), **Theodor Svedberg** (1884–1971)

Blood delivers nutrients and oxygen to the body cells and transports carbon dioxide, a waste product of **metabolism**, from the cells. The oxygen enters the body through the lungs or gills and then this gas is released to burn energy-yielding nutrients. This gas is poorly soluble in water and blood and, therefore, needs an additional component to enhance its oxygen-carrying capacity. This component is satisfied by respiratory pigments: the metal-containing proteins hemoglobin and hemocyanin, which are red and blue, respectively.

Hemoglobin (also spelled haemoglobin), found in virtually all vertebrates and most invertebrates, is the major component of erythrocytes (red **blood cells**). An iron-containing protein, hemoglobin was discovered by Friedrich Ludwig Hünefeld in 1840. The reversible binding of oxygen to erythrocytes, first reported in 1866 by Felix Hoppe-Seyler, a pioneer in the science of biochemistry, is dependent upon the presence of hemoglobin. Vertebrate blood is bright red when hemoglobin is saturated with oxygen. When oxygen binds to hemoglobin, oxygen's solubility increases seventy-fold in mammals.

By contrast, the copper-containing hemocyanin carries oxygen in most mollusks (such as slugs, snails) and **arthropods** (crustaceans, horseshoe crabs, scorpions, centipedes, but rarely in insects). Copper imparts a blue color to oxygenated hemocyanin. Hemocyanin was first discovered in snails in 1927 by the Swedish chemist Theodor Svedberg, better known for his research on analytical ultracentrifugation.

Whereas hemoglobin and hemocyanin share the same respiratory function, differences are seen in how they are transported in circulatory fluids. Hemoglobin is bound to erythrocytes in a closed circulatory system of blood vessels; oxygen diffuses across the walls of the capillaries, the smallest blood vessels, and into the interstitial fluid surrounding cells. By contrast, oxygen-transported hemocyanin is not bound to blood cells but suspended in hemolymph; this fluid is contained in an open circulatory system that directly bathes the body cells. Some insects have a still-simpler open circulatory system in which their hemolymph does not contain oxygen-carrying molecules.

SEE ALSO: Arthropods (c. 570 Million BCE), Pulmonary Circulation (1242), Metabolism (1614), Harvey's *De motu cordis* (1628), Blood Cells (1658), Blood Types (1901), Blood Clotting (1905).

A heme is a chemical compound consisting of an iron ion surrounded by a large organic ring called a porphyrin. Heme B, represented here by a space-filling three-dimensional model, is an important component of hemoglobin and myoglobin.

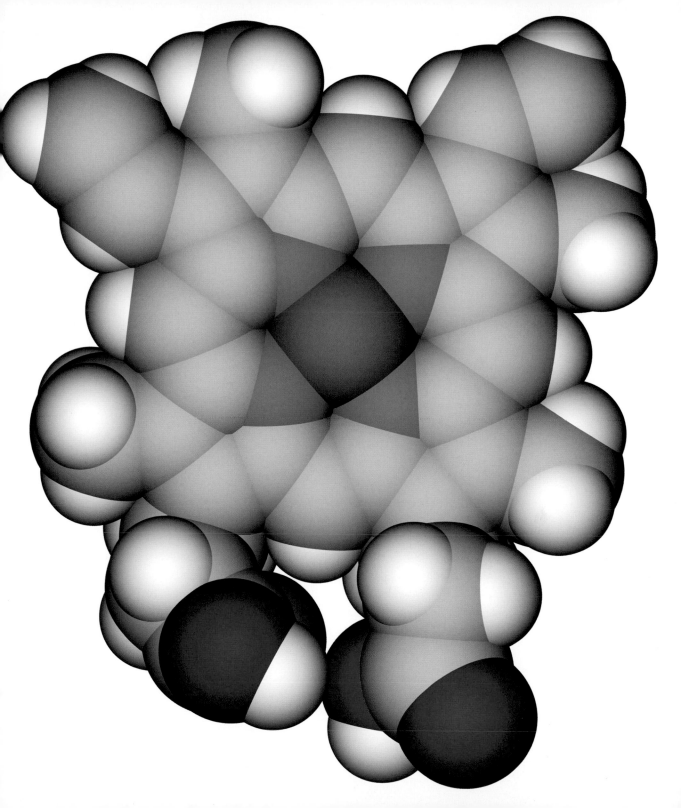

Deoxyribonucleic Acid (DNA)

Friedrich Miescher (1844–1895), **Albrecht Kossel** (1853–1927),
Phoebus Levene (1869–1940), **Oswald T. Avery** (1877–1955),
Erwin Chargoff (1905–2002), **Francis Crick** (1916–2004),
Rosalind Franklin (1920–1958), **James D. Watson** (b. 1928)

DNA, undoubtedly the most familiar of all chemicals in biology, had its beginnings in 1869. Friedrich Miescher, a Swiss physician-biologist in Germany, was interested in the chemistry of the **cell nucleus** and used lymph cells present in pus taken from patient bandages obtained at local hospitals. Chemical analysis failed to reveal the presence of protein that Miescher was anticipating, and he termed the new unknown substance "nuclein."

In the final decade of the nineteenth century, the German biochemist Albrecht Kossel isolated and described Miescher's "nuclein," which he renamed *nucleic acid*. Further analysis revealed the presence of five organic bases: adenine (A), cyosine (C), guanine (G), thymine (T), and uracil (U), which he collectively termed nucleobases—discoveries for which Kossel was awarded the 1910 Nobel Prize. It was subsequently determined that there were actually two nucleic acids, deoxyribonucleic acid (DNA) and ribonucleic acid (RNA).

Phoebus Levene was medically trained in his native Russia, but religious persecution impelled him to move to the United States in 1893 to practice medicine and study biochemistry. During a medical leave, while recovering from tuberculosis, he worked with Kossel. After several decades, in the mid-1930s, Levene, at the Rockefeller Institute, correctly determined that the nucleic acids were linked to sugars (deoxyribose and ribose) and a phosphate group—he called this combination a *nucleotide*—but incorrectly postulated how they were linked together. By the late 1940s, based primarily upon work by Oswald Avery, it was generally understood that DNA was involved in the hereditary process. But its chemical structure continued to be enclosed in a shroud.

Austrian biochemist Erwin Chargaff left Nazi Germany during the 1930s and, at Columbia University, analyzed the chemical neurobases of DNA. In 1950, he discovered that different organisms have different amounts of DNA, but that A and T, as well as G and C, were always present in approximately equal quantities to each other. The final chapter in correctly assembling the components of DNA, **the double helix**, would be disclosed by Franklin, Watson, and Crick in 1953.

SEE ALSO: Cell Nucleus (1831), DNA as Carrier of Genetic Information (1944), The Double Helix (1953).

A simplified illustration of DNA, showing two biopolymer chains. Each chain is twisted about the other in a double helix formation, with nitrogen bases linking them.

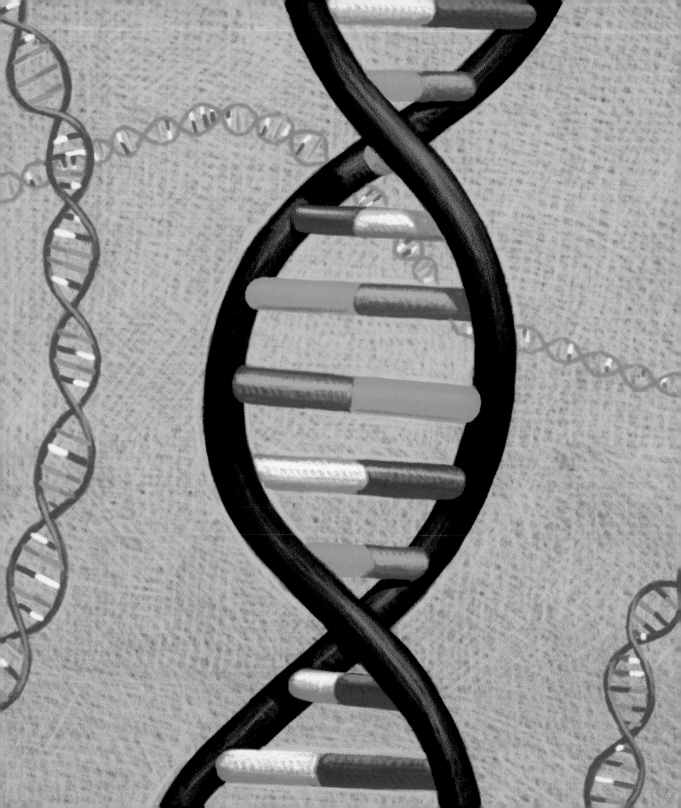

Sexual Selection

Charles Darwin (1809–1882)

Charles Darwin, in *Origin of Species* (1859), proposed that evolution was based on natural selection. Whereas the book provided a mere fleeting allusion to human evolution, hinting that it would be considered at a later day—this was sufficient to ignite passionate debate. It implied that humans had evolved from lower life forms and directly challenged the Book of Genesis. Twelve years later, in his *The Descent of Man, and Selection in Relation to Sex* (1871), Darwin specifically extended his theory to humans.

In the first of the two-volume, 900-page *Descent*, Darwin sought to provide evidence that all humans are a single species that have descended from a common apelike ancestor and had developed by evolution, as had other species; in 1871, no human fossil evidence had been discovered. He noted the similarities between humans and other **primates** and argued that human mental and emotional capabilities are not uniquely human traits, but that they differed in degree but not in kind from other higher animals.

Darwin then argued, from an evolutionary perspective, for the commonality and equality of the human race. He rejected the theory of polygenesis (a concept embraced by a number of prominent biologists) that the human races are derived from different lineages, created separately, with some races being inferior. Rather, he supported the theory of monogenesis, that all humans evolved from a common origin, and that the differences between the races—skin color, hair type—were all superficial; when taken in its totality, all humans closely resemble one another.

The theory underlying sexual selection was first briefly proposed by Darwin in *Origins* and later examined in extensive detail as it relates to humans and animals in *Descent*. Whereas the driving force underlying natural selection is survival, sexual selection is based on the need to reproduce. Darwin envisioned two kinds of "sexual struggles": between members of the same sex each vying for a member of the opposite sex, and between members of the opposite sex seeking to attract them. In the latter case, the object of the attention, usually the female, selects the more desirable partner.

SEE ALSO: Primates (c. 65 Million BCE), Anatomically Modern Humans (c. 200,000 BCE), Darwin's Theory of Natural Selection (1859), Parental Investment and Sexual Selection (1972), Oldest DNA and Human Evolution (2013).

"A Venerable Orang-outang," an editorial cartoon of Charles Darwin depicted as an ape, published in 1871 in The Hornet, *a British satirical magazine.*

Coevolution

Charles Darwin (1809–1882), Hermann Müller (1829–1883)

Survival of an organism is dependent upon its ability to thrive in an environment with other organisms. In a predator-prey relationship, the hunter is favored by evolving traits that aid in killing its prey. To counter these advantages, the prey must evolve traits to avoid detection and to successfully escape, sometimes by utilizing physical or chemical defenses. Examples of such one-upmanship have been aptly referred to as an "evolutionary arms race."

Based on the **theory of natural selection**, if the predator evolves enhanced offensive capabilities, the prey's survival necessitates the evolution of a commensurately improved defense. Examples of evolved traits exist in plant-insect relationships. Plants may employ chemical defenses to ward off culinary advances by insect herbivores. Insects, in turn, may evolve metabolic capabilities that neutralize the noxious plant chemical. The plant reciprocates by evolving a more effective chemical deterrent.

By contrast, some of the classic examples of coevolution result from mutually beneficial specialized relationships between plants and pollinator insects (as bees) and between a number of species of flowering plants with specific pollinators, such as bats and insects. Moth-pollinated plants and moths have coevolved such that the plant tubes are the exact length of the moths' "tongue." Upon examining the size and shape of a Madagascar orchid, Darwin predicted the existence of a pollinating moth with an 11-inch-long (28 centimeter) proboscis. Some forty years later, decades after Darwin's death, such a moth was discovered.

Coevolution refers to reciprocal evolutionary changes that occur between pairs of species as they interact with and are dependent upon one another. Charles Darwin commented briefly on this phenomenon in *Origin of Species* (1859) and far more extensively in his *The Descent of Man* (1871). In the latter work, Darwin quoted the German biologist Hermann Müller, a pioneer in the study of coevolution, and his studies on bees and the evolution of flowers. These were described in Müller's work, *The Fertilization of Flowers*, appearing first in German in 1873 and in an English translation a decade later.

SEE ALSO: Plant Defenses against Herbivores (c. 400 Million BCE), Angiosperms (c. 125 Million BCE), Darwin's Theory of Natural Selection (1859), Ecological Interactions (1859), Sexual Selection (1871).

Moth-pollinated plants and moths (such as this hummingbird hawk moth) have coevolved such that the length of the plant tubes exactly matches the moth's pollinating proboscis.

Nature versus Nurture

John Locke (1632–1704), **Francis Galton** (1822–1911)

Certain physical characteristics, such as eye color and **blood type**, are genetically determined, as are perfect pitch and the ability to recall musical notes from memory. But the relative influence that nature versus nurture plays in the development of human traits is a debate that goes back to ancient Greece and has continued to the present. The seventeenth-century philosopher John Locke argued that at birth the human mind was a *tabula rasa*—a blank slate, lacking in mental content—and that traits, such as personality, social and emotional behavior, and intelligence, were acquired from environmental influences. Our modern understanding of "nature versus nurture" was popularized by Francis Galton in 1874, who argued that intelligence was largely inherited and advocated **eugenics** to improve the genetic stock of the human population.

Prior to attempting to determine whether a middle ground exists between these dichotomous positions, we might consider our contemporary understanding of nature and nurture. "Nature" refers to the influence of our genetic makeup and most greatly affects our physical characteristics. "Nurture" was previously limited to environmental influences but is now redefined to also include the influence of prenatal, parental, extended family and peers, and socio-economic status. If the environment did not play a role in determining individual traits and behaviors, identical twins, even when reared apart, should be the same in all respects—but this is not the case. A spirited debate currently rages as to whether sexual orientation is inherited or a learned behavior.

A genetic link exists for many common disorders—diabetes, heart disease, cancers, alcoholism, schizophrenia, and bipolar disorder—and their occurrence can be positively or negatively affected by such influences as diet, exercise, and smoking. **Epigenetics** is the study of the intersection of these two influences: how environmental inputs affect the expression of genes. A primary goal of the **Human Genome Project** is to identify those genes associated with diseases and determine what environmental factors might contribute to their occurrence.

SEE ALSO: Mendelian Inheritance (1866), Eugenics (1883), Genetics Rediscovered (1900), Blood Types (1901), Inborn Errors of Metabolism (1923), Human Genome Project (2003), Epigenetics (2012).

The relative influence of genetic factors versus nongenetic and environmental factors on the development of human traits continues to be the subject of debate. Studies of twins—especially monozygotic twins who have identical genetic material but have been reared apart—shed light on the link between heredity and various outcomes, with differing outcomes suggesting the influence of environmental factors and the lifestyle pursued by each twin.

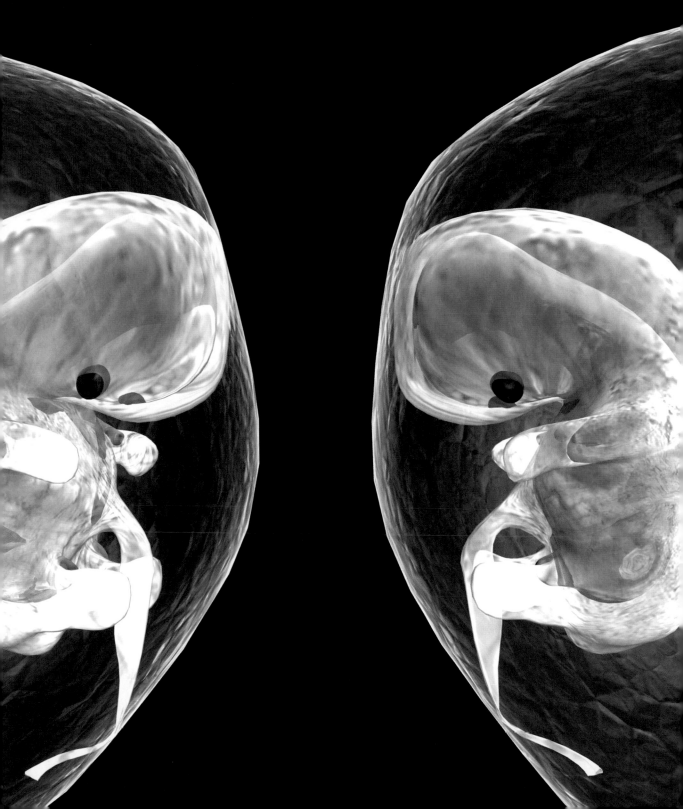

Biosphere

Eduard Suess (1831–1914), Vladimir I. Vernadsky (1863–1945)

The famed Austrian geologist Eduard Suess first introduced the concept of the biosphere in 1875, using it to refer to the "place on the Earth's surface where life dwells." This concept was built upon and very significantly expanded by the Russian mineralogist-geochemist Vladimir Vernadsky, who defined it in terms that combined elements of geology, chemistry, and biology in his 1926 book *La biosphere*. Vernadsky envisioned that the biosphere contained two types of matter: living matter in all its forms and "inert" (nonliving) matter, such as minerals, which were preserved over time. Vernadsky argued that just as nonliving matter was transformed by living organisms, the biosphere was transformed by human cognition, and that life and human cognition were essential components in the evolution of the Earth.

The contemporary concept of the biosphere is the space on or near the earth's surface that contains or supports living organisms as well as dead matter produced by living organisms. The biosphere is a core concept in ecology and biology; it represents the highest level of biological organization and includes all biodiversity on Earth, from simple molecules to structures within a cell (organelles), organisms, populations, communities, and terrestrial and aquatic ecosystems.

Certain environmental conditions must be met for organisms to live, including the proper temperature and moisture, but in addition, they require energy and nutrition. Nutrients are contained in dead organisms or in the waste products of living cells and are recycled and transformed into compounds that other organisms can use as food. Vernadsky was the first scientist to recognize that the oxygen, nitrogen, and carbon dioxide present in the Earth's atmosphere was the result of biological processes. The biosphere has evolved since the initial appearance of the first single-celled organism some 3.9 billion years ago, at a time when the carbon dioxide-rich atmospheric conditions resembled that of our celestial neighbors, Venus and Mars. The plants caused a breakdown and release of oxygen from the carbon dioxide, giving rise to an oxygen-rich (O_2) atmosphere for breathing and stratospheric ozone (O_3), which protects the Earth's inhabitants against ultraviolet radiation.

SEE ALSO: Origin of Life (c. 4 Billion BCE), Prokaryotes (c. 3.9 Billion BCE), Land Plants (c. 450 Million BCE), Plant Nutrition (1840), Photosynthesis (1845), Ecological Interactions (1859), Global Warming (1896).

Miles above Earth's biosphere, astronaut Robert C. Stewart tests a hand-controlled manned maneuvering unit, which allows astronauts to move freely in space without a tether, during the 1984 flight of the space shuttle Challenger.

Meiosis

August Weismann (1834–1914), Oscar Hertwig (1849–1922)

Meiosis, which evolved 1.4 billion years ago in **eukaryotes**, both reduces the number of chromosomes needed for sexual reproduction, and contributes to genetic variation, leading to the process of evolution. In 1876, Oscar Herwtig was the first to recognize the role of the **cell nucleus** and chromosomal reduction in sea urchin eggs during meiosis, and noted that the fusion of the egg and sperm and their nuclei contributed to the inherited traits of the offspring. August Weismann extended these findings in 1890 when he found that meiosis required two cycles of cell division if the number of chromosomes was to remain stable. The word *meiosis* (Greek = "lessen") refers to the reduction in the number of chromosomes by one half in the daughter cells after sexual reproduction.

Hereditary information is passed from parents to their offspring in genes written in the language of DNA. In asexual reproduction, which occurs in **prokaryotes** (bacteria) and a few eukaryotes, the organism simply divides, with the resulting offspring an exact genetic replica of its single parent, inheriting both its strengths and weaknesses; in the absence of a mutation, no evolution is possible. By contrast, in sexual reproduction, which occurs in most eukaryotes, each parent contributes genes. The genome of one diploid germ cell, composed of DNA wrapped in a chromosome, undergoes DNA replication, followed by two rounds of division (process of reduction division—called meiosis), resulting in haploid cells called *gametes*. Each gamete, containing a complete set of chromosomes, fuses with a gamete of the opposite sex during fertilization to form a new diploid cell or zygote.

ENABLING EVOLUTION TO OCCUR. As a direct result of the process of crossing over during meiosis, there is a recombination of genes, with a scrambling of alleles (alternate forms of each gene). The resulting offspring has a unique combination of genes contributed by each parent that is genetically different than either parent. This genetic diversity creates an opportunity for changes to occur in the offspring through natural selection. Natural selection is the basis for evolution and the opportunity for organisms to successfully meet the demands associated with a changing and often challenging environment.

SEE ALSO: Prokaryotes (c. 3.9 Billion BCE), Eukaryotes (c. 2 Billion BCE), Cell Nucleus (1831), Darwin's Theory of Natural Selection (1859), Mitosis (1882), Genes on Chromosomes (1910), DNA as Carrier of Genetic Information (1944).

During the process of meiosis, there is a recombination of genes contributed by each parent. The result is an offspring that has a unique combination of genes, providing the opportunity for evolution to occur.

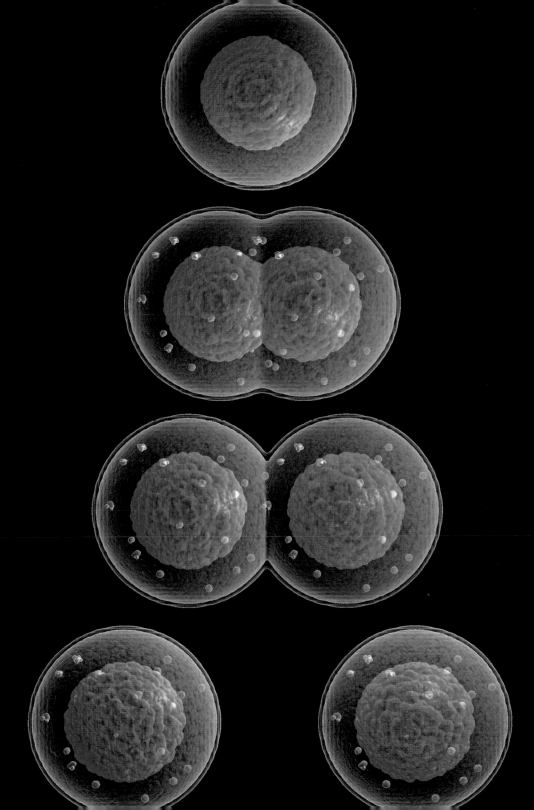

Biogeography

Charles Darwin (1809–1882), Alfred Russel Wallace (1823–1913)

During the nineteenth century, Alfred Russel Wallace was lionized as being among the greatest naturalist explorers and biologists. He was a widely published and acclaimed author of twenty-two books and hundreds of scientific papers, as well as being a pioneer in the study of *biogeography*—the geographical distribution of plants and animals. Sadly, during the following century, his reputation had been largely relegated to Charles Darwin's understudy in the theory of evolution. Unlike Darwin, whose family inheritance permitted him to devote full-time effort to study and writing, Wallace was obliged to sell many of his collected biological specimens, give lectures, and write books to support his family.

Having an early thirst for exploring the natural world, from 1848 to 1852 Wallace traveled to the **Amazon rainforest** and collected a wide array of specimens, which were lost in a ship fire on his return to England. In 1854 he again set sail, this time for the Malay Archipelago, where he was to remain for eight years studying hundreds of thousands of its animals and plants. This led him to independently embrace the concept of evolution through natural selection—far from mainstream thinking in the 1850s. While still in Malaysia, his paper on evolution was jointly presented with Darwin's before a scientific audience in 1859.

As he traveled throughout the Malay Archipelago, he noted that while the terrain and climate were similar, the distribution of animal species differed in the northwest and in the southeast. Animals in Sumatra and Java were more similar to those in Asia, while those in New Guinea bore similarity to the animals of Australia. There was a clear boundary between the islands—later called the Wallace line—that separated the Oriental and Australian biogeographic regions. In 1874, he divided the world into six geographic regions based on their geography and their animal inhabitants, and these appeared in his 1876 classic *Geographical Distribution of Animals*. This work served as a virtual tour guidebook of animals and where they could be found. His 1880 book, *Island Life*, examined plant and animal species on three separate and distinct island types.

SEE ALSO: Amazon Rainforest (c. 55 Million BCE), Darwin and the Voyages of the *Beagle* (1831), Darwin's Theory of Natural Selection (1859).

During the mid-1800s, Wallace traveled throughout the Malay Archipelago, an area shown on this century-old atlas map.

China
Yangtse
Hang-chow
Foochow
Canton
Yun-nan
Hong Kong
Formosa
Mariana Is.
Guam
Tongking
Hanoi
Hai-nan
Luzon
Philippine Is. (U.S.)
Hue
Anam
Siam
Bangkok
Mekong
S. China Sea
Manila
Palawan
Mindanao
Pelew Is.
Caroline Is.
Saigon
Malay Penina.
Brunei
Sarawak
Singapore
Malacca Str.
Sumatra
Borneo
Celebes
Gilolo
140°
120°
Bismarck Arch.
New Guinea
Batavia
Java
Timor
Torres Str.
C. York
Christmas
Keeling Is.
Melville I.
G. of Carpentaria
Kimberley
North Australia
COMMONWEALTH OF AUSTRALIA
Western Australia
Cen. Aust.
Queensland
South Australia
North West C.
Sharks B.
Coolgardie
Perth
New South Wales
C. Leeuwin
Albany
Adelaide
Murray
Victoria
Sydney
Canberra
Melbourne
Bass Str.

Marine Biology

Aristotle (384–322 BCE), **Charles Darwin** (1809–1882),
Charles Wyville Thompson (1830–1882)

Until the late nineteenth century, knowledge of marine biology was limited to the upper few fathoms of the ocean and shallow waters. Within these rather circumscribed limits, Aristotle had described many forms of marine life, and Charles Darwin, in his voyage on the *Beagle* in 1831, noted **coral reefs**, plankton, and barnacles.

This state of knowledge was dramatically altered after the *Challenger* expedition in 1872–1876, the first voyage exclusively dedicated to a study of the marine sciences. (Moreover, there were practical considerations for the voyage, namely, growing demands for transcontinental telegraph communication utilizing ocean cables.) Charles Wyville Thompson, a Scottish marine biologist and professor at the University of Edinburgh, who had established a reputation for his invertebrate marine studies in the late 1860s, was selected scientific director for the expedition. The around-the-world voyage on the *Challenger*, a Royal Navy vessel refitted for scientific applications, was almost 70,000 nautical miles (30,000 kilometers). The data collected included identification of some 4,700 new species of marine life and disproof of the belief that life could not exist at depths below 1,800 feet (550 meters). Ocean currents and temperatures were systematically plotted, maps of the bottom deposits were prepared, and the underwater Mid-Atlantic Ridge—the longest mountain range in the world—was discovered.

In 1873, Wyville Thompson authored an early marine biology book, *The Depth of the Sea*, based on his initial findings. Upon his return to great honors and a knighthood in 1877, he worked to prepare a report of the voyage that filled fifty volumes and almost 30,000 pages, appearing in his 1880 work, *The Voyage of the Challenger*. His journey focused upon the collection, description, and cataloging of marine organisms, using newly developed methods for capturing and preserving specimens for study.

Contemporary study of marine biology investigates such questions as how particular organisms adapt to the chemical and physical properties of seawater and how ocean phenomena control the distribution of marine life. Of particular interest is study of marine ecosystems, namely, understanding food chains and webs and predator-prey relationships.

SEE ALSO: Coral Reefs (c. 8,000 BCE), Aristotle's *The History of Animals* (c. 330 BCE), Darwin and the Voyages of the *Beagle* (1831), Food Webs (1927).

A July 1874 photograph of a Tongan seaman, identified as a crewmember during the Challenger *expedition, which is thought to be the first marine expedition to carry an official photographer and an official artist.*

Enzymes

Wilhelm Kühne (1837–1900), Eduard Buchner (1860–1917), James B. Summer (1887–1955)

Life cannot exist without enzymes. Thousands of chemical reactions occur in living cells: old cells are being replaced by new ones; simple molecules link to form complex ones; food is digested and converted to energy; waste materials are disposed of; and cells reproduce. These reactions, involving buildup and breakdown, are collectively referred to as **metabolism**. For each of these reactions to occur, a certain degree of energy is required (activation energy) and in the absence of such energy, these reactions would not occur spontaneously. The presence of these enzymes—which are usually proteins or RNA enzymes—reduces the amount of activation energy required for these reactions to occur and increases the rate of these reactions by millions. In the process, enzymes are neither consumed nor chemically changed.

Each of the chemical reactions in the body is a component of a pathway or cycle, and most enzymes are highly specific and act on only a single substrate (reactant) in the pathway to produce a product in the metabolic sequence. Most of the more than 4,000 enzymes in living cells are proteins, with a unique three-dimensional configuration, the shape of which accounts for their specificity. An enzyme is commonly named by adding the suffix *ase* to the root name of the substrate on which it acts, although more specific (and descriptive) names are used in chemically oriented literature.

It was known in the late seventeenth and early eighteenth centuries that meat was digested by secretions in the stomach and starch could be broken down to simple sugars by saliva and plant extracts. Wilhelm Kühne, a German physiologist, was the first to coin the name *enzyme* in 1878 to refer to trypsin, a protein-digesting enzyme he had discovered, and, in 1897, Eduard Buchner at the University of Berlin first demonstrated that enzymes could function outside living cells. In 1926, working with the jack bean, James Summer at Cornell University isolated and crystallized the first enzyme, urease, and provided conclusive proof that it was a protein. Summer was the co-recipient of the 1946 Nobel Prize in Chemistry.

SEE ALSO: Metabolism (1614), Human Digestion (1833), Inborn Errors of Metabolism (1923), Protein Structures and Folding (1957).

Certain anticancer and immunosuppressive drugs target purine nucleoside phosphorylase (PNP), an enzyme that carries out housekeeping functions by clearing away certain waste molecules that are formed when DNA is broken down. The image depicts a computer-generated model of PNP.

Phototropism

Charles Darwin (1809–1882), Francis Darwin (1848–1925), Nikolai Cholodny (1882–1953), Frits Warmolt Went (1903–1990)

Charles Darwin, aided by his son Francis, was intrigued by the movement of plants toward light, a phenomenon called *phototropism*. In these studies, he tested the canary grass coleoptiles, the hollow sheath surrounding the stem of grasses. The Darwins found that when their tips were covered, the phototropic response was absent. Further study revealed that the coleoptile tips were most responsive to light, while bending occurred in their middle section. The elder Darwin described these results in *The Power of Movement in Plants* (1880), which laid the groundwork for discovery of auxin, the first plant hormone.

While still a graduate student, the Dutch American biologist Frits Went extended Darwin's findings. Went concluded that the tip contained a phototropic chemical, which he called *auxin*, and later chemically identified as indoleacetic acid (IAA). In 1927, Went and Nikolai Choladny, at the University of Kiev, independently observed that auxin is a plant growth hormone, concentrated in plants furthest from the light source, on the dark side of the stem. Auxin activates **enzymes**, the expansins, which weaken the cells in the wall of the stem. The dark-side cells grow faster than those on the light-exposed side causing the stem to move upward toward the light; the Choladny-Went Theory, explaining phototropism, remains controversial. When plants spread out their leaves in daytime, it makes **photosynthesis** possible.

If a plant is placed in the ground on its side, it realigns itself such that its shoots point upward and roots downward. In addition to phototropism, auxin influences geotropism (aka gravitropism) by more selectively building up on the lower side of the coleoptile than on the upper side. This causes the downward growth of plants.

Auxin also exerts other growth-promoting effects that influence the amount and type of plant growth. Auxin is produced in the plant tips and moves downward toward the base, causing an elongation of plant cells along the shoots and influencing the branching process. A reduced flow of auxin from the branch tips signals that the branch is not productive, which results in growth resources being redirected to more fruitful branches.

SEE ALSO: Land Plants (c. 450 Million BCE), Photosynthesis (1845), Enzymes (1878), Secretin: The First Hormone (1902).

Sunflower buds exhibit the phenomenon of phototropism. In the morning, they face the east and follow the position of the sun throughout the day, returning to an eastward orientation the next morning. This painting, Farm Garden with Sunflowers *(1905–1906), was created by Austrian artist Gustav Klimt (1862–1918).*

Mitosis

Gregor Mendel (1822–1884), Walther Flemming (1843–1905)

A basic tenet of the **cell theory** is that all cells arise from pre-existing cells, and one hallmark of all living organisms is its ability to reproduce. The German anatomist Walther Flemming, a leader in the science of *cytogenetics*—the study of the cell's hereditary material, the chromosome—played a pivotal role in our basic understanding of these phenomena.

In 1879, he developed and used an aniline dye to visualize the structure of the nucleus from salamander embryo cells. Within the nucleus was a coiled mass of threadlike material, which he called *chromatin*. He observed that these paired threads—later named chromosomes—split longitudinally into two halves, with each unpaired thread half moving to the opposite side of the cell. He named this process of chromosomal splitting *mitosis* (Greek = "thread") and described it in his 1882 book, *Cell-Substance, Nucleus, and Cell-Division*. Later scientists discovered that immediately after mitosis—involving the separation of chromosomes in the nucleus, and consisting of six distinct phases—the parent cell divides into two daughter cells, each identical in cellular content to its parent, in a process called *cytokinesis*.

Flemming was not aware of Gregor Mendel's work and his rules of heredity, nor was he aware that traits are transmitted by genes contained in chromosomes. Thus, the import of the discoveries by Mendel and Flemming were not appreciated until the early 1900s, when genes were recognized to be the functional unit of heredity.

Mitosis is among the most fundamental of all biological processes in all living organisms: The number of cells increases, and the organism grows by mitosis—the process by which all single-celled organisms reproduce. Mitosis repairs damaged or worn out cells and tissues. Moreover, the applied study of mitosis has led to stem cell technology in which undifferentiated stem cells can differentiate into specialized cells. Errors in mitosis can lead to cancer. Hence, we can readily understand that the discovery of mitosis and chromosomes is considered to be one of the ten most important in cell biology and one of the one hundred most significant of all scientific discoveries.

SEE ALSO: Cell Theory (1838), Mendelian Inheritance (1866), Meiosis (1876), Genetics Rediscovered (1900), Genes on Chromosomes (1910), Induced Pluripotent Stem Cells (2006).

Mitosis, one of the most important processes in biology, refers to the division of a "parent" cell into two identical "daughter" cells.

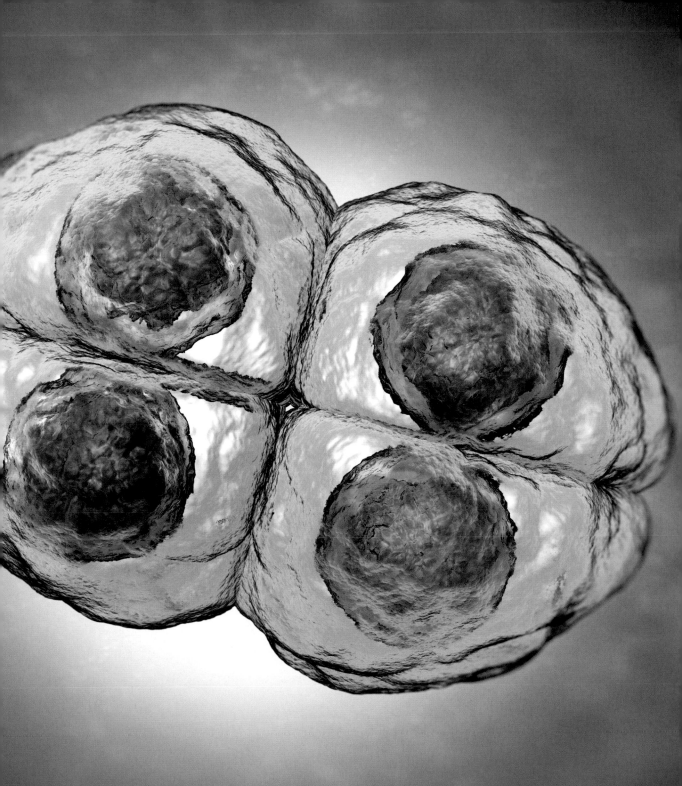

Thermoreception

Johann Wilhelm Ritter (1776–1810)

An animal's survival depends upon its ability to maintain internal body temperature within a critical range required for biochemical and physiological processes to occur. Thermoreception is a sense that enables an animal to detect the temperature of its external environment and its internal environment (body or core temperature). Endotherms (commonly referred to as "warm-blooded animals"), such as **birds** and **mammals**, require stable internal temperatures for survival and create most of their body heat by metabolic processes. By contrast, ectotherms (so-called "cold-blooded animals"), including some fish, **amphibians**, and **reptiles**, have a variable body temperature that is dependent upon external sources of heat; they adjust their behavior in ways that control their temperature. While most insects are ectotherms, flying insects generate considerable body heat that must be lost in order to maintain their normal body temperature.

In 1801, the German chemist and physicist Johann Ritter provided the first evidence that warmth and cold were sensory qualities, and these represent two of the four types of touch sensations. Several groups of researchers in the early 1880s noted that sensory spots on the skin were selectively sensitive to thermal sensations; these were thermoreceptors. In response to warm and cold stimuli, electrical signals were detected from single nerve fibers in a cat's tongue (1936) and from human skin (1960).

The general nature of thermoreceptors in external parts of the body is similar in almost all animal species and, based upon the species, is selectively sensitive to specific ranges of temperatures, as well as to the rate at which temperature changes. In birds and mammals, thermoreceptors in the hypothalamus activate processes that promote heat production and loss, maintaining inner body temperature within a normal range.

Below and in front of the eyes, pit vipers, including rattlesnakes, have thermosensitive pits that detect body heat from potential prey and also serve to locate its direction and distance from the snake. Most insects have thermoreceptors located in their antennae. Blood-sucking insects, such as mosquitoes and lice, use the warmth of their victim's body as the primary influence to stimulate such behavior and to guide their blood feeders.

SEE ALSO: Amphibians (c. 360 Million BCE), Reptiles (c. 320 Million BCE), Mammals (c. 200 Million BCE), Birds (c. 150 Million BCE), Metabolism (1614), Homeostasis (1854), Negative Feedback (1885).

Reptiles, such as this Jackson's chameleon (Trioceros jacksonii), *found in Hawaii and East Africa, have body temperatures that vary with external sources of heat.*

Innate Immunity

Élie Metchnikoff (1845–1916)

In 1882, while strolling down the beach in Messina, Sicily, where he maintained a private laboratory, the Russian zoologist Élie Metchnikoff found a live starfish that he poked with a rose thorn. The following morning, to his surprise, he found cells covering the thorn as though trying to engulf and destroy it. Metchnikoff recognized that this phenomenon of *phagocytosis*—an early and immediate line of defense utilized by most vertebrates, invertebrates, microbes, and plants to combat pathogenic organisms or foreign cells—had broader biological significance not limited to the starfish. This primitive invertebrate had remained unchanged for 600 million years, and it could provide insights into the evolution of the immune system, which defends the body against disease. Metchnikoff was the first to recognize innate or natural immunity, a discovery for which he was awarded the 1908 Nobel Prize.

Innate immunity is a rapid nonspecific response that does not require prior exposure to the foreign organism or cell and utilizes a range of several modes of defense: The first line response to disease-causing microbes is the presence of anatomical or physical barriers, such as the skin or shell, mucus, and cells of the gastrointestinal and respiratory systems. Phagocytosis is carried out by neutrophils, white blood cells (leukocytes) in mammalian blood, and macrophages in tissues. An inflammatory response is activated by chemicals released at the site of injury that wall off and prevent the spread of infection. Then a complement system—in which more than thirty proteins (such as natural killer cells and interferon) are activated and mobilized—destroys and eliminates invaders.

Both innate immunity and adaptive (or acquired) immunity—the latter only present in vertebrates and only activated after prior exposure to the microbe or cell—are based on the premise that the invaded animal can recognize which are its own cells (self) and which are foreign (nonself). In innate immunity, detection of nonself is signaled by the presence of pattern recognition molecules, which are present in foreign microbes but absent from animals. Identification of these molecules triggers the range of immune responses. Such was the response by Metchnikoff's starfish.

SEE ALSO: Lymphatic System (1652), Adaptive Immunity (1897), Ehrlich's Side-Chain Theory (1897), Acquired Immunological Tolerance and Organ Transplantation (1953).

This red starfish (Fromia elegans), *which has not evolved in over 600 million years, exhibits the phenomenon of* phagocytosis, *a primitive innate immune response, to remove pathogens.*

Germ Plasm Theory of Heredity

Jean-Baptiste Lamarck (1744–1829), **August Weismann** (1834–1914)

The concept that individuals inherited traits acquired through use and disuse by their parents dates back to the ancient Greeks and was formally stated by Jean-Baptiste Lamarck in the early 1800s. Throughout most of the nineteenth century, Lamarckism and the blending of the characteristics of the father and mother were the favored theories of heredity. Among the greatest of all biologists was the German August Weismann who, in a dramatic experiment, greatly weakened support for Lamarckism. He amputated the tails of 901 mice for five successive generations; no mouse was ever born without a tail and the tail lengths in the fifth generation were as long as those in the first. Rather than refuting **Lamarckian inheritance** by logic, Weismann did so by experimentation.

In his germ plasm theory, proposed in 1883 and detailed in his 1893 book, *The Germ-Plasm: A Theory of Heredity*, Weismann emphasized the stability of germ plasm (hereditary material, now called *genes*) that was transmitted, without change, from generation to generation. In his theory, the environment had little, if any, effect on the germ plasm, even if the environment altered external body characteristics. Weismann drew a clear distinction between body material (soma) and hereditary material (germ plasm). He postulated that in multicellular organisms, germ plasm is independent of other body cells; that somatic cells (non-sex cells) are involved in bodily activities but do not function in heredity; and, most importantly, that germ plasm is the essential element of germ cells or gametes (sperm cells and egg cells).

Weismann's germ plasm theory had a great influence on biological thinking— namely, the rediscovery of Mendel's rules and the role of chromosomes in inheritance— but a number of recent findings have severely undermined its validity. Weismann's view that there is only a set quantity of germ plasm, which is reduced with successive generations after successive somatic cell divisions, was undermined when the clone ewe, Dolly, produced by somatic cell nuclear transfer, was found to possess a full complement of germ plasm. In addition, Lamarckism has been recently resurrected as **epigenetics**.

SEE ALSO: Lamarckian Inheritance (1809), Mendelian Inheritance (1866), Genetics Rediscovered (1900), Genes on Chromosomes (1910), Cloning (Nuclear Transfer) (1952), Epigenetics (2012).

This illustration from "Thumbelina," in a 1913 edition of Hans Andersen's Fairy Tales, shows an old field mouse giving shelter to a tiny girl. An experiment in which five successive generations of mice had their tails amputated challenged the Lamarkian view of inheritance, as the tail lengths of the fifth generation matched those of the first.

Eugenics

Charles Darwin (1809–1882), Francis Galton (1822–1911)

Francis Galton, a man of many intellectual talents, made significant contributions to such diverse areas as meteorology (weather maps), statistics (correlation and regression analysis), and criminology (fingerprinting). Upon reading his cousin Charles Darwin's *Origin of Species*, he became inspired by the notion that if natural selection enables the fittest organisms to survive and pass on their traits, it must also apply to humans—human ability and intelligence must be hereditary.

In 1883, Galton initiated a social movement, which he called *eugenics* ("good birth"), intended to improve the genetic composition of the human population. Eugenics, called "social Darwinism" by some, enjoyed its greatest popularity during the early decades of the twentieth century. It was practiced throughout the world and actively promoted by governments and some of society's most influential and respected individuals. While its advocates argued that the results would lead to more intelligent and healthier people by eliminating such hereditary diseases as hemophilia and Huntington's disease, its opponents viewed eugenics as a justification for state-sponsored discrimination and human rights violations.

Practices arising from the eugenics movement varied among countries. Great Britain sought to decrease the birth rate among the urban poor. In the United States, many states enacted laws prohibiting the marriage of epileptics, the "feebleminded," and mixed-race individuals. Thirty-two states had eugenics programs that resulted in the sterilization of 60,000 individuals from 1909 to the 1960s.

By far, the most egregious interpretation of eugenics was responsible for the racial policies of Nazi Germany seeking to promote a pure and superior "Nordic race" and eliminate the less fit and undesirable, which led to the annihilation of millions of Jews, Romani (Gypsies), and homosexuals. By the end of World War II, because of its association with Nazi Germany and concerns that what is improved or beneficial is highly subjective and often based on prejudice, the active pursuit of eugenic programs fell into disfavor. More recently, some have argued that medical genetics, with *in utero* testing for mutations leading to diseases or fetal gene manipulation, is the new eugenics. These are decisions made by the individual, however—not the state.

SEE ALSO: Darwin's Theory of Natural Selection (1859), Mendelian Inheritance (1866), Genetics Rediscovered (1900), Sociobiology (1975).

Ultrasound, commonly performed during the eigteenth to twentieth weeks of pregnancy, can be used to detect birth defects such as spina bifida and Down syndrome.

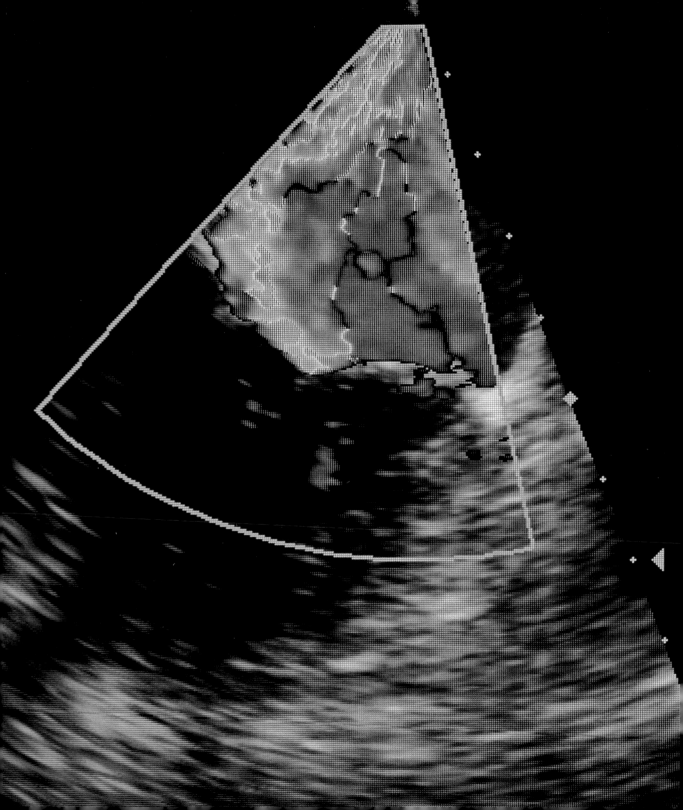

Gram Stain

Hans Christian Gram (1853–1938)

One year after graduating from medical school, in 1884, while working in a Berlin mortuary, the Danish scientist Hans Christian Gram developed a stain that permitted him to visualize some but not all bacteria in lung tissue. This simple but major discovery subsequently led to the finding that many bacteria can be differentiated into two broad categories based on the thickness of their cell walls, aiding in the diagnosis and treatment of bacterial infections.

Cell walls, found in bacteria, plants, and fungi, but not in animals or protozoa, provide protection and support of the cell, and perhaps most important, prevent bursting if excess water enters the cell. It is the cell wall that traps certain dyes, permitting their bacteria to be visualized.

In the Gram stain procedure, Gentian (crystal) violet is poured over a slide containing bacteria and Lugol's (iodide) solution is added to fix the dye. The slide is then washed with ethanol. Certain bacteria (such as the pneumonia-causing *Streptococcus pneumoniae*) retain the dye and appear purple; these are Gram-positive bacteria. Other microbes (the typhus- and syphilis-causing bacteria, for example) become decolorized by the alcohol and assume a red or pink color—Gram-negative bacteria. Gram-positive bacteria have thick cell walls and trap the purple stain in their cytoplasm, while Gram-negative bacteria have much thinner cell walls from which the dye is readily washed. The Gram stain is routinely used in medicine as a diagnostic tool to differentiate infections caused by Gram-positive or Gram-negative bacteria and provides a rational basis for the selection of **antibiotics**.

Most antibiotics are preferentially effective against either Gram-positive or Gram-negative bacteria. For example, penicillin combats many Gram-positive bacteria by interfering with their ability to synthesize cell walls that are essential for their survival. (Animal cells lack cell walls and, therefore, penicillin is not toxic to them.) The thick outer membrane of Gram-negative bacteria protects it against the body's defenses and also impedes the passage of many antibiotics into the cell. The aminoglycosides are a class of antibiotics used for the treatment of these bacteria.

SEE ALSO: Germ Theory of Disease (1890), Endotoxins (1892), Antibiotics (1928).

After applying the Gram stain, Gram-positive Bacillus cereus *are dyed violet and appear in a series of chains, while Gram-negative* Escherichia coli *are the small pink clusters in the background.*

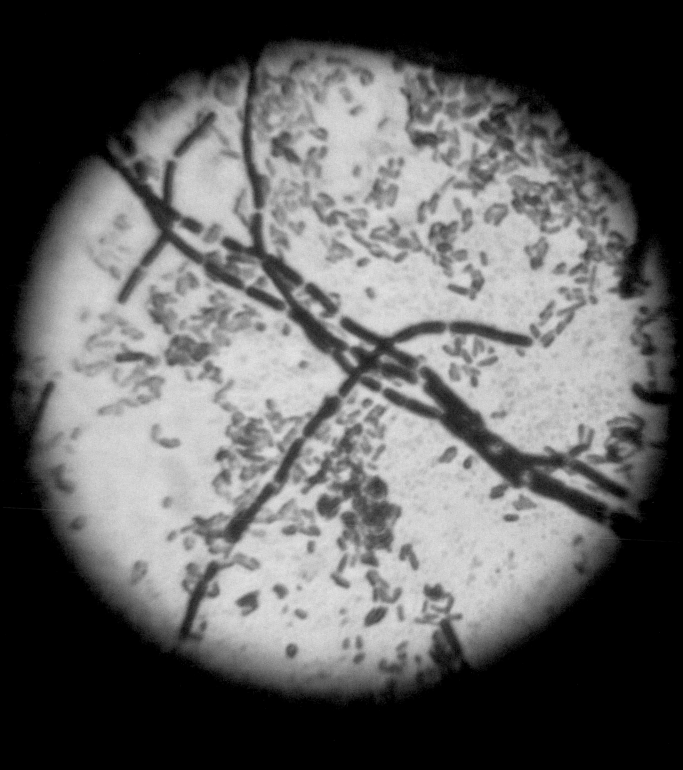

Negative Feedback

Claude Bernard (1813–1878), **Albert Butz** (1849–1905),
Walter B. Cannon (1871–1945)

Homeostasis, a fundamental principle in biology, is a concept that was developed by Claude Bernard during the 1850s and expanded upon and popularized by W. B. Cannon during the 1920s to1930s. It is the process by which a living being maintains a constant internal environment when its external environment is changing. Negative feedback control systems, whether in biological or nonliving systems, consist of three integral components: a receptor that detects changes in the system; a control center that compares the change with a set or reference point, which in biological systems are the normal values; and an effector, which initiates appropriate action to return the system to its reference point. By analogy, consider the home furnace and thermostat. In 1885, Albert Butz invented the earliest functional thermostat. The furnace continues to heat the facility until a set temperature is detected by the thermostat, which shuts the furnace down and then turns it on when the facility's temperature falls below the set temperature.

Many endocrine systems, such as blood glucose levels, are linked to control centers by homeostatic negative feedback mechanisms that operate in a cyclical and continuous manner. After eating a carbohydrate-rich meal, blood glucose levels rise, stimulating the release of **insulin** from the beta cells of the pancreas. Glucose enters body cells, and the liver takes up the excess sugar, which it stores as glycogen. Blood glucose levels are detected and compared with set levels (70–110 mg glucose/100 ml blood). If levels are too low, insulin secretion stops and glucagon is released from the alpha cells of the pancreas stimulating the breakdown of liver glycogen to glucose, which is released in the blood.

Negative feedback inhibition also controls the amount of final product that is synthesized in many enzyme-catalyzed biochemical pathway reactions. After an optimal amount of end product is formed, the end product reacts with an enzyme in the pathway, interfering with synthesis of additional compounds.

SEE ALSO: Metabolism (1614), Homeostasis (1854), Enzymes (1878), Thyroid Gland and Metamorphosis (1912), Insulin (1921), Progesterone (1929), Second Messengers (1956), Energy Balance (1960), Hypothalamic-Pituitary Axis (1968), Cholesterol Metabolism (1974).

An illustration of a complex machine with a steam boiler, gears, levers, pipes, meters, furnace, flue, and presumably a thermostat to provide a negative feedback loop that will keep the temperature of the machine at a reasonable level.

Germ Theory of Disease

Antonie van Leeuwenhoek (1632–1723), **Louis Pasteur** (1822–1895),
Robert Koch (1843–1910)

Prior to the latter half of the nineteenth century, and dating back to ancient times in China, India, and Europe, it was widely accepted that such infectious diseases as cholera and the Black Death were caused by "bad air" or miasma. Infections were thought to be spread by contact with poisonous vapors filled with decomposed or rotting matter.

The "germ theory of disease" may be the most important contribution of microbiology to modern medical science and practice, and it has served as the basis for the use of **antibiotics** for the treatment of infectious diseases. The concept that microbes were the cause of some diseases evolved over several hundred years, with multiple scientists providing evidence that culminated in the theory and its acceptance by the medical and scientific community.

Using a simple microscope, microbes were first seen and described in the 1670s by the Dutch lens maker Antonie van Leeuwenhoek. Almost two centuries later, in 1862, Louis Pasteur conducted decisive experiments that refuted another long-held theory—namely, spontaneous generation—that living organisms could arise from nonliving matter. Pasteur demonstrated that microbes were present in the air but were not created by air.

Robert Koch transformed from a simple practicing German physician to one of the pioneer founders of microbiology (as was Pasteur) after he received a late-twenties birthday gift of a microscope from his wife. From 1876 to 1883, he discovered the bacterial causes of anthrax, tuberculosis, and cholera, and devised methods for isolating pure cultures of disease-causing microbes. In 1890, he devised rules that are still used (with some modification) to determine whether a given microbe causes a disease. These postulates state that: (a) the microbe must be present in every case of the disease; (b) the microbe must be isolated and grown in pure culture; (c) the disease must be produced when the microbe is administered to a healthy individual; and (d) the microbe must then be reisolated from the individual. Koch was awarded the 1905 Nobel Prize for his work on tuberculosis, a disease that was responsible for one of every seven deaths in the mid-nineteenth century.

SEE ALSO: Scientific Method (1620), Refuting Spontaneous Generation (1668), Leeuwenhoek's Microscopic World (1674), Miasma Theory (1717), Gram Stain (1884), Endotoxins (1892), Antibiotics (1928).

When developing his germ theory of disease, Koch used Bacillus anthracis (*shown in this digital illustration*), *employing purified cultures of the microbe that had been isolated from anthrax-diseased animals.*

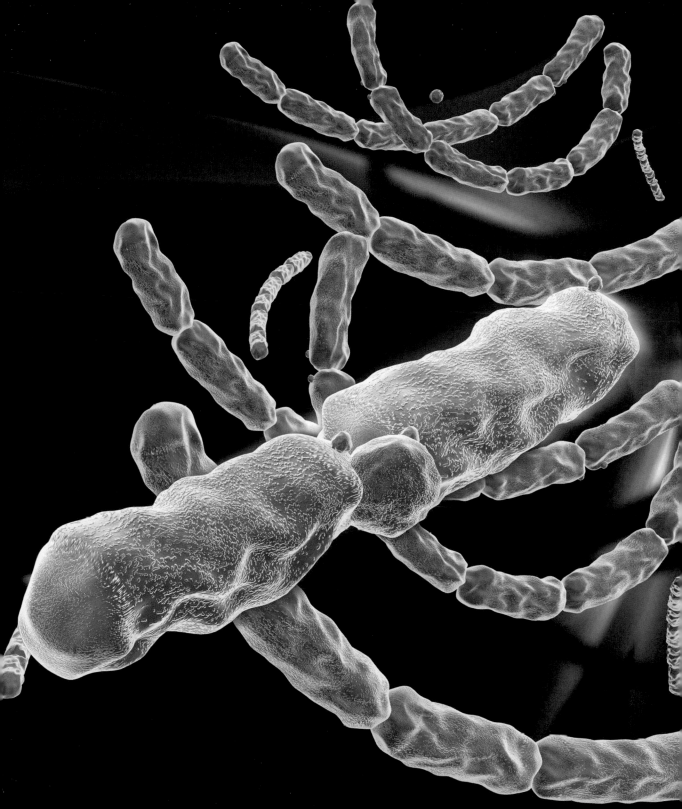

Animal Coloration

Robert Hooke (1635–1703), **Charles Darwin** (1809–1882),
Fritz Müller (1821–1897), **Henry Walter Bates** (1825–1892),
Edward Bagnall Poulton (1856–1943)

One cannot help but be impressed by the diversity of colors seen in animals. In his *Colour of Animals*, written in 1890, Edward Poulton, an evolutionary biologist and Oxford professor of zoology, provided the first comprehensive text on animal coloration. As a subtext, the book was intended to actively support Charles **Darwin's theory of natural selection**, which was then besieged by many of his contemporaries.

Poulton was not the first to comment on coloration in animals. Robert Hooke, a pioneer microscopist, first described the structure and brilliant colors of a peacock's feathers in his classic 1665 work *Micrographie*. In *Descent of Man, and Selection in Relation to Sex* (1871), Darwin proposed that conspicuous coloration evolved to provide individual animals, in particular, male **birds**, with a reproductive advantage in attracting females. Moreover, duller colors provided birds and **insects** with camouflage to conceal themselves from the covetous eyes of predators, a concept elaborated upon by Poulton.

In the *Colour of Animals* and findings by others, coloration was observed to provide animals with a diverse array of survival benefits. Poulton was first to emphasize that camouflage coloration enabled prey to avoid potential predators but also enabled predators to conceal themselves or to lure unsuspecting prey. He acknowledged the work of Henry Bates (1862) on the use of coloration by butterflies to resemble another species and thereby deceive predators; and by Fritz Müller who, in 1878, introduced the concept that coloration served as a warning signal (aposematism) to an approaching predator that the would-be prey was prepared and capable of defending itself.

Coloration provides animals other survival benefits: Some use flashes of light, bold patterns, or motion to divert attacks by predators. Coloration can protect others against sunburn, while certain frogs lighten or darken their skin to control their body temperature. Male monkeys use coloration to assess the social status of their peers. Poulton concluded that pigments in animal tissues produced coloration and that the brilliant colors seen in some birds were the result of consuming carotenoid-containing plants.

SEE ALSO: Insects (c. 400 Million BCE), Birds (c. 150 Million BCE), Darwin's Theory of Natural Selection (1859), Biological Mimicry (1862), Sexual Selection (1871).

Feather from the male Indian peafowl (Pavo cristatus), *the national bird of India. Male birds are generally more colorful or ornamented than females, perhaps because it confers a reproductive advantage, while females can more easily conceal themselves from predators while raising their young.*

Neuron Doctrine

Joseph von Gerlach (1820–1896), **Wilhelm Waldeyer** (1836–1921), **Camillo Golgi** (1843–1926), **Santiago Ramón y Cajal** (1852–1934)

One of greatest scientific battles of the late nineteenth century involved the fundamental nature of the nervous system's structure. This debate pitted two preeminent rival neuroanatomists—Camillo Golgi, an Italian, and Santiago Ramón y Cajal, a Spaniard—who harbored hostility toward each other even as they were jointly awarded the 1906 Nobel Prize.

The **cell theory** of 1838 proposed that the cell is the fundamental unit of life, a concept not extended to the nervous system, which is far more complex in its structural organization. In 1873, Golgi announced that when he used his newly formulated silver stain, a reazione nera (a "black reaction"), it permitted him a clear full view of single nerve cells on a yellow background. He described the cells as a branching network or reticulum that was the anatomical and functional unit for **nervous system communication**. This description supported the reticulum theory proposed in 1872 by the German histologist Joseph von Gerlach and became the prevailing view toward the end of the nineteenth century. Nerve cells were viewed an exception to the cell theory.

Cajal, working in virtual scientific isolation in Spain in the late 1880s, used the same stain as Golgi but reached a diametrically different conclusion. His microscopic analysis revealed that each neuron (nerve cell) was a distinct entity, not contiguous with other cells. Cajal reported his initial findings in 1891 in Spanish, a little-read language in science, so Wilhelm Waldeyer formally consolidated Cajal's findings and proposed the neuron doctrine in a widely read German publication. The doctrine, which stated that the neuron was the structural and functional unit of the nervous system, was later conclusively established with the **electron microscope**. This view is now considered to be the foundation of neuroscience.

Waldeyer's name is closely associated with the neuron doctrine, although he provided no original observations in its formulation. In 1892, Cajal hypothesized the law of dynamic polarization, which stated that electrical impulses in neurons traveled in only one direction from dendrite → cell body → axon → dendrite in another cell.

SEE ALSO: Leeuwenhoek's Microscopic World (1674), Nervous System Communication (1791), Cell Theory (1838), Neurotransmitters (1920), Electron Microscope (1931).

The neuron (nerve cell) is the structural and functional unit of the nervous system and is physically separated from adjacent neurons by a synapse (gap). Neurons communicate with one another via chemicals called neurotransmitters.

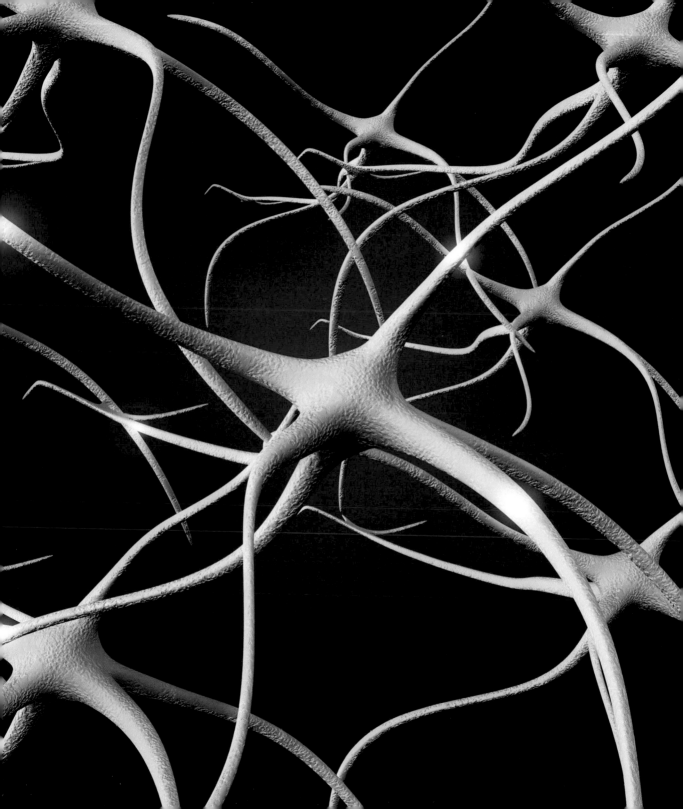

Endotoxins

Filippo Pacini (1812–1883), **Robert Koch** (1843–1910), **Richard Friedrich Johannes Pfeiffer** (1858–1945), **Eugenio Centanni** (1863–1942)

DEATH BY CHOLERA. Likely originating on the Indian subcontinent in ancient times, cholera was one of the most widespread and deadly nineteenth-century diseases, killing tens of millions in Asia and Europe. Cholera victims experienced high fever, extreme diarrhea and vomiting, rapidly became dehydrated, and often died. The bacterial cause of an 1854 cholera pandemic in Florence was first identified by the Italian anatomist Filippo Pacini, but his findings were ignored by the medical community, which preferred subscribing to the traditional **miasma ("bad air") theory** of disease. In 1883, the bacterium *Vibrio cholera* was rediscovered by the German bacteriologist Robert Koch, who established the **germ theory of disease** in 1890, despite being unaware of Pacini's earlier finding. The bacterial cause of cholera was now accepted.

In 1892 while working with the cholera-causing bacterium, Richard Pfeiffer, a protégé of Koch's at the Institute of Hygiene in Berlin, first conceived and then proved the concept of an endotoxin. Pfeiffer injected experimental animals with a mixture of cells that had ruptured after being exposed to the cholera bacterium, causing the animals to go into shock and die. Pfeiffer postulated that a substance was released when the envelope surrounding certain bacteria was broken down. It was subsequently determined that the observed endotoxic response is the consequence of an inflammatory reaction mounted by the host (patient), intended to combat localized infections. However, when confronted by a severe, body-wide infection, such as cholera, the inflammatory response becomes excessive, leading to septic shock, in which blood pressure precipitously falls and death may ensue. (Pfeiffer differentiated an endotoxin from an exotoxin—the latter, a toxin released by bacteria into the environment.)

Italian pathologist Eugenio Centanni showed that while this endotoxic substance was released from some Gram-negative microbes, it was never associated with Gram-positive bacteria. In 1935, lipopolysaccharide (LPS), part of the outer cell membrane of Gram-negative microbes, was found to be the trigger for the endotoxic effects in such infectious disorders as cholera, salmonella, and bacterial meningitis. LPS is now used synonymously with the more historic designation *endotoxin*.

SEE ALSO: Miasma Theory (1717), Gram Stain (1884), Germ Theory of Disease (1890).

During the First Balkan War (1912–1913) between Turkey and the Balkan league, the Turkish army was ravaged by a cholera epidemic that caused 100 deaths per day. This image from a 1912 French magazine shows the grim reaper decimating a column of Turkish soldiers.

Le Petit Journal

ADMINISTRATION
61, RUE LAFAYETTE, 61

Les manuscrits ne sont pas rendus

On s'abonne sans frais dans tous les bureaux de poste

5 CENT. **SUPPLÉMENT ILLUSTRÉ** **5** CENT.

23me Année ✱✱ Numéro 1.150

DIMANCHE 1er DÉCEMBRE 1912

ABONNEMENTS

	SIX MOIS	UN AN
SEINE et SEINE-ET-OISE	2 fr.	3 fr. 50
DÉPARTEMENTS	2 fr.	4 fr. »
ÉTRANGER	2 50	5 fr. »

LE CHOLÉRA

Global Warming

Jean Baptiste Joseph Fourier (1768–1830), **John Tyndall** (1820–1893), **Svante Arrhenius** (1859–1927)

The concept of the *greenhouse effect*, which underlies global warming, evolved from a series of observations dating back to the early nineteenth century. In 1826, Joseph Fourier calculated that the Earth's temperature, if it were warmed only by the sun, would be 60°F (15.5°C) cooler and further surmised that the atmosphere served as an insulator that prevented heat loss. Later, in 1859, John Tyndall discovered that water vapor and carbon dioxide (CO_2) in the atmosphere were responsible for this heat trapping. In 1896, Svante Arrhenius noted a quantitative relationship between the concentration of atmospheric carbon dioxide (CO_2) and the average surface temperature of the Earth. He called this phenomenon *hothouse*, which a decade later was renamed "the greenhouse effect."

Overwhelmingly, the scientific community has attributed this temperature rise to an increase in greenhouse gases (GHG). The most important GHG are water vapors and CO_2, with human activity responsible for most of the CO_2. Sources of this gas are fossil fuels used by cars, factories, electricity production, and deforestation. The Intergovernmental Panel on Climate Change (IPCC) projects a further average rise of 2–5.2°F (1.1–2.9°C) during this century, to be most extreme in the Arctic, leading to glacial melting. Other predicted consequences of global warming include more extreme weather (heat waves, drought, heavy rain), decreased crop yields, changes in migratory patterns of animals, a decrease in biodiversity, and plant and animal species extinction.

The IPCC, a United Nations group with representatives from all major industrialized nations, and virtually all national scientific academies agree that both the Earth's surface temperature and the rate of warming of the atmosphere and oceans have been rising faster in recent decades. Some scientists and members of the general population, however, question whether these temperature changes are within normal climatic variation, if human activities are responsible for their occurrence, and what the appropriate measures for its remediation may be. Some of these questions undoubtedly arise from sincere differences in the interpretation of the data, while others are motivated by political, philosophic, or economic considerations.

SEE ALSO: Amazon Rainforest (c. 55 Million BCE), Depletion of the Ozone Layer (1987).

Some scientists predict that global warming will cause more extreme weather, including abnormally persistent periods of drought and heavier rainfall.

Adaptive Immunity

Thucydides (c. 460–c. 395 BCE), **Edward Jenner** (1749–1823),
Hans Buchner (1850–1902), **Paul Ehrlich** (1854–1915)

Those who recovered from a previous exposure to the plague could safely care for patients who were ill without contracting the disease a second time. So observed the Greek historian Thucydides during the plague of Athens in 430 BCE. This same principle applied to Edward Jenner's 1796 observation that milkmaids who had contracted cowpox did not get smallpox. A century later, the mechanism was uncovered: In 1890, Hans Buchner discovered a "protective substance" in blood serum capable of destroying bacteria, and in 1897, Paul Ehrlich identified these as antibodies responsible for conferring immunity.

Invertebrates and vertebrates initiate an immediate defensive mechanism when confronted by a pathogenic microbe or foreign tissue. This primitive, nonspecific response is called **innate immunity**. Vertebrates have an additional and far more powerful layer of immunological protection, called *adaptive* or *acquired immunity*, which develops over weeks. Adaptive immunity is characterized by molecular specificity directed against that pathogen and an immunological memory that specifically targets that pathogen even when reexposure occurs at a distant future time, when the immunologic response is rapid and enhanced.

Two types of adaptive immunity occur, both arising from lymphocytes (a type of white blood cell): Humoral immunity or B-cell immunity results in antibody formation in the blood that attacks the microbe or foreign cell. The second, cell-mediated immunity or T-cell immunity, causes the formation of a large number of activated lymphocytes specifically also intended to kill the foreign cell. An antigen, any protein substance that stimulates a response from either B- or T-cells, binds to a specific antigen receptor on either cell type. The binding of an antigen to a B-cell antigen receptor leads to the formation of an antibody or immunoglobulin that eliminates the pathogens in the blood, while activated T-cells promote the production of antibodies or kill infected cells.

Adaptive immunity can also be conferred artificially via vaccinations, such as those against polio, measles, and hepatitis. Tissue and organ transplants, containing foreign cells, may initiate an immune response that can result in their rejection.

SEE ALSO: Innate Immunity (1882), Acquired Immunological Tolerance and Organ Transplantation (1953).

A scanning electron micrograph of a human T-cell. After recognizing molecules on the surface of a virus, the host's T-cells are activated to mobilize an attack on the foreign invaders, ideally resulting in their destruction.

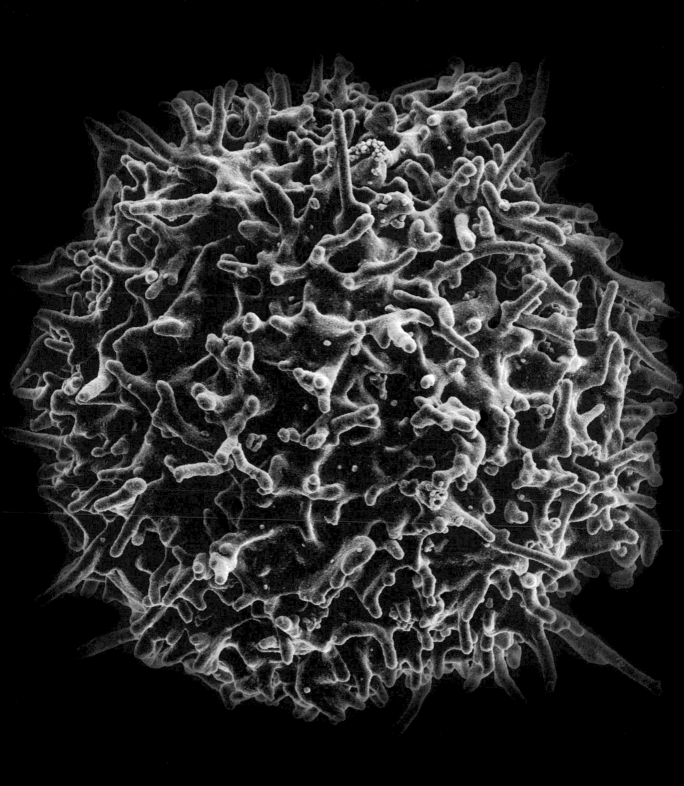

Associative Learning

Ivan Pavlov (1849–1936), **Edward L. Thorndike** (1874–1949),
B(urrhus) F(rederic) Skinner (1904–1990)

When owners show their "best friends" their leash, dogs commonly exhibit unrestrained enthusiasm barking and running about. What we are witnessing is *associative learning*, which is any learned process associated with a specific stimulus, and includes classical and operant conditioning. In 1905, the Columbia University psychologist Edward Thorndike postulated that a behavioral response (R) was most likely to reoccur if the subject was presented with the same stimulus (S).

Thorndike's Law of Effect was dramatically demonstrated by the Russian physiologist Ivan Pavlov and described in his 1897 monograph *The Work of the Digestive Glands*. Pavlov was interested in studying stomach digestion in dogs and measured secretions from the salivary gland in response to food placed in their mouths. Initially, the dogs salivated when food (S) was in their mouth but, after a number of trials, they began to salivate (R) prior to being given food; Pavlov termed this "psychic secretion," and it became the focus of his studies. The dogs continued to salivate when a bell was paired with the food, and, with repeated sessions, they came to associate the bell with the food, salivating even in the absence of food. This stimulus-response learning is referred to as *classical* or *Pavlovian conditioning*, and its formulation earned Pavlov the 1904 Nobel Prize. In *Clockwork Orange* (1962), a novella by Anthony Burgess adapted to film in 1971 by Stanley Kubrick, Alex, the antihero, undergoes the Ludovico technique to cure his antisocial behavior. He is given a nausea-eliciting drink, which is paired with violent acts portrayed on the screen; the procedure renders him totally averse to committing acts of violence even without experiencing extreme nausea.

Operant (or *instrumental*) *conditioning* was championed from the late 1940s to 1970s by Harvard psychologist, B.F. Skinner, who was named the most influential psychologist of the twentieth century in a 2002 survey of psychologists. In this procedure, a test subject (pigeon or rat) is provided a food reward ("reinforcement") or can avoid a noxious foot shock after completing predetermined learned responses. This type of learning serves as the basis for rewards that teachers bestow upon students after completing their work meritoriously or payouts from slot machines at gambling casinos.

SEE ALSO: Imprinting (1935).

While held in the highest esteem by Vladimir Lenin, Pavlov, whose likeness appears on this bronze bust, was highly vocal in his contempt for Soviet persecution of intellectuals during the 1920s and 1930s.

Ehrlich's Side-Chain Theory

Carl Weigert (1845–1904), **Paul Ehrlich** (1854–1915)

Paul Ehrlich, born in Prussia in 1854, was a German medical scientist who conducted pioneering work in hematology, immunology, and chemotherapy, and discovered the first effective drug treatment for syphilis. Introduced to the staining of cells by his cousin Carl Weigert, a renowned neuropathologist, this interest—perhaps even obsession— persisted and influenced Ehrlich's conceptual thinking for much of his scientific career. During his medical school days, while continuing to experiment with chemical dyes, Ehrlich observed that some cells and tissues selectively took up and bound chemical dyes and were stained, yet others did not. After completing medical school, he developed a dye that permitted the differentiation of the numerous **blood cells**, and this served as the basis for the study of hematology.

In 1893, while working on an antiserum for the treatment of diphtheria, Ehrlich began formulating his side-chain theory, which described how antibodies—proteins produced by the immune system—are formed, and how they interact with foreign substances (antigens). Based on an analogy of a lock and key, he postulated that the surface of each cell contains distinct receptors or "side chains" that specifically bind to disease-causing toxins produced by the infectious agent. The binding of the toxin to the side chain (key-to-lock) is an irreversible interaction, and prevents any additional binding of toxin molecules.

The body responds by producing an excess number of side chains (antibodies), but the cell lacks the capacity to accommodate all the side chains on its cell surface. The excess side chains are released, where they remain in circulation, prepared to protect the individual against subsequent attacks by the disease-causing toxins. Ehrlich's first paper describing his side-chain theory appeared in 1897. The theory was publically presented at a Royal Society meeting in London in 1900, where it was enthusiastically received, and for which he was a co-recipient of the 1908 Nobel Prize. By 1915, the year of Ehrlich's death, exceptions to his theory were identified, and many details were found to be incorrect. The theory fell into disfavor, but his concepts on antigens and antibodies serve as the basis for immunology.

SEE ALSO: Blood Cells (1658), Adaptive Immunity (1897), Protein Structures and Folding (1957), Monoclonal Antibodies (1975).

The "lock and key" analogy Ehrlich used to formulate his side-chain theory was also the basis for his attempt to develop a "magic bullet": a drug that selectively kills a disease-causing microbe without harming the patient. This led to his 1910 discovery of Salvarsan, the first effective treatment for syphilis.

Malaria-Causing Protozoan Parasite

Charles-Louis-Alphonse Laveran (1845–1922), Ronald Ross (1857–1932)

Although the parasite responsible for causing malaria has been in existence for at least 50,000 to 100,000 years, its population significantly increased around 10,000 years ago, a time concurrent with the beginnings of **agriculture** and human settlements. Malaria was once common in most of Europe and North America and was only declared eliminated from the United States in 1951. The World Health Organization estimated that in 2010 there were 219 million cases of malaria and 600,000 deaths, 90 percent of them in Africa.

The lifecycle of the malaria parasite, which involves both an insect and a human vector, was determined during the final decades of the nineteenth century. In 1880, the French army doctor Charles-Louis-Alphonse Laveran observed the presence of protozoa (single-celled microbes) within the red **blood cells** of malaria-infected patients and speculated that it might be responsible for its cause. Working in Calcutta, in 1898, the British physician Ronald Ross determined the complete lifecycle of the malaria parasite in mosquitoes and established that the mosquito was the vector carrying the *Plasmodium* parasite to humans. Ross and Laveran were awarded the 1902 and 1907 Nobel Prizes, respectively.

The female *Anopheles* mosquito, carrying the *Plasmodium*, feeds on human blood. In the process, it injects the parasite into the bloodstream, through which it invades liver cells and produces tens of thousands of merozoites per liver cell. The merozoites enter the bloodstream (where they cause malaria-characteristic periodic chills and fever), penetrate red blood cells, and reproduce. When a mosquito bites an infected human, it ingests sporocytes, which travel from the mosquito's gut to its salivary gland, restarting the cycle when it bites another victim.

One type of genetic resistance to malaria has been attributed to changes in red blood cells, deforming the cells into a sickle shape, which interferes with the parasite's ability to invade and reproduce in these cells. Sickle-cell disease, the most commonly inherited disease among individuals of African descent, reduces the frequency and severity of malaria attacks, particularly in young children who are most affected by malaria. Sickle-cell disease may, therefore, confer an evolutionary advantage for those residing in Africa where malaria is prevalent.

SEE ALSO: Agriculture (c. 10,000 BCE), Blood Cells (1658), Inborn Errors of Metabolism (1923).

An Israeli postage stamp illustrates both the Anopheles *mosquito and the sharp decline in malaria in that country. There are some 484 species of* Anopheles, *but only 30–40 transmit the* Plasmodium, *which causes malaria in endemic areas.*

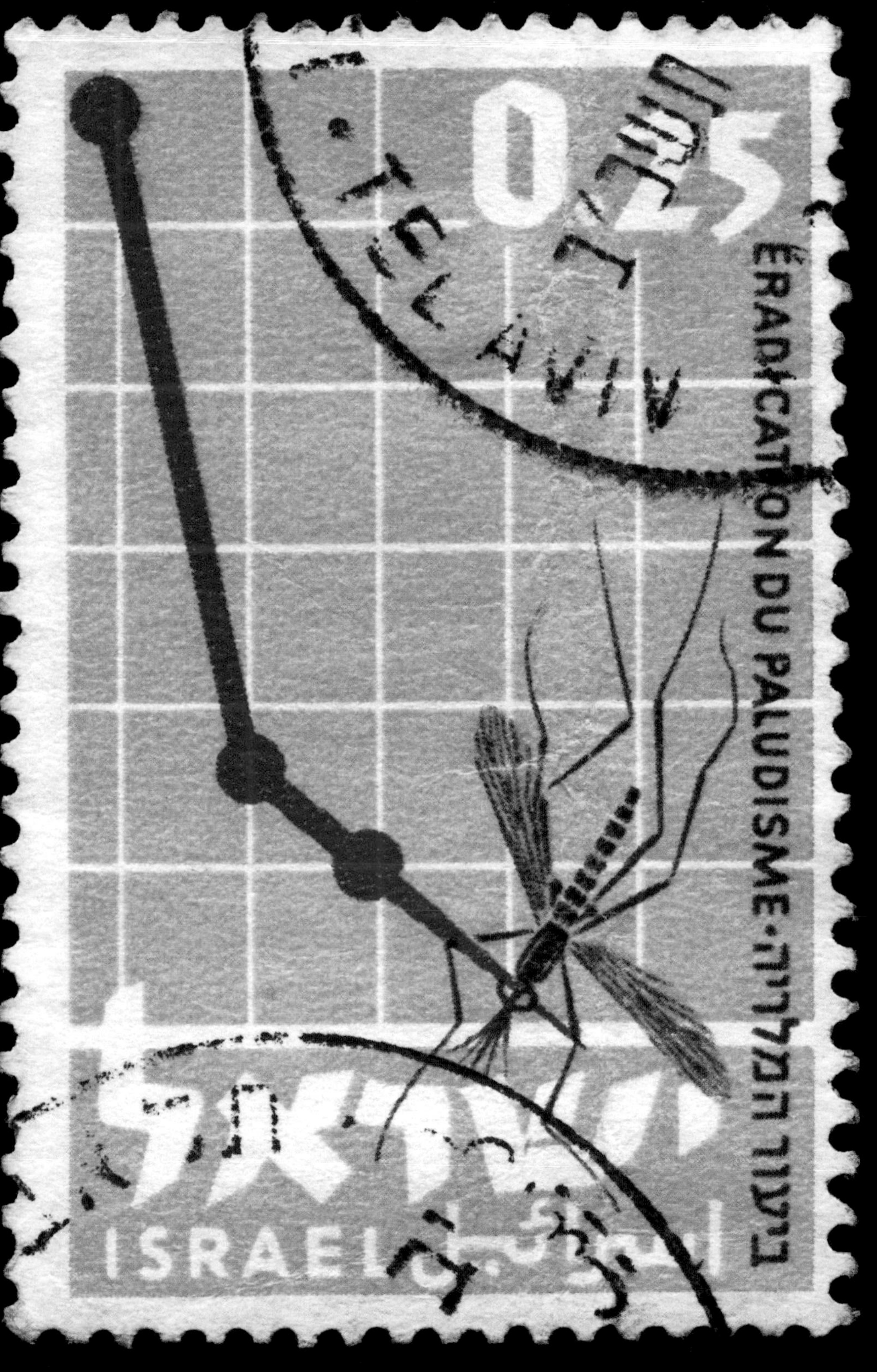
0.25
ÉRADICATION DU PALUDISME
ISRAEL
TEL AVIV

Viruses

Adolf Mayer (1843–1942), **Martinus Beijerinck** (1851–1931),
Dmitri Ivanovsky (1864–1920), **Max Knoll** (1897–1969),
Wendell M. Stanley (1904–1971), **Ernst Ruska** (1906–1988)

The **tobacco** plant was brought to Europe from the New World during the early sixteenth century, and by the middle of the nineteenth century, it was a major Dutch crop. In 1879, Adolf Mayer was asked investigate a disease in Holland that stunted the growth of the tobacco plant and mottled its leaves; he reproduced the condition (which he called *tobacco mosaic disease* — TMD) in healthy plants by rubbing them with the sap from infected ones. About a decade later, Dmitri Ivanovsky investigated this same plant disease in Ukraine and Crimea and, in 1892, reported that the cause of TMD was able to pass through a fine porcelain filter that traps bacteria.

Martinus Beijerinck, a Dutch microbiologist, repeated Ivanovsky's study, also finding that whatever caused TMD passed through the porcelain filter, and in 1898 he concluded that it was smaller than a bacterium. But, while capable of reproducing in living plants (unlike bacteria), it was unable to grow in culture media; he called it a *virus* (Latin for "poison"). During the first three decades of the twentieth century, researchers were able to grow viruses in suspensions of animal tissues, and then, in 1931, in fertilized chicken eggs, which proved invaluable for research and vaccine production.

The structure and chemistry of viruses were next studied. In 1931, the newly invented **electron microscope** by Ernst Ruska and Max Knoll made it possible for them to actually create images of a virus. Four years later, the American biochemist Wendell Stanley crystallized and described the molecular structure of the tobacco mosaic virus (TMV), the first virus ever to be discovered. He was a co-recipient of the 1946 Nobel Prize in chemistry for this achievement.

Stanley found that viruses share the properties of living and nonliving matter: When not in contact with living cells, they are dormant — no more than a large chemical. Viruses are nucleic acids (DNA or RNA) surrounded by a protein coat. However, when in contact with an appropriate living plant or animal cell, they become active and reproduce. In short, they reside in a gray area between lifeforms and chemicals.

SEE ALSO: Tobacco (1611), Deoxyribonucleic Acid (DNA) (1869), Tissue Culture (1902), Bacteriophages (1917), Electron Microscope (1931).

The tobacco mosaic virus (TMV) was the first virus ever discovered. The image depicts the TMV capsid, the protein shell of the virus that surrounds its core of genetic material.

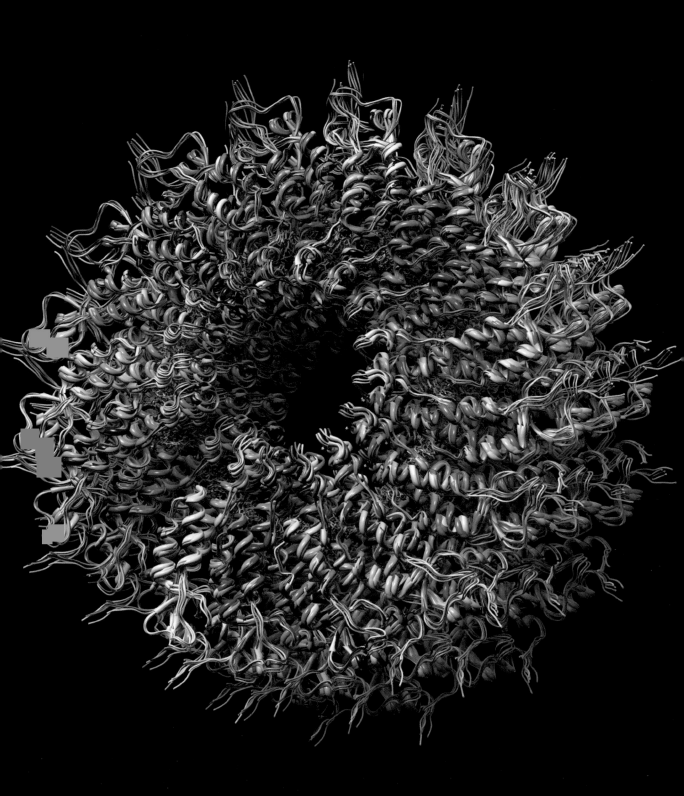

Ecological Succession

Adolphe Dureau de la Malle (1777–1857), Henry Chandler Cowles (1869–1939), Frederic Clements (1874–1945)

In 1825, the French naturalist Adolphe Dureau de la Malle first used the term *ecological succession* to describe the growth of vegetation after forest clear-cutting. Henry Chandler Cowles reintroduced this term in 1899 in his doctoral dissertation at the University of Chicago to describe the succession of vegetation and soil on the Indiana sand dunes, located on the southern end of Lake Michigan. Cowles, who became an early leader in the emerging study of ecology, wrote about ecological succession as the history of the ecosystem from its birth to maturity: changes occur in the species community, with some becoming less abundant over time, and replaced by more plentiful species.

Frederic Clements was a contemporary of Cowles, classmate of the writer Willa Cather at the University of Nebraska, and a leading early plant ecologist. Based on his studies of vegetation in his native state of Nebraska and the western United States, in 1916 he proposed that vegetation gradually changes over time in a sequence of predictable and deterministic stages—which he likened to the development of an individual organism—toward a mature climax stage. This concept was a dominant theory through much of the twentieth century.

Primary succession refers to plant communities occupying sites that had not been previously vegetated and include sandy or rocky surfaces or those covered by lava flow. The earliest signs of vegetation include lichens and grass (pioneer plants), which require few nutrients and use the minerals from rocky surfaces to survive. These are succeeded by small shrubs, trees, and then animals in the climax stage, forming a fully functional ecosystem. By contrast, secondary succession occurs in an area that has undergone a major disturbance or removal of a preexisting vegetative community because of fire, flood, hurricane, or such human factors as logging or agricultural activities. Secondary succession proceeds much more rapidly to the climax stage. Cowles referred to these intermediate stages progressing toward the climax community as *sere*. As each new plant species grows, it modifies the habitat to permit and facilitate the growth of the subsequent species.

SEE ALSO: Land Plants (c. 450 Million BCE), Agriculture (c. 10,000 BCE), Ecological Interactions (1859).

Located on the northeastern shore of the Big Island of Hawaii, Hi'ilawe Falls drops more than 1,400 feet down a steep cliffside of sacred Waipi'o Valley. The moss-covered lava rocks surrounding the waterfall exhibit primary succession.

Animal Locomotion

Eadweard Muybridge (1830–1904)

In 1872, former California governor and business tycoon Leland Stanford hired the English-born photographer Eadweard Muybridge to settle a bet as to whether all four feet of a horse were off the ground when trotting. (They are.) Just between 1883 and 1886, Muybridge made more than 100,000 images analyzing animal and human movement, utilizing multiple cameras, at speeds that were imperceptible to the human eye. His 1899 classic, *Animals in Motion*, remains in print.

Locomotion is not the same as movement. All animals move but locomotion is the progression from one place to another. Locomotion enhances the animal's success in finding food, reproducing, escaping from predators, or departing from unfavorable habitats. Locomotion can be either passive or active: In passive locomotion, the simplest and most energy efficient type, the wind and waves provide the transportation. Active locomotion requires an energy expenditure to overcome such negative forces as friction, drag (resistance), and gravity, and animal body designs have evolved to expend the least energy in active movement on land, air, or water.

Terrestrial locomotion includes walking, running, hopping, and crawling, in which the animal expends energy overcoming inertia, opposing gravity, and maintaining balance. To maintain balance when walking, bipedal animals keep one foot on the ground, while mammalian quadrupeds keep three feet grounded at any one time. Aerial locomotion— flying and gliding—is utilized by insects, **birds**, and bats, and was used by the pterosaurs (flying reptiles extinct for millions of years). The challenge for flying animals is overcoming gravity and air resistance; energy expenditures are minimized by the shape of the wings that maximize utilization of air currents to remain aloft. Aquatic locomotion— swimming and floating—requires overcoming water resistance. Fast animal swimmers are benefitted by their streamlined fusiform bodies, which are tapered at each end.

The efficiency of each of these modes of locomotion has been analyzed by comparing their relative energy expenditure. Swimming was found to be most energy efficient, with running least so and flying intermediate. Regardless of the mode of locomotion, small animals expend more energy/unit of body weight than large ones.

SEE ALSO: Fish (c. 530 Million BCE), Birds (c. 150 Million BCE), Animal Migration (c. 330 BCE), Sliding Filament Theory of Muscle Contraction (1954), Energy Balance (1960).

A phenakistoscope was an early animation device that, when spun, gave the illusion of depicting motion. In about 1893, Muybridge prepared this phenakistoscopic disc, which provided the very realistic illusion of a couple waltzing.

Genetics Rediscovered

Gregor Mendel (1822–1884), **Hugo de Vries** (1848–1935),
William Bateson (1861–1926), **Carl Correns** (1864–1933),
Erich von Tschermak (1871–1962)

In 1866, Gregor Mendel, an obscure Augustinian monk, published the paper "Experiments on Plant Hybrids," in German, in an obscure journal. As the paper's title denoted, it dealt with plant hybridization, not heredity or inheritance. It provided evidence that inherited traits were not passed to successive generations by blending or being the average of the traits from the two parents, as was commonly believed; rather, traits were independently transmitted, and the dominant trait was expressed in the outward appearance of the progeny. (The scientific world, including Mendel, was unaware that such traits were transmitted by genes.) Mendel's paper did not proclaim, nor even hint, that these findings were revolutionary.

Many have speculated why this paper remained unnoticed for one-third of a century, but Mendel was an unknown amateur scientist, with no connections or scholarly affiliations, working in a humble monastery—not a distinguished laboratory or university. As such, his paper failed to arouse scientific interest until 1900. In that year, Hugo de Vries, Erich von Tschermak (whose grandfather was Mendel's botany professor), and Carl Correns—Dutch, Austrian, and German, respectively—working with three different plant hybrids, independently concluded studies similar to those of Mendel. They each found his paper only in the final stages of their research while preparing their own findings for publication. De Vries was the first to publish and relegated mention of Mendel to a footnote; it remains problematic whether his conclusions were independently derived or "borrowed" from Mendel. Von Tschermak had little apparent understanding of Mendel's results. Only Correns fully acknowledged Mendel's earlier findings and their significance. It has been advanced that Mendel's paper achieved fame because of the public dispute amongst the three for "rediscovering" the paper and the science of genetics.

The English botanist William Bateson read Mendel's paper and was so enthralled that he translated it into English. Widely publicizing its findings to the scientific world, he was the first to refer to the science of heredity and biological inheritance as *genetics* in 1905.

SEE ALSO: Darwin's Theory of Natural Selection (1859), Mendelian Inheritance (1866), Genes on Chromosomes (1910), Evolutionary Genetics (1937), The Double Helix (1953).

Around 1900, Carl Correns used the Mirabilis jalapa *(four o'clock flower) as a model plant to rediscover Mendelian genetics. The plant was exported from the Peruvian Andes in 1540.*

Ovaries and Female Reproduction

Aristotle (384–322 BCE), **Antonie van Leeuwenhoek** (1632–1723), **Emil Knauer** (1867–1935), **Josef von Halban** (1870–1937), **Siegfried W. Loewe** (1884–1963), **Edgar Allen** (1892–1943), **Edward A. Doisy** (1893–1986)

The earliest indirect reference to the ovary appeared in **Aristotle's** *The History of Animals*. Although not recognizing the existence of the ovaries, Aristotle did describe the process of spaying sows, then a common agricultural practice. Camels were also spayed to "quench in them sexual appetites and stimulate in them growth and fatness." Aristotle's understanding of the ovaries in the reproductive process was limited by the prevailing belief in the "seed and soil" concept of reproduction. The male provided the "seed," with the female playing a passive role, providing the "soil" in which the seed would grow. Aristotle saw a connection between the "seed" and the male's semen. However, the existence of **spermatozoa** only became evident in 1677, when first visualized under Leeuwenhoek's microscope. Interest in the ovary and female reproductive system was resumed during the sixteenth through the nineteenth centuries and was focused upon its anatomy and, later, various disorders afflicting it.

The loss of the ovaries was known for some time to result in atrophy of the uterus and loss of sexual function. In 1900, Emil Knauer showed that the ovaries exerted control over the female reproductive system. After transplanting ovaries to experimental animals, he prevented the symptoms associated with their removal. Josef von Halban extended these studies the same year when he demonstrated that transplantation of ovaries into an immature spayed guinea pig permitted the animal to attain normal puberty. Thus, the ovaries were found to not only maintain the female genital tract but were also responsible for its development.

On the basis of these findings, the existence of an internal secretion from the ovary—a single hormone—was postulated. Identification of this substance required the development of a sensitive assay, which Edgar Allen and Edward Doisy perfected in 1923, that was capable of detecting estrogen, the female sex hormone, in the blood and urine of pregnant and nonpregnant females. In 1926, Siegfried Loewe detected the presence of a female sex hormone in the urine of menstruating females, noting that the concentration of this hormone varied with the phase of the menstrual cycle.

SEE ALSO: Aristotle's *The History of Animals* (c. 330 BCE), Spermatozoa (1677), Progesterone (1929).

In this image of the female reproductive system, the ovaries are depicted as blue-green structures. At the time of ovulation, the egg travels through the Fallopian tube to the uterus, where it is potentially fertilized by a sperm cell.

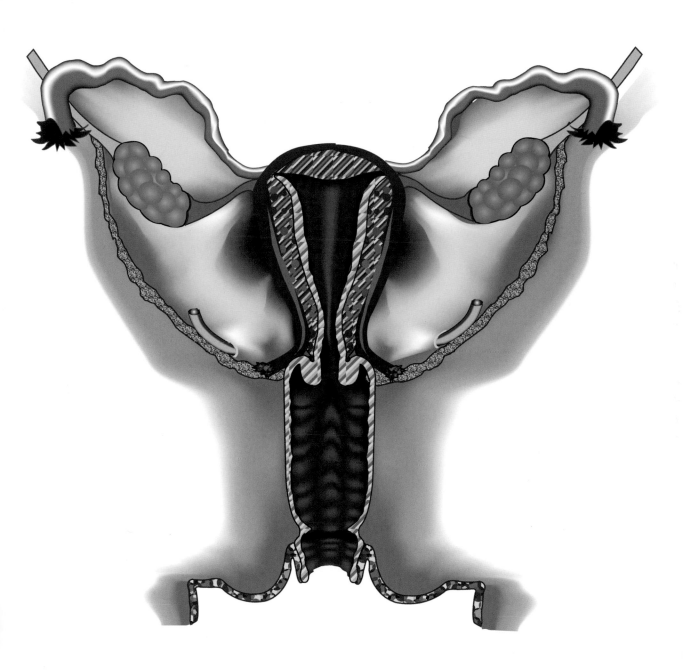

Blood Types

William Harvey (1578–1657), **Richard Lower** (1631–1691), **Jean-Baptiste Denys** (1643–1704), **James Blundell** (1791–1878), **Karl Landsteiner** (1868–1943)

Several decades after William Harvey demonstrated how blood circulated, attempts were made to transfuse blood. In 1665, Richard Lower successfully kept dogs alive by transfusing blood from other dogs, and two years later, Jean-Baptiste Denys performed the first documented blood transfusion into a human subject. Whereas single transfusions were sometimes successful, the recipients often succumbed after receiving their second or third transfusion. By the end of the seventeenth century, blood transfusions were banned and fell into obscurity until 1818, when James Blundell, an English obstetrician, performed the first successful transfusion of human blood for the treatment of postpartum bleeding. Between 1825 and 1830, he performed ten transfusions—five of which were beneficial. Not only was Blundell medically successful but also financially, earning the present-day equivalent of $50 million from his invention of blood transfusion instruments.

The discovery of blood types by Karl Landsteiner, an Austrian-born American immunologist, transformed transfusions into a routine medical procedure. In 1901, at Vienna's Institute of Pathology, he reported that when the blood of some individuals was brought into contact with some other's, it could cause agglutination (clumping) of the red **blood cells** with fatal consequences, and this resulted from an immunological (antigen-antibody) reaction. He identified three human blood groups, A, B, and C (later renamed O), and subsequently a fourth group, AB.

Landsteiner's blood typing served as the basis for the first successful transfusions of compatible blood into recipients in 1907 at Mt. Sinai Hospital in New York and on a large-scale basis on the World War I battlefields. In 1927, ABO blood types were introduced for use in paternity suits to establish the biological parents of a child. Landsteiner was awarded the 1930 Nobel Prize for his discovery of human blood groups and the ABO system of blood typing. In 1940, while working at the Rockefeller Institute (now Rockefeller University), he discovered another blood factor in Rhesus monkeys. This Rh factor was responsible for the potentially life-threatening hemolytic disease of the newborn, which occurs when a mother and her fetus have incompatible blood types.

SEE ALSO: Harvey's *De motu cordis* (1628), Blood Cells (1658).

In 1901, Landsteiner identified the four human blood types and, four decades later, found the Rh factor. In 1968, the drug RhoGAM was introduced to suppress the Rh– mother's normal production of antibodies to her fetus's Rh+ red blood cells, effectively preventing hemolytic disease of the newborn.

Tissue Culture

Gottlieb Haberlandt (1854–1945), **Ross Granville Harrison** (1870–1959), **George Otto Gey** (1899–1970)

Animal and plant tissue cultures have proven to be valuable tools with commercial, scientific, and medical applications. *Tissue culture*, a term used interchangeably with organ culture and cell culture, involves the growth of fragments of plant or animal tissue in an artificial sterile external environment where they can be easily manipulated and studied. Such testing includes examination of their biochemical or genetic activity, their metabolic, nutritional, or specialized function, and for the effects caused by physical, chemical, and biological agents, including drugs.

In 1902, the Austrian botanist Gottlieb Haberlandt was the first to conceive of tissue culture in plants. He was able to maintain plant cells in a living state for several weeks, but they failed to reproduce because of the absence of growth hormones in the culture media. With advances in research, plant tissue culture (micropropagation) has been used to develop more hardy and pest-resistant crops for the preparation of pharmaceuticals—such as the anticancer drug Taxol—and in genetic engineering.

Ross Granville Harrison at Yale, in 1907, developed a new tissue culture technique and used it to settle the long-standing debate as to how nerve fibers originate. This approach, the "hanging drop" culture, became a major tool for viral research during the 1940s and 1950s and was used for the manufacture of vaccines for the prevention of polio, measles, mumps, rubella, and chicken pox, and later for the production of **monoclonal antibodies**.

The oldest and most commonly used cell line is HeLa, which was brought to the public consciousness in Rebecca Skloot's nonfiction bestseller *The Immortal Life of Henrietta Lacks* (2010). The HeLa cell line, derived from cervical cancer cells taken in 1951 from Ms. Lacks, who died six months later, are immortal—that is, they can divide an unlimited number of times in cell culture, as long as a suitable culture environment is maintained. The HeLa cells were propagated by George Otto Gey, who shared them with scientific colleagues. The cells have been used for significant scientific research but without permission of the family, who did not profit from their highly successful commercialization.

SEE ALSO: Biotechnology (1919), The Immortal HeLa Cells (1951), Monoclonal Antibodies (1975), Genetically Modified Crops (1982), Induced Pluripotent Stem Cells (2006).

Laboratory cultures permit researchers to conduct studies involving a large number of samples under very carefully controlled experimental conditions.

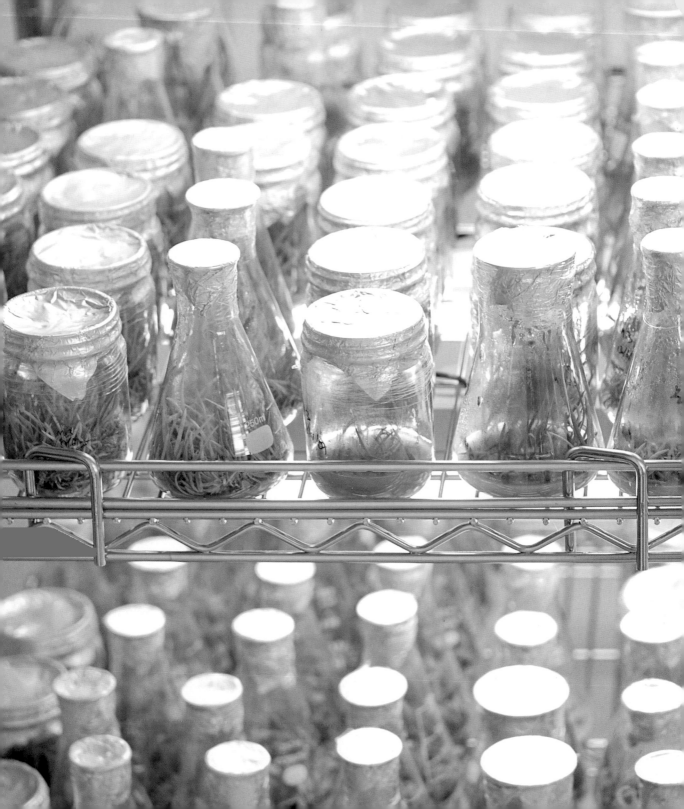

Secretin: The First Hormone

Claude Bernard (1813–1878), **William M. Bayliss** (1860–1924),
Ernest H. Starling (1866–1927)

During the 1840s, Claude Bernard determined that pancreas secretions are responsible for digesting dietary fats—a process occurring in the small intestine, not the stomach. In 1902, the English physiologists William Bayliss and Ernest Starling sought to better understand this pancreatic secretion. Were nervous system signals responsible for the flow of digestive juices, or perhaps some chemical? They severed all nerves to the pancreas, and the juices continued to flow, ruling out direct nervous system involvement.

Bayliss and Starling then focused upon chemical factors responsible for the release of pancreatic digestive juices. They grinded pieces of a dog's duodenum in stomach acid (hydrochloric acid) and injected this extract into another dog. Finding that pancreatic juice was released in the dog, as it would during normal feeding, they surmised that a chemical—which they called *secretin*—must have been secreted from the lining of the intestine, carried into the bloodstream, and transported to the pancreas, triggering its release of digestive fluid. In 1905, Starling called this chemical messenger a *hormone* (from Greek "to excite").

To be designated a hormone, the chemical must be directly released into the blood from a ductless (endocrine) gland and transported in the blood to a distant target site. Secretin was the first of many hormones to be discovered and would be joined in the coming decades by adrenaline, thyroid, **insulin**, testosterone, and estradiol. The endocrine system is one of the two major systems responsible for communication in the body and for maintaining **homeostasis**. (The other is the nervous system.) Hormones are released from such glands as the thyroid, adrenal glands, and ovaries, regulating such functions as growth and development, reproduction, energy **metabolism**, and behavior.

Vertebrates have no monopoly on hormones. Invertebrates have endocrine systems, and these are particularly well developed in insects. In different invertebrates, their endocrine systems regulate such processes as reproduction, development, and water balance. Phytohormones (plant hormones) are fewer in number and are primarily involved with growth and development.

SEE ALSO: Metabolism (1614), Nervous System Communication (1791), Homeostasis (1854), Phototropism (1880), Ovaries and Female Reproduction (1900), Thyroid Gland and Metamorphosis (1912), Neurotransmitters (1920), Insulin (1921), Progesterone (1929), Hypothalamic-Pituitary Axis (1968).

Secretin, released by the cells lining the intestine (pink), increases secretions of digestive juices from the pancreas (golden).

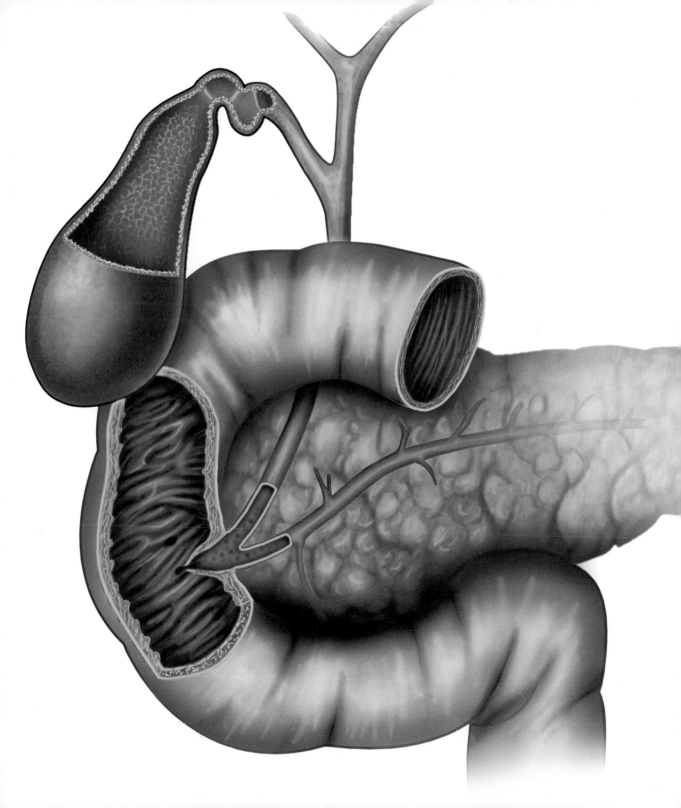

Dendrochronology

Leonardo da Vinci (1452–1519), **Percival Lowell** (1855–1916), **A(ndrew) E(llicott) Douglass** (1867–1962), **Clark Wissler** (1870–1947)

In 1894, the astronomer Percival Lowell sent A. E. Douglass to Flagstaff, Arizona, to build an observatory. There, Douglass noted similar ring widths in the trees cut for the observatory construction. As an astronomer, he observed that the sunspot cycle affected climatic changes and that there was a correlation between the climate and tree ring widths; moreover, all trees within a given region exhibited the same relative tree ring growth. (Douglass was not the first to study tree rings; in about 1500, Leonardo da Vinci commented that tree ring number corresponded to the tree's age and ring thickness to climatic dryness.)

In 1904, Douglass began his scientific study of tree rings (also called "growth rings") or *dendrochronology* (dendro = "tree"). In 1914, Douglass was approached by Clark Wissler of the American Museum of Natural History about using his tree ring timing to date Native American ruins in the southwestern United States, a successful project that continued for fifteen years. In addition to studying patterns in climate change and the age of archeological ruins, dendrochronology has been used to date glacial movements and volcanic eruptions.

When the trunk of a tree is cut horizontally, we see tree rings, with each ring marking the passage of one year of tree life. These rings are the result of new growth in the vascular cambium layer of cells nearest to the bark. Early in the growing season, the cells are thin-walled (earlywood), and thicker cells (latewood) are produced by the tree later in the season. One annual ring is marked from the beginning of earlywood to the end of latewood.

Rings represent new vascular tissue that transports water and nutrients up the tree into the leaves. During the growth season, large tubes open to permit the flow of greater volumes of water, while during the dormant and drier season, the new ring growth is reduced and the tight tubes reflect a lesser amount of water transported. Tree growth is dependent upon such climatic factors as weather, rainfall, temperature, **plant nutrition**, soil activity, and carbon dioxide concentration.

SEE ALSO: Land Plants (c. 450 Million BCE), Devonian Period (c. 417 Million BCE), Gymnosperms (c. 300 Million BCE), Plant Nutrition (1840).

An image of the middle of a polished slice of a petrified tree from Arizona, showing its tree rings. An enlargement of the image enables one to see insect borings in the wood that date back 230 million years.

Blood Clotting

Giulio Bizzozero (1846–1901), Paul Morawitz (1879–1936)

Coagulation of blood is an important defensive mechanism in both vertebrates and invertebrates that prevents the loss of blood and introduction of disease-causing microbes into the body. While blood clotting follows the same basic sequence in all vertebrates, from the primitive fishlike jawless lamprey to mammals, the number of components associated with the clotting process increases and becomes far more complex as we move up the evolutionary scale.

Blood contains three types of cells: red **blood cells** (erythrocytes) that transport oxygen; white blood cells (leucocytes), involved with combating infection; and platelets (thrombocytes), involved in blood clotting. Injury to a blood vessel in mammals initially causes a spasm and constriction of the blood vessel followed by activation of platelets, forming a plug to stem blood loss. The platelets also activate a cascade of multiple clotting factors leading to the generation of thrombin and the formation of a fibrin clot that stabilizes the platelet plug and arrests blood loss.

The multiple roles of the platelets in the clotting process were first described in 1882 by Giulio Bizzozero. In 1905, Paul Morawitz assembled the then-known clotting factors (including the four he had discovered) that lead to the formation of thrombin and the fibrin clot. This compilation continues to serve as the basis for following the process by which clotting occurs. From the 1940s into the 1970s, additional clotting factors were identified—now numbering thirteen and designated by Roman numerals—as well as additional cofactors and regulators required for normal coagulation to occur. A deficiency of clotting factor IX is responsible for hemophilia B, a genetically determined disorder that afflicted members of European royalty who were descendants of Queen Victoria. Factor IX, or Christmas factor, was discovered in 1962 and named after Stephen Christmas, who was lacking this factor.

In invertebrates, clotting factors have been identified in such **arthropods** as the horseshoe crab and crayfish. In some invertebrates, spasms of blood vessels are sufficient to stop the flow of hemolymph—a fluid analogous to blood and interstitial fluid in vertebrates that directly bathes the cells in arthropods and most molluscs—from wounds.

SEE ALSO: Arthropods (c. 570 Million BCE), Blood Cells (1658), Hemoglobin and Hemocyanin (1866), Blood Types (1901).

An underside view of an Atlantic horseshoe crab (Limulus polyphemus). *Their blood contains hemocyanin, giving it a blue color, as well as amebocytes, which can be used to detect bacterial endotoxins in medical devices, vaccines, and drugs.*

Radiometric Dating

Henri Becquerel (1852–1908), **Bertram Borden Boltwood** (1870–1927), **Ernest Rutherford** (1871–1937), **Frederick Soddy** (1877–1956)

The determination of the absolute age of fossils has been invaluable to biologists as they trace the history of life and the evolution of living organisms. One of the most common techniques for determining these ages is radiometric dating, which is based on the observations of physicists and chemists at the turn of the twentieth century and their studies on the decay of radioisotopes.

During the course of his work on uranium salts, in 1896, Henri Becquerel, teacher of Marie Curie, serendipitously found that uranium spontaneously emitted radiation, which led to his discovery of radioactivity. Building upon Becquerel's findings, Ernest Rutherford, father of nuclear physics, and his student Frederick Soddy, working at McGill University in 1902, found that the process of radioactive decay changes atoms from one element (parent isotope) to another (daughter isotope) at a constant rate. (Isotopes have the same number of protons in their nucleus but a different number of neutrons, giving rise to the same element but with different atomic masses.) Rutherford and Soddy predicted the time it would take for one-half the atoms in an isotope to decay, and this half-life ($t_{1/2}$) decay is unique for each isotope. Carbon–14 has a $t_{1/2}$ of 5,730 years and has been used in more recent times to determine the age of organic materials, such as wood, bone, shells, and fabrics that are up to 75,000–80,000 years old. By contrast, uranium-238, with a $t_{1/2} = 4.5$ billion years, is used to determine the age of older fossil samples.

Bertram Boltwood was a Yale University radiochemist and pioneer in the study of radioisotopes. Initially a correspondent and later a close friend of Rutherford, in 1907 he was the first to apply Rutherford's principles to what would become radiometric dating. Using the ratio of the half-life decays of uranium-238/lead-206, Boltwood estimated the age of the Earth to be 2.2 billion years, ten times older than had previously been determined, but one-half the present age calculation.

SEE ALSO: Origin of Life (c. 4 Billion BCE), Fossil Record and Evolution (1836), Dendrochronology (1904), X-Ray Crystallography (1912), Lucy (1974).

An archeologist cleans a mammoth's tooth. The earliest mammoths appeared about 5 million years ago, but while most became extinct around 10,000 years ago, some dwarf mammoths lived as recently as 4,000 years ago.

Probiotics

Élie Metchnikoff (1845–1916), Minoru Shirota (1899–1982)

The human body is home to more than 100 trillion bacteria, with a total weight of five pounds; the mouth alone houses several hundred bacterial species. In the past, bacteria were exclusively associated with infectious diseases, often with fatal consequences, and in cases of food poisoning. Modern medicine has compounded this problem with the overuse of **antibiotics** that nonselectively rid the body of harmful microbes, as well as those that provide benefit. This is particularly true in the intestinal tract, where the use of broad-spectrum antibiotics disrupts the normal microbial balance in the gut, causing diarrhea.

THE GOOD BACTERIA. In 1907, the Russian biologist Élie Metchnikoff, who was a co-recipient of the 1907 Nobel Prize for immunity research, conceived of the notion that it was possible to modify the gut flora and replace harmful microbes with beneficial ones. More specifically, fermented milk could "seed" the intestines with lactobacillus, which would acidify the intestines and suppress the growth of these proteolytic bacteria (bacteria that break down proteins). Metchnikoff attributed the aging process to the accumulation of waste materials in the lower segment of the large intestines that empties into the rectum, and the leaking back of these toxic substances from the rectum into the bloodstream (known as *autointoxication*). Metchnikoff noted that the rural population of Bulgaria, who subsisted primarily on milk fermented with lactobaccilus (lactic acid bacteria), was exceptionally long-lived.

During the 1930s, Minoru Shirota in Japan, inspired by Metchnikoff's concepts, developed Yakult, a yogurt-like drink containing a stronger strain of lactobacillus that was intended to destroy harmful intestinal bacteria. This product, and others like it that have been introduced in recent years, are referred to as *probiotics*, which are simply living microbes used as food supplements. They benefit the user by improving the intestinal microbial balance.

Although many health claims have been made for probiotics, from curing intestinal disorders to a very wide range of major systemic disorders, to date no claim for medical use has been approved by the Food and Drug Administration or the European Food Safety Authority.

SEE ALSO: Prokaryotes (c. 3.9 Billion BCE), Microbial Fermentation (1857), Antibiotics (1928).

Balkan yogurt differs from Greek and Swiss styles because it is allowed to ferment inside individual containers rather than in large vats. The image depicts Balkan homemade yogurt—a great source of probiotics—served in a traditional ceramic dish.

Why Does the Heart Beat?

Galen (c. 130–c. 200), **Jan Evangelista Purkinje** (1787–1869), **Walter Gaskell** (1847–1914), **Wilhelm His, Jr.** (1863–1934), **Arthur Keith** (1866–1955), **Sunao Tawara** (1873–1952), **Martin Flack** (1882–1931)

As far back as the second century, Galen noted that a heart, removed from the body and its nerves, continued to beat. The answer to what triggered the heartbeat was the subject of debate, with successive anatomical pieces being assembled to help solve this puzzle, until the first decade of the twentieth century.

The first pieces were laid down in 1839 by Jan Purkinje, a Bohemian physiologist-anatomist, who spoke thirteen languages, was a poet, and translated the poetry of Goethe and Schiller. Among his many significant discoveries was a series of fibers—the Purkinje fibers—in the ventricles (lower chamber) of the heart, but he failed to recognize their function. Working at Cambridge University in the 1880s, Walter Gaskell studied the formation of heart impulses and conduction of a heartbeat proceeding as a wave from the atria (upper chambers) to the ventricles. He noted that after surgically separating the atrial and ventricular chambers, the ventricles ceased to beat. In 1893, the Swiss-born cardiologist-anatomist Wilhelm His, Jr., described a bridge connecting the upper and lower chambers, but not the function of this muscle branch ("bundle of His").

The year 1868 marked the end of the shogunate and the start of Japan's path from feudal state to modernized nation. Japan opened its doors to Western culture, adopting a German system of education. Sunao Tawara, a Japanese medical graduate sent to Germany in 1903, studied the conduction system of the heart and discovered the atrioventricular (AV) conducting system and AV node. He recognized that electrical impulses traveled from the bundle of His to the Purkinje fibers, which were part of an electrical conducting system. The final puzzle piece was laid in place in 1907 when the Scottish anthropologist-anatomist Arthur Keith and the medical student Martin Flack microscopically discovered the sino-atrial (SA) node—the "pacemaker"—the site at which the impulse driving the heartbeat originated, and its conduction from the AV node through the ventricles. Keith gained notoriety, however, as a racist and collaborator in the Piltdown Man hoax.

SEE ALSO: Harvey's *De motu cordis* (1628), Blood Pressure (1733).

An electrocardiographic (ECG or EKG) recording traces the electrical current moving through the heart. Each heartbeat begins with an electrical impulse arising from the heart's main pacemaker, the sinoatrial node. This impulse first activates the upper chambers (atria), and then flows down to activate the lower chambers (ventricles).

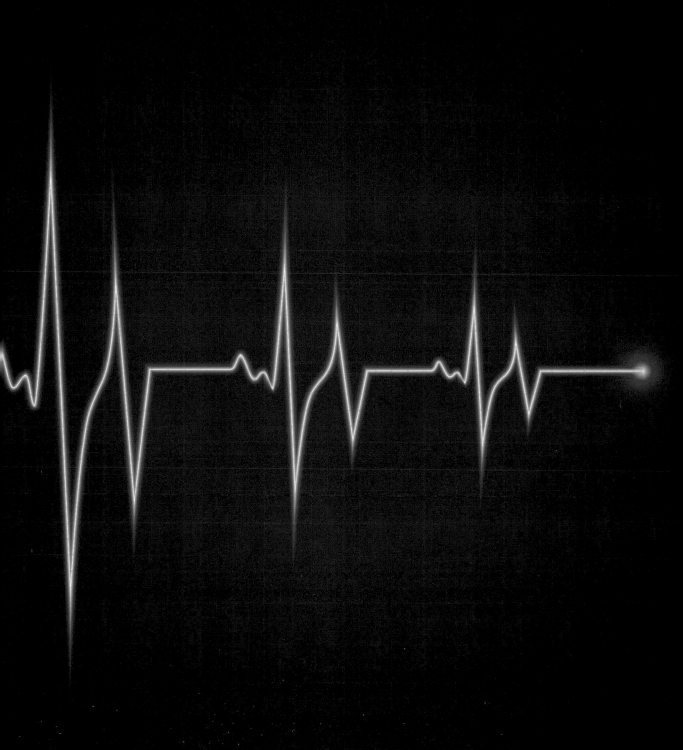

Hardy-Weinberg Equilibrium

Charles Darwin (1809–1882), **Gregor Mendel** (1822–1884), **Wilhelm Weinberg** (1862–1937), **Godfrey Hardy** (1877–1947)

In *Origin of Species* (1859), Darwin provided evidence that natural selection was the basis for evolution and, while recognizing that heritable traits were being passed from parents to offspring, was unaware of how this occurred. In 1866, Mendel presented a model whereby heritable units (now called genes) were transmitted, and that alternate forms of this unit (alleles) were present on the chromosome that would be expressed in offspring as a trait. Scientists were faced with the dilemma of reconciling Mendel's evidence for large changes passed from parent to offspring with Darwin's theory involving gradual generational changes.

In 1908, the English mathematician Godfrey Hardy and the German physician Wilhelm Weinberg independently developed a model to determine whether evolution has occurred and detect whether any changes in gene frequencies in the population took place. If no evolution was occurring, an equilibrium of allele frequencies would remain in effect in each generation of reproducing individuals. But in order for equilibrium to occur, each of the following five conditions had to be satisfied: the population must be infinitely large to prevent genetic drift (the random chance that allele frequencies would change); mating must be random, with individuals pairing by chance; no mutations can occur, thereby preventing the introduction of new alleles into the population; no individuals can move into or out of the population; and no natural selection can occur, thereby certain alleles are not preferred or excluded.

In real life, these conditions would never be met and, therefore, evolution would occur. The formulas made possible the detection of allele frequencies changing from generation to generation proving that evolution is occurring while allowing scientists to estimate the percentage of a population carrying an allele for an inherited disease.

Throughout much of the twentieth century, the study of genetics was dominated by English speakers who were unaware of Weinberg's German paper, which had introduced the concept six months earlier in 1908. Therefore, until 1943, the Hardy-Weinberg Equilibrium was solely attributed to Hardy.

SEE ALSO: Darwin's Theory of Natural Selection (1859), Mendelian Inheritance (1866), Genetics Rediscovered (1900), Genes on Chromosomes (1910).

The Hardy-Weinberg equilibrium provides a mathematical model to detect changes in population gene frequencies. This collection of Donax variabilis *shells shows diverse coloration and patterning based on differences in their genotypes.*

Genes on Chromosomes

Charles Darwin (1809–1882), **Gregor Mendel** (1822–1884),
Hugo de Vries (1848–1935), **Thomas Hunt Morgan** (1866–1945),
Alfred H. Sturtevant (1891–1970)

Gregor Mendel's hereditary studies on garden peas were "rediscovered" in 1900 and, with them, the basis for genetics. Thomas Hunt Morgan was one of many biologists at the turn of the twentieth century who accepted Darwin's theory of evolution but rejected his notion of natural selection, and also Mendel's findings. One of the three "rediscoverers" of genetics was the Dutch botanist Hugo de Vries, who in 1886, while working with the evening primrose, found evidence for mutations (sudden changes in body form).

Morgan, a zoologist at Columbia University, in 1907 began research on *Drosophila melanogaster*—the common fruit fly—in an attempt to show that mutations, rather than the gradual variation proposed by Darwin, were the basis for natural selection. He selected fruit flies because 1,000 could be housed in a one-quart milk bottle, they produced one generation every twelve days, males and females were readily distinguishable, and mutations were readily detected. After three years of breeding, the first mutant was detected: a white-eyed fly. Subsequent breeding revealed that females were exclusively red-eyed and that only some males had white eyes.

In 1910, he proposed the chromosomal theory of heredity. Each chromosome contained a collection of small units called *genes*, that were arranged on the chromosome as "beads on a string." Moreover, some traits (such as yellow body color or rudimentary wings) are linked to the sex-determining chromosome. In 1913, Morgan's student Alfred H. Sturtevant found that each gene could be assigned a specific position on a chromosome map, which became the basis for mapping the human genome.

The gene was the missing link in Mendel's theory of heredity and Darwin's theory of evolution, and Morgan filled in that gap. (By 1916, Morgan had accepted **Darwin's theory of natural selection**.) Thus, Morgan was awarded the 1933 Nobel Prize for determining the role of chromosomes in heredity. As an additional legacy, of the students who worked with Morgan or one of his students, five were awarded their own Nobel Prizes.

SEE ALSO: Darwin's Theory of Natural Selection (1859), Mendelian Inheritance (1866), Genetics Rediscovered (1900), DNA as Carrier of Genetic Information (1944), The Double Helix (1953), Human Genome Project (2003).

In this illustration of the human karyotype, all twenty-two chromosome pairs are displayed, along with the XX and XY sex chromosome pairs.

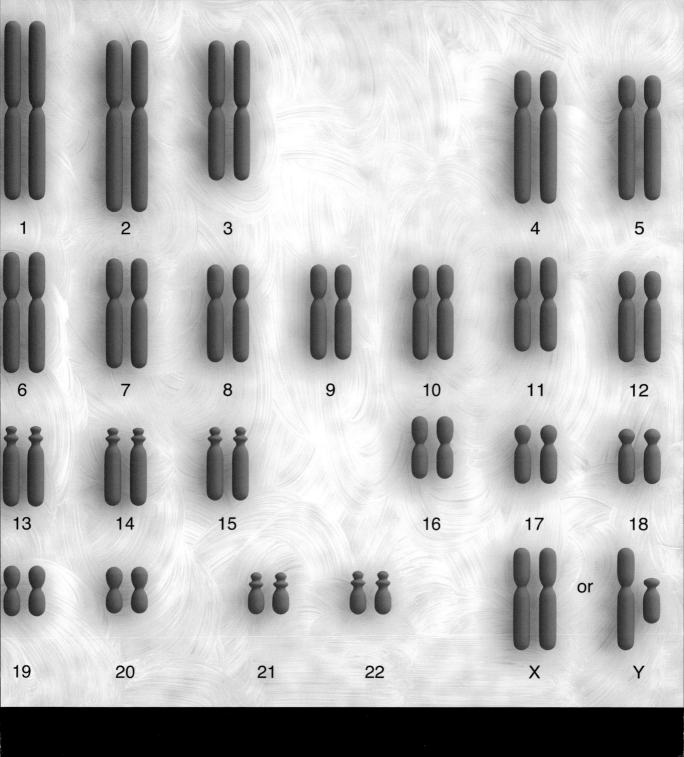

Cancer-Causing Viruses

Francis Peyton Rous (1879–1970), **J. Michael Bishop** (b. 1936), **Harold Varmus** (b. 1939)

In 1911, the American pathologist Peyton Rous had recently graduated from medical school, and with only four years of research experience, had been placed in charge of cancer research at the Rockefeller Institute. For the next twenty years, the results of his studies were sometimes ignored and often derided by the scientific community. Only fifty-five years later—four years before his death—were they fully acknowledged, when he received his Nobel Prize in 1966; this may represent a record for the time elapsed between a discovery and the award. We now know that some 15 to 20 percent of all cancers are caused by **viruses**, most of which occur in animals.

Among Rous's first projects at Rockefeller involved determining the cause of a sarcoma, a large tumor growing on the breast of a Plymouth Rock hen. Recent literature accounts of unusual transmissible growths in animals aroused his interest. He successfully produced tumors in similar healthy hens (but not other birds) after transplanting small samples of the tumor—the first demonstration that a cancer could be transmitted from one bird to another. Later studies revealed that it was unnecessary to transfer intact cells, as they could be produced by injecting a cell-free, bacteria-free filtrate obtained from the tumor. Rous concluded that tumors in hens were caused by a filterable agent. For decades after the discovery of the first virus—the tobacco mosaic virus in 1892—viruses were referred to as filterable agents. Later, the designation *virus* was restricted to filterable agents that could only grow on living cells.

Rous had discovered the Rous sarcoma virus (RSV), a retrovirus, and the first oncogenic (tumor-causing) virus to be described. A retrovirus contains RNA instead of DNA, and transcribes information to DNA once inside its host. An oncogen is a **cancer-causing gene** that is found in retroviruses. In 1976, J. Michael Bishop and Harold Varmus found that normal oncogenic genes in healthy cells could cause cancer if they were picked up by retroviruses—work for which they were awarded the 1989 Nobel Prize.

SEE ALSO: Viruses (1898), Cancer-Causing Genes (1976), HIV and AIDS (1983).

This illustration from 1989 shows how a normal cell becomes a cancer cell after a cancer-causing agent, such as a retrovirus, activates an oncogene on the cell's DNA.

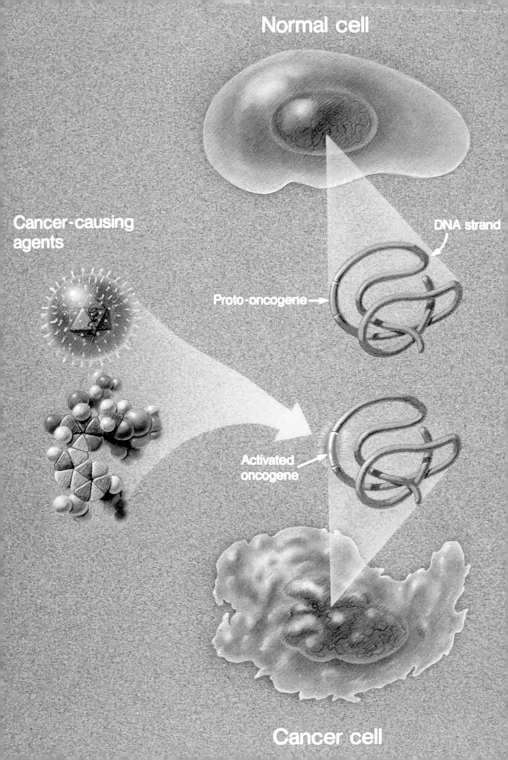

Normal cell

Cancer-causing
agents

DNA strand

Proto-oncogene →

Activated
oncogene

Cancer cell

Continental Drift

Alexander von Humboldt (1769–1859), Alfred Wegener (1880–1930)

Even casual inspection of a map of the Southern Hemisphere suggests that the coastlines of eastern South America and western Africa fit together like the pieces of a jigsaw puzzle. This same thought occurred to the naturalist-explorer Alexander von Humboldt who, during the early 1800s, found similarities between animal and plant fossils in South America and western Africa, and common elements between the mountain ranges in Argentina and South Africa. Subsequent explorers saw similarities between fossils in India and Australia.

In 1912, the German geophysicist-meteorologist and polar explorer Alfred Wegener went a step further and proposed that the present continents were once fused into a single landmass, which he called *Pangaea* ("All-Lands"). Expanding upon this theory in his 1915 book, *The Origin of Continents and Oceans*, Wegener described how Pangaea subsequently split into two supercontinents, Laurasia (corresponding to the present-day Northern Hemisphere) and Gondwanaland, also called Gondwana (Southern Hemisphere)—an event now thought to have occurred 180 to 200 million years ago. Wegener could not provide an explanation for continental drift, and his concept was roundly rejected until after his death in 1930 from heart failure during an expedition to Greenland. The occurrence of continental drift was finally accepted in the 1960s, when the concept of *plate tectonics*—involving plates that are in constant motion relative to each other, sliding under other plates and pulling apart—was established.

Long before continental drift was acknowledged by the scientific community, naturalists were finding ancient fossils of the same or similar plants and animals in continents thousands of miles apart and separated by oceans. Fossil remains of the tropical fern *Glossopteris* were found in South America, Africa, India, and Australia, while those of the family *Kannemeyrid*, a mammal-like reptile, were uncovered in Africa, Asia, and South America. By contrast, some living plants and animals on different continents are very different from one another. For example, all the native mammals in Australia are marsupials and not placental mammals, which suggests that Australia split off from Gondwanaland before placental mammals evolved.

SEE ALSO: Devonian Period (c. 417 Million BCE), Paleontology (1796), Darwin and the Voyages of the *Beagle* (1831), Fossil Record and Evolution (1836), Biogeography (1876).

According to the theory of continental drift, a single giant landmass, Pangaea, split into the two supercontinents, Laurasia (Northern Hemisphere) and Gondwana (Southern Hemisphere).

Pangaea

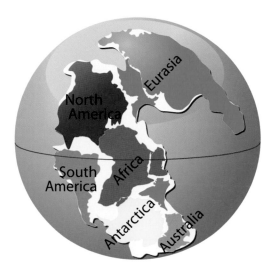

North America

Eurasia

South America

Africa

Antarctica

Australia

PANGAEA
200 million years ago

Laurasia

Gondwana

North America

Eurasia

South America

Africa

Antarctica

Australia

LAURASIA & GONDWANA
120 million years ago

Vitamins and Beriberi

Takaki Kanehiro (1849–1920), **Christiaan Eijkman** (1858–1930),
Frederick Hopkins (1861–1947), **Casimir Funk** (1884–1967)

First described in a Chinese medical book dating back 4,500 years, beriberi has long been endemic in Asia, where polished rice is a staple of the diet. Its symptoms involve the nervous system and heart, and its name is derived from the Sinhalese "weak, weak," duplicated for emphasis and referring to paralysis. White rice is milled and polished rice whose husks, bran, and germ have been removed, as have been important nutrients.

In 1884, Takaki Kanehiro, a Japanese naval physician, had returned home after medical studies in England. He noted that beriberi was uncommon among Western navies and Japanese naval officers who ate a Western diet, but was widespread among ordinary seamen whose diet consisted exclusively of white rice. He conducted an experiment comparing diets and observed that the incidence of beriberi was ten times higher on the rice-only diet when compared with the Western diet. Kanehiro concluded that diet was responsible for beriberi, contrary to the prevailing view that its cause was an infectious disease. The prevailing view dominated, and the diet was not changed. During the 1904–1905 Russo-Japanese War, 27,000 Japanese soldiers died of beriberi, almost half the number succumbing to combat wounds.

The Dutch physician Christiaan Eijkman was sent to the Dutch East Indies (now Indonesia) to study beriberi in 1897. After inducing beriberi in chickens that received a white rice diet, he switched the diet to unpolished rice, and the chickens recovered. Eijkman proposed that polished rice lacked a component present in unpolished rice, an "anti-beriberi factor." In 1911, Casimir Funk, the Polish chemist, discovered this chemical, thiamine, an amine (a nitrogen-containing compound), and called it a vital amine or *vitamine*. When other such compounds were discovered that were not amines, the final *e* was dropped. In 1912, the British biochemist Frederick Hopkins performed experiments to compare the growth of mice receiving only a synthetic diet with one supplemented with milk; the former group stopped growing, but growth resumed with the addition of milk. Hopkins proposed the "Vitamin Hypothesis of Deficiency," that the absence of a vitamin could cause disease. Hopkins and Eijkman were co-recipients of the 1929 Nobel Prize.

SEE ALSO: Rice Cultivation (c. 7000 BCE), Albumin from Rice (2011).

Removal of the husk of brown rice extends its shelf life, but the resulting white rice lacks thiamine (vitamin B1) content.

Thyroid Gland and Metamorphosis

Thomas Wharton (1614–1673), **Theodor Kocher** (1841–1917),
J. Frederick Gudernatsch (1881–1962)

The thyroid gland was first identified in 1656 by the English anatomist Thomas Wharton. For describing the thyroid's functions, the Swiss surgeon Theodor Kocher received the 1909 Nobel Prize. Starting in 1874, Kocher removed the entire thyroid gland from a series of patients with enlarged glands over the next decade. His operations were highly successful, reducing the mortality rate of this previously dangerous operation to negligible levels. But almost uniformly, his patients experienced fatigue, lethargy, and feelings of excessive cold. We now know that the thyroid participates in a variety of essential life cycle events, including energy utilization, growth and development, metamorphosis, reproduction, hibernation, and heat generation. The thyroid is among the largest of the endocrine glands, present in all vertebrates, and located in the neck region in tetrapods (those with limbs having digits).

Thyroid hormone increases the metabolic activity of virtually all body tissues and increases the rate at which foods are utilized as sources of energy. In response to a release of its hormones, there is an increase in the number and size of **mitochondria**—intracellular bodies where adenosine triphosphate (ATP) is generated to provide energy to carry out cellular functions and, in the process, generate body heat. The effect of thyroid hormone on growth is mostly seen in growing children, with hypothyroid children exhibiting mental retardation and growth that is slow and stunted.

During the course of evolution, the thyroid gland has assumed different functions in different species. In 1912, J. Frederick Gudernatsch, working at the Cornell University Medical School, fed tadpoles mammalian thyroid and was able to induce metamorphosis. Tadpoles transformed into adult frogs: the external gills of newly hatched tadpoles disappeared, a large jaw developed, the eyes and legs grew rapidly, and the tail was resorbed. By contrast, when the thyroid gland was removed from tadpoles, metamorphosis failed to occur. Flatfish also undergo metamorphosis. In an early stage in their development, they are bilaterally symmetrical, with eyes on both sides of the body. During metamorphosis, one eye moves to the other side, which becomes the upper side of the fish.

SEE ALSO: Amphibians (c. 360 Million BCE), Metabolism (1614), Secretin: The First Hormone (1902), Mitochondria and Cellular Respiration (1925).

Thyroid hormones play a major role in the development of vertebrates. Perhaps nowhere is this more striking than in the metamorphosis of the tadpole into the adult frog.

X-Ray Crystallography

Wilhelm Conrad Röntgen (1845–1923), **William Henry Bragg** (1862–1942), **Max von Laue** (1879–1960), **William Lawrence Bragg** (1890–1971), **Dorothy Crowfoot Hodgkin** (1910–1994), **Francis Crick** (1916–2004), **Rosalind Franklin** (1920–1958), **James D. Watson** (b. 1928)

Credit for determining the structure of DNA has been attributed to James Watson and Francis Crick. But many researchers believe that equal billing should have been assigned to Rosalind Franklin, their collaborator, whose X-ray diffraction images clearly illustrated that the structure of DNA was a **double helix**. While X-ray crystallography was originally used to determine the size of atoms and the nature of chemical bonds, its applications now include chemistry, mineralogy, metallurgy, and the biomedical sciences. It is a powerful analytical tool used by biologists to determine the structure and function of molecules of biological interest including vitamins, proteins, DNA and RNA, as well as drugs and their rational design.

During the course of his 1895 studies on the passage of an electric current through a gas, Wilhelm Röntgen observed that the emitted rays recorded an image on a photographic plate. Because of the unknown nature of these rays, he called them *X-rays*, a discovery for which he was awarded the first Nobel Prize in Physics in 1901. In 1912, Max von Laue discovered that crystals defracted X-rays. Building upon Laue's work, from 1912 to 1914, William Lawrence Bragg and his father William Henry Bragg conducted research analyzing crystalline structures using X-ray defraction, for which they were jointly awarded the 1915 Nobel Prize. They found that by measuring the angles and intensities of the diffraction pattern of X-rays through the crystalline sample, a detailed three-dimensional picture of a molecule could be determined. Lawrence Bragg, who at twenty-five was (and is) the youngest Nobel laureate, developed the basic law for determining crystalline structures in 1912: the Bragg law of X-ray refraction, which is still used.

Among the foremost biological crystallographers was Dorothy Crowfoot Hodgkin, who determined the structures of cholesterol (1937), penicillin (1946), vitamin B_{12} (1956) for which she received the 1964 Nobel Prize, and **insulin** (1939), which she worked on for thirty years. Hodgkin also studied the three-dimensional nature and structure of biomolecules, such as proteins.

SEE ALSO: Insulin (1921), Amino Acid Sequence of Insulin (1952), The Double Helix (1953).

An illustration of the crystalline structure of manganese tetrafluoride (MnF_4), as determined by X-ray crystallography.

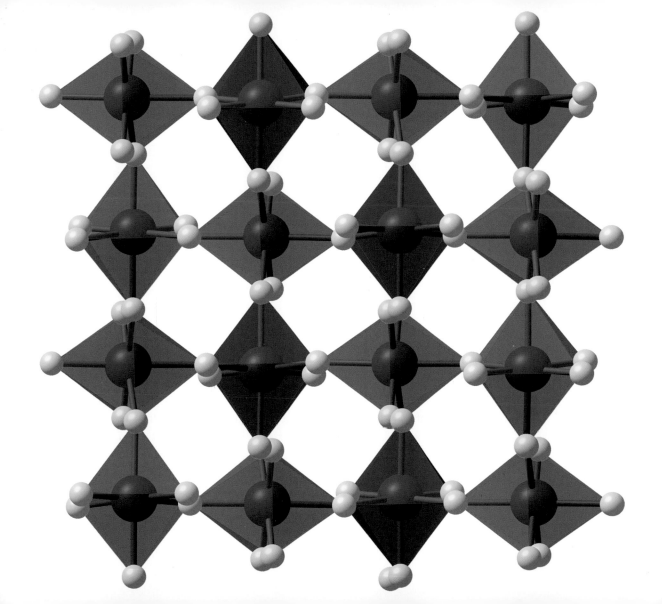

Bacteriophages

Ernest H. Hankin (1865–1935), Félix d'Hérelle (1873–1949), Frederick W. Twort (1877–1950)

Ernest Hankin was a British bacteriologist in India in the 1890s studying malaria and cholera. In 1896, he reported that there was something in the Ganges and Jumma rivers that exerted an antibacterial effect against cholera, an effect that continued even after the river waters were passed through a porcelain filter, which held back bacteria. He theorized that this substance was responsible for limiting the spread of the cholera epidemic but proceeded no further in studying this mysterious, invisible substance.

During the early 1900s, the English bacteriologist Frederick Twort was engaged in experiments growing bacteria in an artificial media and noted that some bacteria were killed by an unknown agent, which he designated an "essential substance." This substance passed through porcelain filters and required bacteria for growth; he thought that it was possibly a virus. His report, published in 1915, was largely ignored for decades. Further work was interrupted by World War I and the lack of funding.

Félix d'Hérelle was a French Canadian microbiologist working at the Pasteur Institute in Paris. As did Twort, d'Hérelle observed the effects of an "invisible antagonist microbe . . . of the dysentery bacillus" that passed through a porcelain filter. He recognized that he had discovered a virus he dubbed a *bacteriophage* ("bacteria eater"), or simply *phage*, which he reported in 1917. Although he appeared to be cognizant of Twort's prior finding, he failed to adequately acknowledge it and largely claimed credit for the discovery. Sensing its antibacterial potential, in 1919, d'Hérelle tested phage in a Paris children's hospital for the treatment of dysentery and was later involved with establishing a commercial laboratory that produced five different phage preparations as treatments against different bacterial infections.

The initial enthusiasm for phages—bacterial **viruses** that selectively attach to one or a few bacterial hosts and kill them by lysis (destruction by dissolution)—as antibacterial agents waned after the introduction of **antibiotics** in the 1940s. Interest was renewed in the 1990s when drug-resistant bacteria emerged. Phages continue to be used to treat bacterial infections in Russia and Eastern Europe and experimentally as model systems to study multiplication of viruses.

SEE ALSO: Viruses (1898), Cancer-Causing Viruses (1911), Antibiotics (1928), Bacterial Genetics (1946).

A bacteriophage (or phage) is a virus that infects bacteria. The phage consists of a capsid head, which encloses its DNA, and a protein tail with fibers by which the phage attaches to a bacterium.

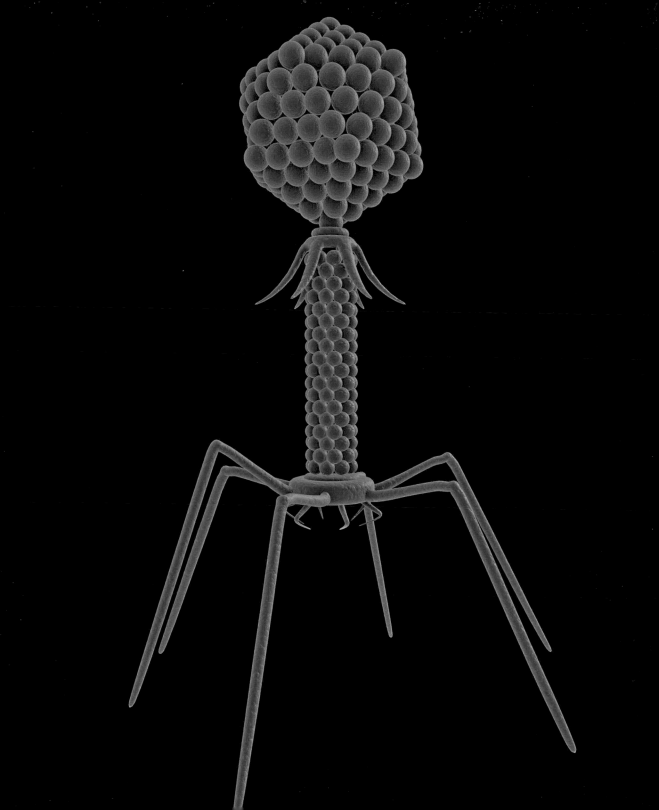

Biotechnology

Eduard Buchner (1860–1917), Károly (Karl) Ereky (1878–1952)

Since the 1970s, biotechnology has entered the common parlance. By biotechnology, we are referring to the use of biological systems for the production of useful products, which in the context of biology includes food, drink, and medicines. Under this umbrella term is included such diverse areas as recombinant DNA, **genetically modified crops**, biopharmaceuticals, and genetic engineering. But the seeds of biotechnology were planted over 10,000 years ago.

After our ancestors evolved from hunter-gatherers to actively producing their own food, they embarked upon the first steps as applied biologists by practicing **artificial selection (selective breeding)** of animals and plants—early biotechnology. Animals were first domesticated and later bred to maximize their utility as coworkers with humans in the field and to provide meat and fur. Plants were selectively bred to improve their nutritional value and to withstand the ravages of adverse climatic conditions and agricultural pests. Over the next several thousand years, cheese and yogurt were prepared from milk, and yeast was used to make beer, wine, and bread. This was the ancient beginning of biotechnology.

Nineteenth century biotechnologists focused their scientific attention on maximizing the process of fermentation—the conversion of sugar and starches in fruits to alcoholic beverages—one of the earliest chemical reactions observed and practiced by humans. In 1896, the German chemist Eduard Buchner showed that the presence of living cells was not essential for fermentation. Fermentation occurred when the products of living cells— *ferments*, now called *enzymes*—were present. This phase in the history of biotechnology was intimately tied to the study and practice of zymology or fermentation, in particular, of beer and wine. Problems of hunger remained to be tackled.

The Hungarian agricultural engineer Karl Ereky was first to coin the term *biotechnology*, which he used in the title of his 1919 book describing how raw materials from pigs could be upgraded to produce socially useful products. In his effort to create an abundance of food in famine-ravished Hungary after World War I, Ereky created one of the largest and most profitable meat- and fat-producing operations in the world.

SEE ALSO: Agriculture (c. 10,000 BCE), Domestication of Animals (c. 10,000 BCE), Artificial Selection (Selective Breeding) (1760), Microbial Fermentation (1857), Enzymes (1878), Green Revolution (1945), Genetically Modified Crops (1982).

During the nineteenth century, applied biotechnologists produced beer by fermentation. This 1897 painting of a monk in a brewery is by the German artist Eduard von Grützner (1846–1925).

Neurotransmitters

John Newport Langley (1852–1925), **Otto Loewi** (1873–1961),
Thomas Renton Elliott (1877–1961)

How do nerves speak to other nerves or muscles? In response to a change within the body or the external environment, nerves are stimulated. An electrical current travels down the nerve and then across a synapse (physical gap) to another nerve or to an effector cell (muscle, the heart, or a gland) that responds. Is this message transmitted across a synapse by an electrical current or a chemical released from the nerve ending?

In 1905, the distinguished British physiologist John Newport Langley at Cambridge University first proposed that a "receptive substance" was the site at which a chemical was released after nerve stimulation. This remarkable concept was based on experimental results obtained by his student T. R. Elliott—results Langley failed to acknowledge. Over the next fifteen years, a number of distinguished scientists conducted experiments, analogous to Elliott's, demonstrating that the response of effector cells after nerve stimulation and the addition of certain chemicals was similar but not always identical.

Otto Loewi, a German-born pharmacology professor at the University of Graz in Austria, long puzzled over how to prove chemical synaptic transmission. On Easter Sunday eve in 1920, the idea for a conclusive experiment came to him during a deep sleep. He scribbled a few notes and resumed sleeping. Upon awakening, he was unable to decipher these notes. At 3:00 a.m. the following morning, while sleeping, the idea reemerged. He rushed to his laboratory, and before the day's end, performed the critical, yet very simple, experiment. He placed two frog hearts in separate tissue baths, and after stimulating the vagus nerve of one, the heartbeat slowed. He then added this bathing fluid to the second bath, and the heartbeat slowed. Thus, he established the release of a chemical, which he called *Vagusstoff* (later identified to be acetylcholine), as the first neurotransmitter. He was a co-recipient of the 1936 Nobel Prize and, two years later, fled Austria after the Nazi invasion.

More than one hundred types of neurotransmitters have been identified in both vertebrates and invertebrates, and many of these have been shown to play a role in normal physiological function, in disease, and in the development of drugs.

SEE ALSO: Animal Electricity (1786), Nervous System Communication (1791), Neuron Doctrine (1891), Action Potential (1939).

The neurotransmitter acetylcholine is released from nerve endings to activate receptor sites on the surface of skeletal (voluntary) muscle fibers, causing their contraction. This microscopic view of nerve cell endings in muscular tissue has been magnified 200 times.

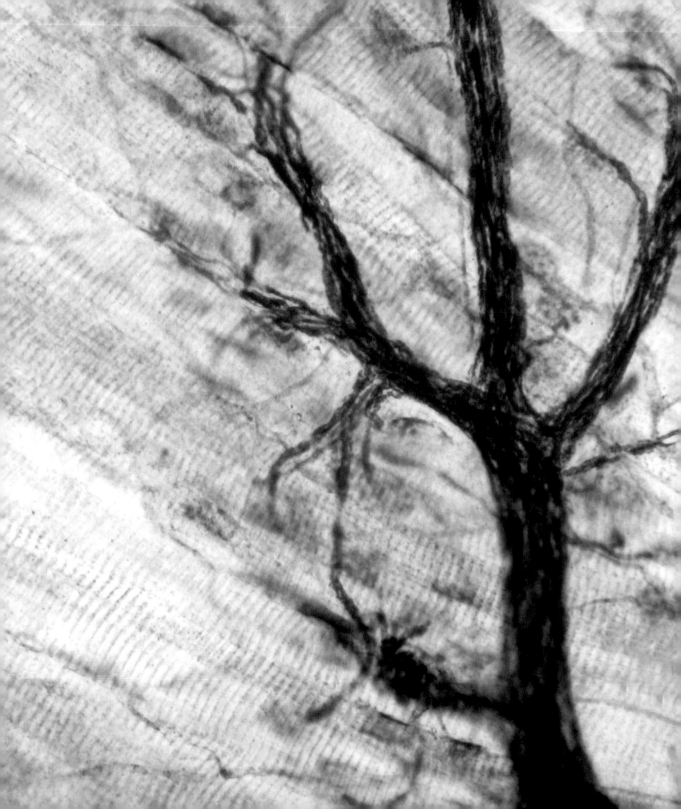

Insulin

1921

Paul Langerhans (1847–1888), **Joseph von Mering** (1849–1908),
Oskar Minkowski (1858–1931), **John J. R. MacLeod** (1876–1936),
Frederick G. Banting (1891–1941), **James B. Collip** (1892–1965),
Charles H. Best (1899–1978)

The body does not waste valuable resources, and insulin is largely responsible for this frugality. Insulin is a hormone that actively conserves and stores energy-rich foods when they are taken in excess: excess carbohydrates are stored in the liver and muscle as glycogen, fat is deposited in adipose tissue, and amino acids are converted to proteins. While our understanding of insulin's effects on the body's chemistry is of relatively recent vintage, descriptions of diabetes, a metabolic disease resulting from insulin resistance or deficiency, appear in ancient Egyptian and Greek manuscripts.

Our modern understanding of diabetes began in 1869 when Paul Langerhans, a medical student, discovered previously unrecognized cells in the pancreas, which were later designated the *islets of Langerhans*. Thirty years later, Joseph von Mering and Oskar Minkowski sought to determine the biological function of the pancreas. After removing this organ from a dog, it exhibited all the signs and symptoms of diabetes, including the telltale presence of sugar in the urine.

In 1921, Canadian surgeon Frederick Banting persuaded John J. R. MacLeod, a physiology professor at the University of Toronto, to allow him to use his laboratory and ten dogs while MacLeod was vacationing. Banting recruited Charles Best, who was waiting to enter medical school, to assist him. They repeated von Mering and Minkowski's experiment and reversed the diabetic symptoms by injecting the dog with a pancreatic extract taken from a healthy dog. In January 1922, a fourteen-year-old diabetic, Leonard Thompson, was the first patient given a pancreatic extract and was successfully treated. Life-saving insulin was discovered! In 1923, the Nobel Prize was awarded to Banting and (an undeserving) MacLeod. Banting shared his prize money with Best, as did MacLeod with Collip, who purified insulin from the pancreatic extract. That same year, Eli Lilly began the commercial distribution of insulin. Prior to the discovery of insulin, a newly diagnosed type 1 diabetic had a life expectancy measured in months or less. Now there is little difference in longevity between well-controlled diabetics and nondiabetics.

SEE ALSO: The Liver and Glucose Metabolism (1856), Secretin: The First Hormone (1902), Amino Acid Sequence of Insulin (1952).

The islets of Langerhans (shown here in a high-power magnification), located in the pancreas, are responsible for the production of insulin.

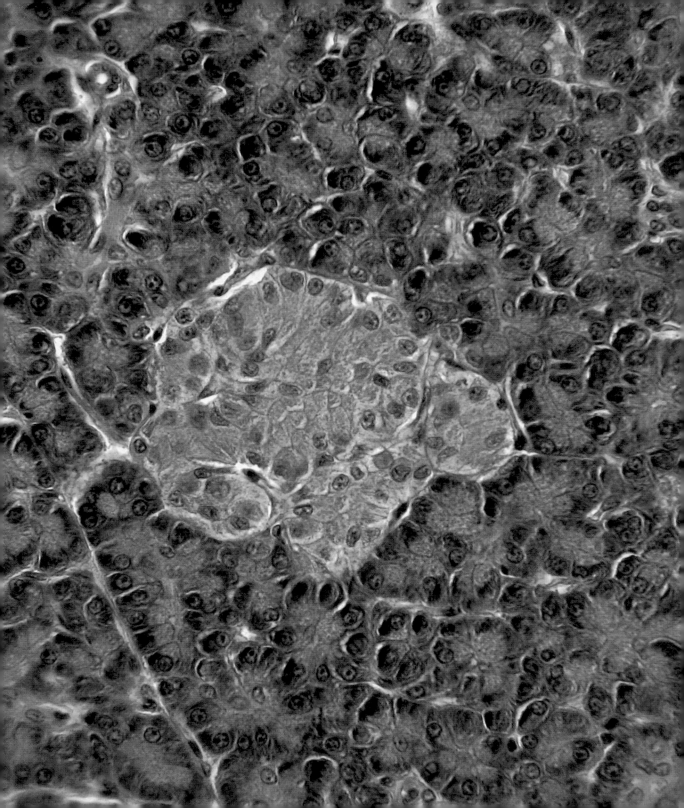

Inborn Errors of Metabolism

Gregor Mendel (1822–1884), Archibald Garrod (1857–1936)

In the 1890s, the English physician Archibald Garrod was called upon to examine Thomas P, a three-month-old boy whose urine was a deep reddish-brown. His diagnosis was alkaptonuria, which is caused by a buildup of homogentisic acid (alkapton), a compound that the body normally breaks down rapidly. At the time, the prevailing thought was that alkaptonuria, a rare disorder, was caused by a bacterial infection; Garrod conceived it to be a disease associated with chemical reactions. Two subsequent siblings of Thomas were born, each with alkaptonuria, from parents who were blood relatives. Further investigation revealed to Garrod that in other families having one or more children with alkaptonia, in every instance, the parents were first cousins. He published these findings in 1902.

Based on his appreciation of Gregor Mendel's rules of inheritance and his knowledge of chemistry, Garrod concluded that some diseases might represent inherited disorders of **metabolism**, and these he described in his classic 1923 text *Inborn Errors of Metabolism*. IEM, also called inherited metabolic diseases, are associated with a single defective gene, resulting in the absence or production of an abnormal specific **enzyme** required to carry out a metabolic reaction. Most abnormal genes are inherited in an autosomal recessive manner—that is, the child must inherit copies of the defective from each parent. IEM represent a large group of disorders in which the body cannot convert food into energy or other essential compounds. As a consequence, the body builds up substances that are toxic or interfere with normal body function or that reduce the body's ability to synthesize essential compounds. The outcome may range from harmless to severe, even fatal.

IEM, of which over two hundred have been identified, are traditionally classified based on the type of metabolism involved—namely, carbohydrate, amino acid, fats, or complex molecules. Each IEM is individually rare, but collectively, they occur in roughly 1 of every 4,000 live births, with the incidence varying among different ethnic and racial groups: sickle cell anemia, 1:600 of African descent; cystic fibrosis, 1:1600 of European descent; and Tay-Sachs, 1:3500 of Ashkenazi Jewish descent.

SEE ALSO: Metabolism (1614), Mendelian Inheritance (1866), Enzymes (1878), Blood Types (1901), One Gene-One Enzyme Hypothesis (1941), Protein Structures and Folding (1957).

Albino kangaroos are variants of kangaroos that are usually red or gray. Occurring in many vertebrates, albinism is an IEM characterized by a defect in tyrosine metabolism that results in deficient melanin production.

Embryonic Induction

Carl Ernst von Baer (1792–1876), **Wilhelm Roux** (1850–1924),
Hans Driesch (1867–1941), **Hans Spemann** (1869–1941),
Hilde Mangold (1898–1924)

In 1828, Carl Ernst von Baer determined that all vertebrate tissues and organs arise from three primary germ layers. Two theories of embryo development sought to explain these findings: Preformation theory argued that, at the time of conception, each embryo had a complete set of organs that increased in size with age. The competing theory of epigenesis postulated that each individual started as an undifferentiated mass and gradually differentiated. In 1888, Wilhelm Roux, an experimental embryologist, killed one of the two cells that arose from the first cleavage of a fertilized frog egg. When only a half embryo developed, he concluded that preformation was correct. When repeated by Hans Driesch four years later, using a sea urchin egg and an improved experimental design, complete sea urchins were formed, thereby refuting preformation.

The German embryologist Hans Spemann, interested in the embryologic development of an organism (morphogenesis), performed experiments involving the grafting of cells from one organism to another. Among the first studies of this master microsurgeon was transplanting the eyecup (from which the eyeball develops) of a new newt embryo to the outermost surface of the abdomen of another newt. This resulted in the formation of an eye lens, the cover of an eyecup.

Spemann's most significant studies were carried out at the Zoological Institute in Freiburg, Germany, by his graduate student Hilde Mangold, who used newt embryos in her doctoral dissertation. Mangold transplanted the upper lip of one embryo (donor) to a distant location on the flank of another embryo (host). Three days later, a nearly complete secondary embryo formed on the flank of the host. Thus, the donor cells in an early embryo were acting as "organizers" and were able to induce or dictate the fate of other cells, establishing that there are no foreordained organs in early-stage embryos. Spemann was awarded the 1935 Nobel Prize; Mangold, killed in a heater explosion in 1924, never witnessed the publication of her full paper. Her dissertation was one of the few in biology to directly lead to a Nobel Prize.

SEE ALSO: Regeneration (1744), Theories of Germination (1759), Germ-Layer Theory of Development (1828), Cloning (Nuclear Transfer) (1952), Induced Pluripotent Stem Cells (2006).

The newt was the test animal used to settle the long-standing controversy of whether, at the time of conception, embryos had a complete set of organs that increased in size (preformation theory) or whether each organism started afresh as an undifferentiated mass of cells (epigenesis theory).

Timing Fertility

Theodoor Hendrik van de Velde (1873–1937), **Kyusaku Ogino** (1882–1975), Hermann Knaus (1892–1970)

Unlike female humans and other **primates** that have menstrual cycles, most other **mammals** have estrus cycles. (The word estrous refers to a "frenzied passion" or, more commonly in animals, to being "in heat.") These mammals are sexually receptive and have the ability to reproduce only during limited times of the year. That time, corresponding to a change in season, is typically one in which there is an adequate source of food and a climate that is hospitable to newborns' survival. It is only at this time that ovulation—that is, the release of a mature egg by a female—occurs. By contrast, animals with menstrual cycles can be sexually active at any time during their cycle throughout the year, quite independent of ovulation. Recent evidence suggests that women are more receptive to sexual activity during their most fertile times, which are the six days prior to ovulation.

Interest in identifying the woman's most fertile time can be traced back to the ancient Greeks, Hebrews, and Chinese, with the prevailing view, until the twentieth century, that these were the days immediately after menstruation. In 1905, the Dutch gynecologist Theodoor Hendrik van de Velde established that women ovulate only once during their menstrual cycle. In the 1920s, two gynecologists—Kyusaku Ogino in Japan (1924) and Hermann Knaus in Austria (1928)—working independently and unaware of each other derived essentially the same formula: each determined that ovulation occurs about fourteen days prior to the next menstrual period. Whereas previous estimations were calculated from the first day of menstruation, the Knaus-Ogino method counted backwards and determined that the most fertile period, based on ovulation and the viability of sperm, was the twentieth to the twelfth day before the start of the next period.

This calendar-based method was initially intended to optimize timing to assist women in becoming pregnant, but it was more commonly used by Catholics as a natural method of birth control condoned by the Church. The Knaus-Ogino, or rhythm, method is far from ideal and, even when used perfectly, has a failure rate of 9 percent.

SEE ALSO: Mammals (c. 200 Million BCE), Primates (c. 65 Million BCE), Ovaries and Female Reproduction (1900), Progesterone (1929).

After ovulation has occurred, a mature egg remains alive in the fallopian tube for only about twenty-four hours.

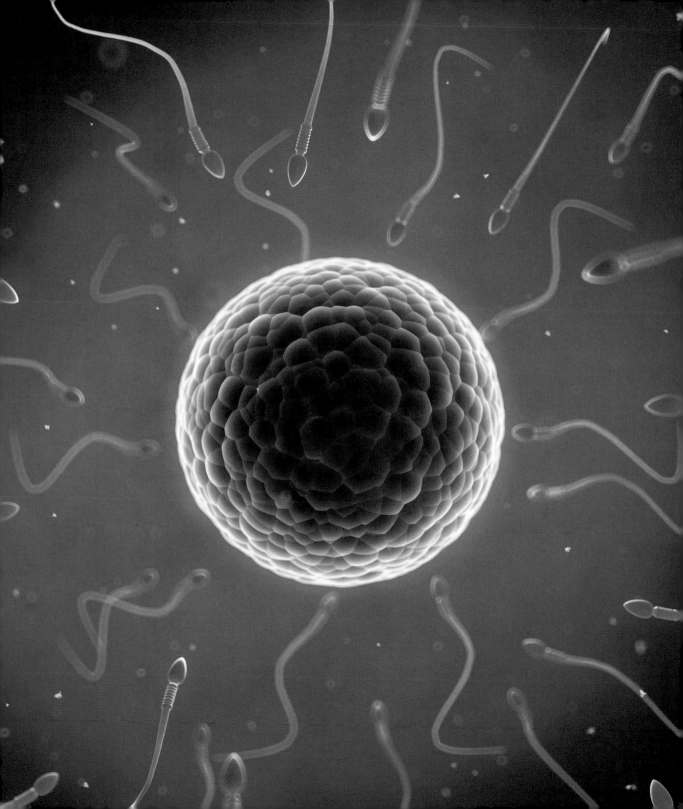

Mitochondria and Cellular Respiration

Otto H. Warburg (1883–1970), **David Keilin** (1887–1963), **Albert Claude** (1898–1983), **Fritz A. Lipmann** (1899–1986), **Hans A. Krebs** (1900–1981)

Mitochondria are the powerhouses of animal cells responsible for converting the energy in our foods to a chemical—adenosine triphosphate (ATP)—that cells can use to carry out their functions. According to the theory of endosymbiosis, several billion years ago mitochondria were free-living organisms that were aerobic (i.e., they used oxygen to produce energy). Large anaerobic (oxygen-lacking) organisms, which were far less efficient in producing energy, engulfed and incorporated the aerobic mitochondria.

The process of cellular respiration—whereby sugar, in the presence of oxygen, is broken down to generate a total of about thirty-six molecules of ATP—takes place in three stages. Stage 1 (Glycolysis): In the absence of oxygen, glucose, a six-carbon sugar, is split into two molecules of pyruvate, a three-carbon sugar, and two molecules of ATP. Stage 2 (Citric Acid Cycle): In the presence of oxygen in the mitochondrion, acetate (derived from carbohydrates, fats, and proteins) is broken down to carbon dioxide, water, and two additional ATP molecules. Stage 3 (Electron Transport Chain or Oxidative Phosphorylation): Electrons from hydrogen are carried down the respiratory chain in the mitochondrion, through a sequence of steps, to produce about thirty-two molecules of ATP.

During the twentieth century, many eminent researchers sought to understand the sequence of events in cellular respiration. Otto Warburg postulated the presence of an intracellular respiratory **enzyme** in mitochondria in 1912. In 1925, David Keilin discovered the cytochrome enzymes and the concept of a respiratory chain. Hans Krebs elucidated the citric acid (Krebs) cycle in 1937. Fritz Lipmann, in 1945, uncovered Coenzyme A, an essential component for the **metabolism** of carbohydrates, fats, and amino acids. Albert Claude separated mitochondria and other organelles using cell fractionation—a process he developed in 1930, permitting biochemical analysis of organelles—and then characterized them using an **electron microscope**.

SEE ALSO: Plant Defenses Against Herbivores (c. 400 Million BCE), Metabolism (1614), Gas Exchange (1789), Photosynthesis (1845), Enzymes (1878), Electron Microscope (1931), Bioenergetics (1957), Energy Balance (1960), Endosymbiont Theory (1967).

The mitochondrion is the site of respiration within the cells of plants and animals. Oxygen is used in the breakdown of organic molecules and the synthesis of ATP, which is used to power the chemical reactions of the cell.

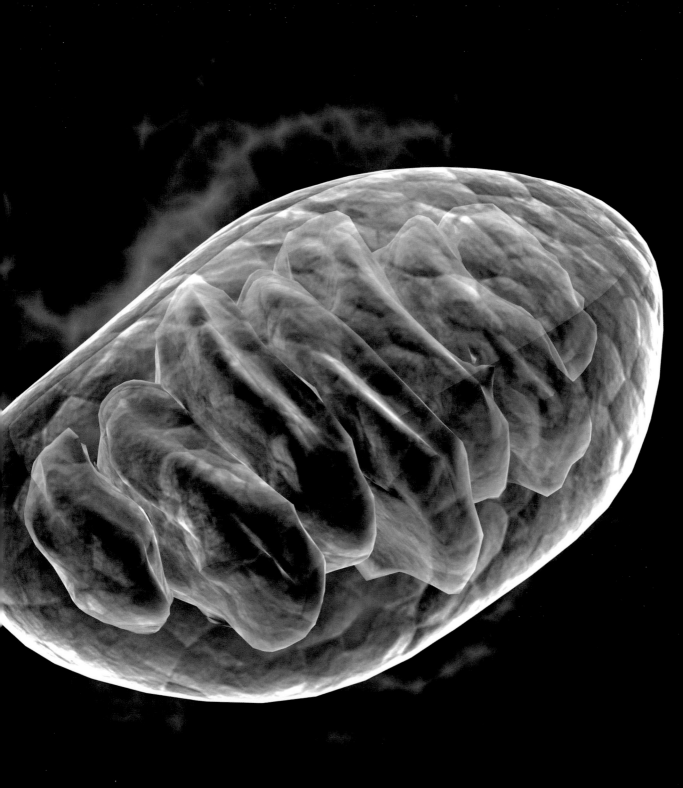

"The Monkey Trial"

Clarence Darrow (1857–1938), **William Jennings Bryan** (1860–1925),
John Thomas Scopes (1900–1970)

In 1925, the right to teach evolution went on trial in Dayton, Tennessee. This trial was broadcast nationally on radio and covered by two hundred newspaper reporters. John Scopes was a 24-year-old high school biology teacher accused of violating the Butler Act (1925), a law that prohibited teachers in any Tennessee-funded public school and university from denying the Biblical account of man's origin, thereby making the teaching of evolution unlawful. Scopes used the textbook *Civic Biology*, which described and was sympathetic to evolution; however, it is unclear whether he actually taught such a lesson or merely admitted doing so to publicize the issues in the case.

The American Civil Liberties Union supported Scopes in defense of the constitutional right of free speech, as well as his academic freedom. Representing Scopes was Clarence Darrow, a nationally acclaimed defense attorney, a leading member of the ACLU, and a self-proclaimed agnostic. On the prosecution team was William Jennings Bryan, a three-time presidential candidate, known for his oratorical skills and fundamentalist beliefs. Bryan was fiercely opposed to evolution and its teaching because it contradicted the word of God, as revealed in the Bible, which he believed superseded human knowledge.

The judge instructed the jury not to consider the merits of evolution but only whether Scopes had violated the law. The outcome of the eight-day trial was preordained, and the jury reached its decision after only nine minutes of deliberation. Scopes was found guilty as charged and was fined $100. The guilty verdict was reversed on appeal, not because it violated the right of free speech as the ACLU had argued, but rather on a subtle technicality. Bryan died in his sleep five days after the trial's end; Scopes went to graduate school and became a geologist studying oil reserves. A highly fictionalized account of the trial was portrayed in Lawrence and Lee's *Inherit the Wind* (1955), and the Butler Act was repealed over forty years later in 1967. Nine decades after the trial, the subject of faith versus science and creationism versus evolution continues. For many, the jury is still out.

SEE ALSO: Darwin's Theory of Natural Selection (1859).

The human pedigree, as envisioned by Ernst Haeckel in his 1874 work The Evolution of Man. *Evolution is more accurately illustrated with a phylogenetic tree, rather than this modified "Great Chain of Being" that links the prosimian (lemurs, etc.) directly to the kangaroo.*

THE MODERN THEORY OF THE DESCENT OF MAN.

Population Ecology

Thomas Malthus (1766–1834), Alfred J. Lotka (1880–1949)

In 1798, the English political economist and demographer Thomas Malthus authored an essay projecting that the human population was growing at such a rapid rate that, if not brought under control, it would lead to mass starvation and poverty. Happily, the agricultural revolution produced food far faster than the production of mouths to feed. Nevertheless, the matter of a rising population became a major issue in the United States in the early 1920s. In response to a severely restrictive immigration law, Alfred Lotka, a Polish immigrant to the United States, brought his mathematical insights to biology. In an influential 1925 article, he demonstrated that the observed rise in population resulted from a disproportionately large number of individuals who had immigrated in previous decades and were now in their peak reproductive years. Restricting the number of immigrants, he argued, would lead to a population decline.

A population refers to members of the same species in a particular geographic area, and population ecology examines factors that influence populations and how these populations interact with their environment. In his 1925 book, *Elements of Physical Biology*, Lotka noted that four variables affected population—death, birth, immigration, and emigration—and that a dynamic equilibrium occurred when losses and gains were equal. The population size is affected by interactions with abiotic and biotic environmental factors: among the abiotic factors are climate and food supply, while biotic factors include predation and intraspecies and interspecies competition. A number of dynamic processes influence population dispersion, population density, and demographic trends, which describe how populations change over time.

Scientists estimate that 99 percent of all species that have ever lived are now extinct. In addition to negative abiotic and biotic factors, humans have contributed to extinction and marked population declines. Human contributions include: pollution, such as runoffs of industrial wastes and agricultural fertilizers; **global warming**; the introduction of **invasive species**, such as zebra mussels in Lake Erie and kudzu in the southern United States; and the removal of competing species or predators.

SEE ALSO: Agriculture (c. 10,000 BCE), Artificial Selection (Selective Breeding) (1760), Population Growth and Food Supply (1798), Invasive Species (1859), Global Warming (1896), Factors Affecting Population Growth (1935), De-Extinction (2013).

The zebra mussel (Dreissena polymorpha), native to the Black Sea and Caspian Sea, were first detected in the Great Lakes in 1988. They feed aggressively on phytoplankton (microscopic plants) but also zooplankton (animal life), which larval fish and native mussels depend upon to survive.

Food Webs

Al-Jahiz (781–868/869), **Charles Elton** (1900–1991), **Raymond Lindeman** (1915–1942)

The concept of a food chain originated with Al-Jahiz, a ninth-century Arabic author of some two hundred books on a wide range of subjects including grammar, poetry, and zoology. In his zoology work, he discussed a struggle for existence among animals who hunt to obtain food and who are, in turn, hunted. Charles Elton, an Oxford faculty member, was among the most important animal ecologists of the twentieth century. In his classic 1927 text *Animal Ecology*, Elton laid out the basic principles of modern ecology, including, rather explicitly, food chains and food webs, which are now central themes in ecology.

At its simplest level, a food cycle follows a linear relationship from the base of the food chain—a species that eats no other (typically, a plant)—to the final predator or ultimate consumer, which is typically three to six feeding levels in length. Elton recognized that this simple food chain depiction was a gross oversimplification of "who eats whom." The food chain failed to account for real ecosystems, in which there are multiple predators and multiple preys, and the reality that a given animal might consume other animals if the preferred prey was not available. Moreover, some carnivores also eat plant material and are omnivores; conversely, herbivores occasionally eat meat. The food web, a concept now preferred to food chain, represents these highly complex interrelationships.

In 1942, Raymond Lindeman postulated that the number of levels in a food chain is limited by *trophic dynamics*, or the effective transfer of energy from one part of the ecosystem to another. After food is consumed, energy is stored in the body of the consumer, and it travels in only one direction. Much of that energy is lost as heat (when the food is being utilized for basic needs), and the remainder eliminated as waste material. In general, only about 10 percent of the energy consumed is available at the next higher trophic (feeding) level. Thus, with each successive level up the chain, less energy is transmitted and, therefore, food chains rarely exceed four to five feeding levels.

SEE ALSO: Agriculture (c. 10,000 BCE), Population Growth and Food Supply (1798), Population Ecology (1925), Energy Balance (1960), Deepwater Horizon (BP) Oil Spill (2010).

The food web is exemplified at Katmai National Park, Alaska, where this grizzly bear is a jawful away from eating a fish, which fed on a smaller fish or microscopic plants or animals floating in the water.

Insect Dance Language

Karl von Frisch (1886–1982)

Animals communicate with one another when seeking to locate food, mate, or signal an alarm in the presence of threats in their environment. A variety of animals use **pheromones** to facilitate various phases of their mating behavior. Not all communication occurs between members of the same animal species, such as the facial expression and body language of our pets. The odor from a skunk's spray is a highly effective defensive weapon used to ward off bears and other potential predators, and it is sufficiently pungent that it can be detected by human noses at downwind distances of one mile (1.6 kilometers).

Animal communication is not limited to vertebrates, with some of the most interesting examples occurring in **insects**. Pioneering studies on insect communications were conducted in the 1920s by the Nobel laureate Karl von Frisch, an Austrian ethnologist at the University of Munich. He observed that a distinctive "dance language" is used by European honeybee (*Apis mellifera*) foragers to inform other bees in the hive about the direction and distance of food. A "round dance," in which the forager executes tight circles, is performed when food is close to the hive—less than 160–320 feet (50–100 meters)—whereas a "waggle dance," resembling a figure-eight movement, signifies food at a distant location.

European honeybees also use a complex chain of communication modes that involve all five senses in a fascinating courtship ritual, with each signaling and triggering a subsequent behavior by the partner: The male visually identifies the female and turns toward her. The female releases a chemical that is detected by the male's olfactory system. He approaches the female and taps her with his limb that, in the process, picks up the chemical. In response, the male extends and vibrates his wings producing a "courtship song," a form of auditory communication. Only after this entire sequence is successively and successfully completed will the female allow the male to perform copulation.

SEE ALSO: Insects (c. 400 Million BCE), Animal Electricity (1786), Neuron Doctrine (1891), Neurotransmitters (1920), Pheromones (1959), Animal Altruism (1964).

Dance language among certain insects—in particular, honeybees—is well developed and has been extensively studied. Here, the Apis cerana japonica *honeybees surround their nest in Japan.*

Antibiotics

Louis Pasteur (1822–1895), Alexander Fleming (1881–1955)

In 1999, *Time* magazine stated that penicillin "was a discovery that would change the course of history." It was the first of many *antibiotics*, or substances derived from microbes (**fungi** or bacteria) that kill or control the growth of other microbes. The earliest accounts of the healing properties of moldy bread appeared from ancient Egypt some 3,500 years ago in the Ebers Papyrus. In 1877, Louis Pasteur demonstrated that one microbe could be used to combat another and termed this *antibiosis*. He inoculated animals with a mixture of the anthrax bacillus and another common bacteria, which protected them against the deadly anthrax infection. Pasteur postulated that microbes released materials that might be used therapeutically, a prediction validated six decades later.

Alexander Fleming served in a battlefield hospital on the Western Front in France during World War I and observed that more soldiers were dying from antiseptics used to treat infected wounds than from the infected wounds. After the war, the Scottish-born Fleming resumed his bacteriological research at St. Mary's Hospital Medical School in London. By 1928, when he turned his attention to the staphylococcus bacterium, he had gained the reputation as a well-respected scientist but one who did not maintain a neat laboratory.

After returning from a month-long family vacation in September 1928, he found that one of his culture dishes was contaminated and that there was no growth of staph colonies surrounding the fungal-contaminated growth. He astutely recognized what countless other scientists had overlooked: the fungus may have released an antibacterial substance. He grew a pure culture of the fungus, *Penicillium notatum*, and found that it selectively killed many, but not all, bacteria. He named this substance *penicillin* and in 1929 authored a paper that described its effects. Minimal interest was shown in his work until war clouds hung ominously over Europe in 1940, when penicillin's potential was recognized and it was isolated and purified. Some sources estimate that penicillin saved the lives of millions of soldiers who would have died from infections of battle wounds. Accordingly, Fleming was a co-recipient of the 1945 Nobel Prize.

SEE ALSO: Prokaryotes (c. 3.9 Billion BCE), Fungi (c. 1.4 Billion BCE), Probiotics (1907), Bacterial Genetics (1946), Bacterial Resistance to Antibiotics (1967).

From a medical perspective, modern military history can be divided into the Infection Era (1775–1918) and the Trauma Era (1919–). On average, during the Infection Era, the ratio of infectious deaths to trauma deaths was 4:1. Thanks in part to penicillin; during World War II that ratio was reduced to 1:1. These photos depict military combatants serving on the Western Front during World War I.

Bundesarchiv, Bild 183-R05148
Foto: o.Ang. | 1916

Progesterone

Gustav J. Born (c. 1850–1900), **John Beard** (1858–1924), **Ludwig Fraenkel** (1870–1951), **George W. Corner** (1889–1981), **Willard M. Allen** (1904–1993)

With the discovery of estrogen and its effects on female reproductive function during the 1920s, many scientists believed that it was the only female sex hormone. But not all researchers were satisfied; doubts had been sowed in 1897. In that year, John Beard theorized that the corpus luteum was an "organ of pregnancy," perhaps even essential for the maintenance of pregnancy. The corpus luteum or "yellow body" is what is left of a follicle after ovulation has occurred.

In 1900, the German researcher Gustav Born observed that the corpus luteum was absent from the ovaries of monotremes, such as the platypus of Australia and New Guinea—the only mammals that feed their babies with milk but lay eggs and lack a placenta. On this basis, Born concluded that the corpus luteum was required for the development of the **placenta**. Furthermore, he speculated that the corpus luteum released an internal secretion that prepared the uterine mucosa (outer layer) for the anticipated arrival of a fertilized egg and its implantation in the uterine wall. After Born's death, his student Ludwig Fraenkel continued this work and, in 1903, showed that destruction of the corpus luteum in pregnant rabbits caused an abortion. Finally, in 1929, George Corner and Willard Allen established that an abortion that normally occurs after removal of the corpus luteum in pregnant rabbits could be prevented by administration of an extract from the corpus luteum. This extract, purified in 1933, was called *progestin* (progesterone).

THE PREGNANCY HORMONE. After ovulation, the corpus luteum springs into action secreting progesterone in anticipation of the potential fertilization of the released egg and ensuing pregnancy. Progesterone stimulates the development of a thick lining of blood vessels in the uterine wall that is needed to sustain the growth of a developing fetus. The corpus luteum continues to produce progesterone for about ten weeks, after which time the placenta assumes this responsibility. Throughout pregnancy, progesterone quiets the uterus and prevents contractions that could lead to a miscarriage. If the egg is not fertilized, the corpus luteum degenerates and no longer secretes progesterone, and a new menstrual cycle ensues.

SEE ALSO: Placenta (1651), Ovaries and Female Reproduction (1900), Secretin: The First Hormone (1902).

In 1803, Nicholai Argunov painted the portrait of a very pregnant Praskovia Kovalyova (1768–1803), who, born into a serf family, became one of the finest opera singers of late-eighteenth-century Russia. Unfortunately, she died weeks after giving birth to her first child.

Osmoregulation in Freshwater and Marine Fish

Claude Bernard (1813–1878), Homer Smith (1895–1962)

As first described by Claude Bernard in 1854, for animals to survive, they must maintain a constant internal environment, including a balance between the gain and loss of water and salts. If excessive water is gained, cells swell and burst; if water loss is too great, cells shrivel up and die. The processes animals use to maintain this balance, osmoregulation, involve two different approaches. Osmoconformers include most marine invertebrates whose internal salt and water concentrations are equivalent to their external environment. They "go with the flow" and have no need to actively control their salt and water balance. By contrast, many marine vertebrates, such as **fish**, have internal salt concentrations different from their external aquatic surroundings and must actively control their salt concentration; they are osmoregulators.

Freshwater fish live in water that is far more dilute than their body fluids and face the problem of salt loss and excessive water gain. They deal with this by drinking almost no water and excreting large volumes of highly dilute urine. Salt stores are built up by eating and by the active uptake of chloride ions across the gills into the body, followed by sodium ions. Marine fish are faced with the reverse problem—living in water that is far more concentrated than their body fluids—and, therefore, they face the loss of body water and the excessive movement of chloride and sodium ions into their body. Their strategy involves drinking large volumes of water and actively transporting chloride ions across the gills and out of the body; once again, sodium ions follow. In 1930, Homer Smith, working at New York University and the Mt. Desert Island Biological Laboratory, determined many patterns of marine fish osmoregulation.

The survival challenges are even greater for anadromous salmon, which, after spending most of their lives in the ocean, then breed in freshwater. After using the osmoregulatory strategies noted above, their acclimation from freshwater to saltwater and back again is not immediate. Salmon remain at the interphase of freshwater and saltwater for days to weeks before moving forward.

SEE ALSO: Fish (c. 530 Million BCE), Urine Formation (1842), Homeostasis (1854).

Small fish scatter as a blue marlin rises to the ocean surface. Marine fish, such as this marlin, live in an environment in which the surrounding water is more concentrated than their body fluids. Their gills and kidneys endeavor to conserve water by actively removing salts from their body.

Electron Microscope

Max Knoll (1897–1969), **Ernst Ruska** (1906–1988),
George Palade (1912–2008)

The electron microscope (EM) is among the most valuable tools used in the study of biology and has revolutionized the discovery and characterization of the subcellular structure of the cell. Viewing these structures was not possible with the standard light microscope (LM) that had been developed at the end of the sixteenth century.

IT'S A VERY SMALL WORLD. The LM magnifies objects up to 2,000 times, while the EM visualizes objects up to 2,000,000 times and at much higher resolution. The two basic EMs are the transmission electron microscope (TEM) and the scanning electron microscope (SEM). The TEM transmits electrons through thin tissue slices, and its two-dimensional images are used to examine the internal structure of cells. In the SEM, the electron beam sweeps across the sample and is used to study the surface detail of solid living specimens. It produces excellent three-dimensional images but is only one-tenth as powerful as the TEM and provides lower resolution.

The extraordinary magnification of EMs comes at a number of costs: EMs are very expensive to purchase and maintain; researchers need considerable training in their use and the preparation of biological specimens; TEM specimens must be stained and visualized in a vacuum that precludes the study of living samples; and EMs are large, and must be housed in vibration-free rooms.

The EM was developed at the University of Berlin by physicist Ernst Ruska and his professor, Max Knoll. Knoll knew that optical resolution (the ability to distinguish between two points, a measure of detail) was dependent upon the wavelength of the source of illumination, and that the wavelength of electrons is 1/100,000 that of light particles. Based on this relationship, using a beam of electrons that was focused on the specimen with electromagnets, they developed the first EM in 1931. It was improved and commercialized in 1939, and Ruska was the recipient of the 1986 Nobel Prize in Physics. In the 1950s, George Palade at the Rockefeller Institute (now Rockefeller University) used the EM to make discoveries on the fundamental organization of cells, for which he was a co-recipient of the 1974 Nobel Prize in Medicine.

SEE ALSO: Leeuwenhoek's Microscopic World (1674).

A scanning electron microscope can produce magnification up to 500,000 times. This SEM image of a flea—which is known to carry a number of diseases transmitted through its bites, including the bubonic plague, caused by the bacterium Yersinia pestis—has been artificially colorized.

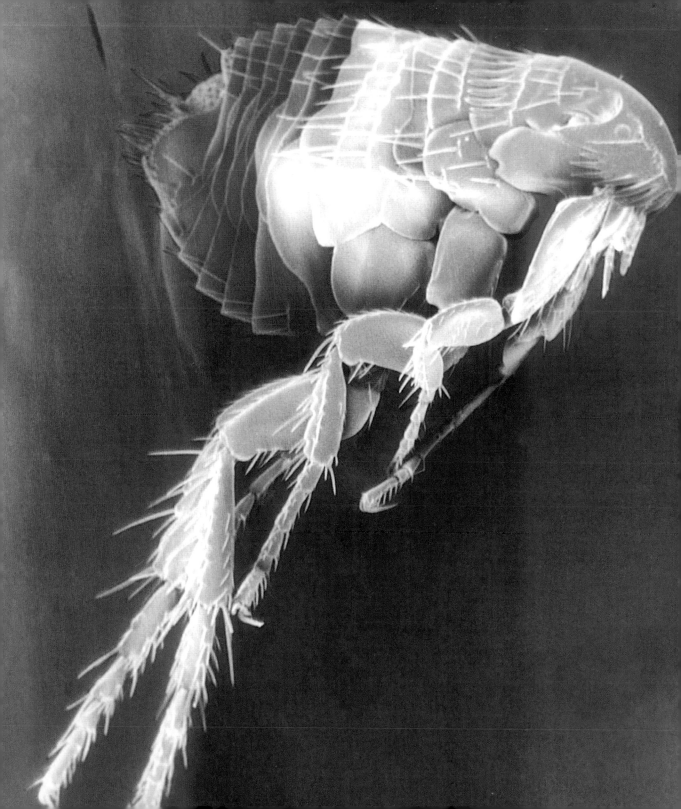

Imprinting

Douglas Spaulding (1841–1877), Oskar Heinroth (1871–1945), Konrad Lorenz (1903–1989)

Within hours after hatching, baby geese or ducks start walking about, and follow their mother. But how do they know they are following *their* mother? In reality, they don't, and studies have shown that they follow the first suitable subject they see within a critical period, measured in hours, and remain attached to this individual for a lifetime. This process is called *imprinting* and was first observed in about 1873 by Douglas Spaulding, an amateur English biologist, and then rediscovered by the German biologist Oskar Heinroth. It was Heinroth's student, Konrad Lorenz, who was the first to study imprinting in detail, and he was the co-recipient of the 1973 Nobel Prize for his work on this topic.

Lorenz, an Austrian zoologist, was one of the founders of modern ethology, the study of animal behavior. In studies conducted in about 1935, he found that after spending their first hours with him, young greylag geese followed him rather than their biological mother and strongly preferred him to a member of their own species. For bird species such as geese, which are feathered and active when hatched, the first thirteen to sixteen hours are critical periods for imprinting to occur. In mallard ducklings and domestic chicks, the imprinting opportunity is lost after thirty hours. By contrast, for birds that are born naked and helpless, the critical period is extended.

Imprinting is an instinctive behavioral act, which, unlike learned associative behaviors such as classical conditioning or operant (instrumental) conditioning, occurs without the need for reinforcement or reward. In nature, the apparent biological function of imprinting is to recognize close relatives and establish an intimate, mutually beneficial social bond between offspring and parent. From the parent's perspective, time, effort, and resources will not be expended for the care of another's offspring. An offspring must recognize its parent or risk being attacked and killed by species members who are not biologically related. Imprinting also extends to sexual preferences, so that a young animal learns the characteristics of a suitable mate and only mates with those individuals that are not too closely related (sibling) or too far removed (another species).

SEE ALSO: Associative Learning (1897), Parental Investment and Sexual Selection (1972).

A number of studies with young birds have shown that the imprinting response is innate and that the young have no inborn recognition of who is their authentic mother.

Factors Affecting Population Growth

Thomas Malthus (1766–1834), Harry S. Smith (1883–1957)

Thomas Malthus noted, "Population, when unchecked, goes on doubling itself every twenty-five years or increases in a geometrical ratio." Under ideal conditions, plant and animal populations continue to grow indefinitely, but this is not the way it is in nature. As resources become limited, birth rates typically decrease and death rates increase, slowing population growth. But does the population density in a given area influence its further decrease or increase?

Density-dependent factors are those that increase the death rate or decrease the birth rate in response to a rise in population. Such high-density pressures are often relieved by outward migrations of animal populations to less populated areas with more plentiful resources. Organisms in close proximity to one another, because of overpopulation, are more likely to be exposed and succumb to highly contagious diseases. Examples include the **American chestnut tree blight**, caused by a fungus, and smallpox and tuberculosis, resulting from a virus and bacterium, respectively. In 1935, Harry Smith, an entomologist at the University of California-Riverside, described the biological control of pest populations using such biotic weapons as predators, pathogens, and parasites. Predators play a major role in controlling population size. The increasing size of the potential prey population provides an incentive for predators to inhabit a geographic area, as with the four-year cycles of increases and decreases seen in the population of lemmings, which are related to their predator's activities.

Abiotic, density-independent factors that occur regardless of population size can rapidly and dramatically reduce and even decimate a population by leaving nutrients short in number and inferior in quality. Recent examples include such catastrophic events as forest fires, hurricane Katrina (2005), and the 1989 Exxon Valdez and 2010 **Deepwater Horizon (BP) oil spill**. Heavy frosts and drought conditions represent some of the climatic factors. Environmental pollutants, such as agricultural pesticides and fertilizers and mining runoffs, have taken their toll on the plant and animal populations, with **amphibians**, fish, and birds at particular risk.

SEE ALSO: Amphibians (c. 360 Million BCE), Population Growth and Food Supply (1798), Population Ecology (1927), Food Webs (1927), Green Revolution (1945), *Silent Spring* (1962), Deepwater Horizon (BP) Oil Spill (2010), American Chestnut Tree Blight (2013).

The population of the Adélie penguin colony on Antartica's Beaufort Island increased 84 percent as ice fields receded from 1958 to 2010. The warmer temperatures increased the ice-free habitat for their breeding.

Stress

Hans Selye (1907–1982)

Hans Selye was a twenty-eight-year-old assistant working in the Biochemistry Department at McGill University in 1934 in search of a new hormone and fame. He was encouraged when, after injecting rats with an extract from ovaries, they developed a wide range of symptoms. His initial elation of a great hormonal discovery turned to dejection when he observed the same findings after injecting a disparate range of other organ extracts. These results led him to recollect back fifteen years to his second-year medical student days at the University of Prague, when patients appeared with a wide range of symptoms that defied a simple diagnosis, that were general in nature, and that had no specific cause.

EUSTRESS OR DISTRESS. In 1936, Selye, a Hungarian heritage Canadian endocrinologist, wrote his first paper in which he formulated the General Adaptation Syndrome (GAS). In it, he introduced, in a biological context, the term *stress*. Selye found this word difficult to succinctly define, nor could it be readily translated into other languages; when used by non-English speakers, it was commonly prefixed by *el, il, lo, der,* etc. Selye referred to stress as a "nonspecific response of the body to any demand, whether it is caused by, or results in, pleasant or unpleasant conditions." Stress, then, could ensue from being terminated from a job or engaging in a hard-fought tennis match with a worthy opponent who is a best friend.

Selye characterized the GAS as having three phases: During the initial phase of alarm, the animal recognizes a challenge and acts by a "fight or flight" response. The "stress hormones" are secreted: cortisol from the adrenal cortex, and adrenaline (epinephrine) from the adrenal medulla. This phase is followed by one of resistance, when the body attempts to restore normality, balance, and **homeostasis**. If the ability to resist is unsuccessful, the final phase of exhaustion ensues, which can lead to the depletion of the body's energy resources and is most hazardous to health; problems such as stomach ulcers, heart disease, hypertension, and depression may be the consequence. Conversely, Selye argued, stress is not necessarily undesirable and should not be avoided.

SEE ALSO: Homeostasis (1854), Negative Feedback (1885), Secretin: the First Hormone (1902), Neurotransmitters (1920), Hypothalamic-Pituitary Axis (1968).

This terrified face, based on a photograph taken by the eminent French neurologist Duchenne de Boulogne (1806–1875), appeared in Charles Darwin's 1872 book The Expressions of Emotions in Animals.

Allometry

Louis Lapicque (1866–1952), **Julian Huxley** (1887–1975),
Max Kleiber (1893–1978), **Georges Tessier** (1900–1976)

BIOLOGICAL SCALING. The smallest microbe and the largest mammal have something in common. Based on their relative body size, they have the same rates of **metabolism**. Similarly, a frog's legs grow in direct size proportion to its body. This simple relationship does not always follow, as small changes in the body size of the Hercules beetle lead to a disproportionately large increase in the dimensions of its legs and antennae. Interest in comparing the relationship between one body part or biological function to the size of the entire body can be traced back to about 1900 when the French physiologist Louis Lapicque compared the brain size in a number of animal species to their body size.

In 1924, the English evolutionary biologist Julian Huxley measured the relative growth rates of the large claw of the fiddler crab (*Uca pugnax*) to its body size at different stages of development, and noted that the claw size was growing at a proportionately greater rate. He developed a mathematical formula to describe this relationship and continued to study biological scaling over the next dozen years. To avoid confusion and bring a state of coherency and mathematical consistency to these studies, in 1936, Huxley and Georges Tessier, who was independently working in this area, published joint papers—one in English, the other in French—in the premier journals in their respective languages. In these papers, they introduced a new neutral term—allometry ("different measure")—to refer to changes in the relative size of one body part to the overall body size.

The term *allometry* has been expanded to now include such relationships as body size and basal metabolic rate (BMR), the metabolic rate of a resting organism. In 1932, the Swiss biologist Max Kleiber determined that elephants had lower absolute BMRs and heart rates than mice but, when their body mass was considered, the BMR was a constant ¾ power of body mass. Using Kleiber's law, this same BMR relationship was subsequently shown to exist from tiny microbes to elephants, suggesting a common evolutionary link.

SEE ALSO: Metabolism (1614), Energy Balance (1960).

Unlike some animals, a frog's legs grow in direct size proportion to its body. This illustration of different species of frogs comes from Ernst Haeckel's Art Forms of Nature *(1904).*

Evolutionary Genetics

Charles Darwin (1809–1882), **Gregor Mendel** (1822–1884),
Theodosius Dobzhansky (1900–1975)

Darwin's theory of natural selection generated debate, then Gregor Mendel's experiments with green peas provided the foundation for genetic research. Thereafter, biologists were faced with the conundrum of reconciling **Mendelian inheritance** with Darwin's theory. Influential Ukrainian-born geneticist Theodosius Dobzhansky provided the linkage in his "modern synthesis." In his first significant studies in 1924, he noted geographic variations in the color and spot pattern in ladybugs, which he attributed to genetic variation resulting from an evolutionary process.

Based on their laboratory studies, most biologists had assumed that all members of a given species of *Drosophila*, the fruit fly, had essentially identical genes. Starting in the early 1930s, Dobzhansky devoted virtually the remainder of his professional career to studying the genetic characteristics of the fruit fly, both in the laboratory and in the field. In the controlled conditions of a laboratory, mutations producing genetic variations could be readily induced, and these flies would successfully breed. Might the same phenomenon occur in nature? In the field Dobzhansky used population cages, which permitted feeding, breeding, and sampling, while preventing escape. His analysis of the chromosomes of different populations of wild fruit flies from various locations revealed that different versions of the same chromosomes predominated, creating new species, which he explained on the basis of mutations.

Spontaneous genetic mutations occur naturally all the time, and many of these are neutral—that is, they do not confer either positive or negative effects on the organism. When the mutated organisms breed within a geographically isolated population, their genetic profile, containing the mutation, spreads within their population until it predominates, forming a new species by natural selection. Thus, Dobzhansky explained, genetic variation is a necessary condition for evolution to occur. In his classic 1937 book *Genetics and the Origin of Species*, Dobzhansky described these experiments and proposed a satisfactory explanation harmonizing natural selection with genetics.

SEE ALSO: Darwin's Theory of Natural Selection (1859), Mendelian Inheritance (1866), Genetics Rediscovered (1900), Genes on Chromosomes (1910).

The common fruit fly (Drosophila melanogaster) *has been a model organism for genetic research because it can be kept in large numbers, it is easy to handle, and it is very inexpensive. Fruit flies have a lifecycle of only two weeks, and its entire genome has been sequenced.*

Coelacanth: "The Living Fossil"

James L. B. Smith (1897–1968), Marjorie Courtenay-Latimer (1907–2004)

In 1938, Marjorie Courtenay-Latimer, the curator of the East London Museum in East London, South Africa, was informed that a pale mauve-blue **fish**, five feet long, had been netted by a trawler in the Indian Ocean off the coast of South Africa. Unable to identify the fish, she contacted her friend J. L. B. Smith, a professor with interests in ichthyology and chemistry at Rhodes University. By the time Smith returned from leave, the fish had been stuffed, but he immediately identified it as a coelacanth, long-thought to be extinct for 65 million years. There were once ninety species of coelacanth, but there are now only two. Coelacanths are classified as being the most endangered order of animals in the world and are designated by the genus of *Latimeria*, in honor of the museum curator.

The coelacanth is not just another old fish but rather a lobe-finned fish more closely related to lungfish, reptiles, and mammals than it is to common ray-finned fish so familiar to us. It is the link between fish and the tetrapods, the first land vertebrate to inhabit the earth, with the two groups diverging some 400 million years ago. The coelacanth is referred to as a "living fossil," as it apparently has not evolved over the millions of years when it made its debut appearance.

Since 1938, about 200 deep-blue coelacanths have been caught in the Indian Ocean, most near Comoro, an island-nation between Mozambique and Madagascar. Brown coelacanths have been caught in the waters off Indonesia. They are up to 6.5 feet (2 meters) in length, have an average weight of about 175 pounds (80 kilograms), and a lifespan of 80 to 100 years. Coelacanths have paired lobed fins used for swimming or walking that extend away from the body and move in an alternating pattern, resembling a trotting horse. The hinged joint in the skull permits the fish to widen its mouth to accommodate large prey, their scales are thick—a characteristic only found in extinct fish—and they have a notocord, which serves as a backbone.

SEE ALSO: Fish (c. 530 Million BCE), Fossil Record and Evolution (1836).

*A model of the Indonesian coelacanth (*Latimeria menadoensis*) is on display at the Tokyo Sea Life Park.*

Action Potential

Emil du Bois-Reymond (1818–1896), **Julius Bernstein** (1839–1917), **John C. Eccles** (1903–1997), **Alan L. Hodgkin** (1914–1998), **Andrew F. Huxley** (1917–2012)

In 1939, Andrew Huxley, a recent Cambridge graduate, joined Alan Hodgkin at the Laboratory of the Marine Biological Association in Plymouth, England, to study nerve conduction in the giant axon of Atlantic squid, which has the largest known neurons (nerve cells). They successfully inserted a fine electrode into the axon and were the first to record the intracellular electrical activity. Within weeks, in September, Germany invaded Poland, and war was declared. Their research was suspended for about seven years, during which time they separately supported the war effort, working on military-related projects.

Hodgkin and Huxley were not the first to study the electrical properties of animal tissues. In 1848, the German physiologist Emil du Bois-Reymond discovered the *action potential*, and in 1912 Julius Bernstein hypothesized that the action potential resulted from changes in the movement of potassium ions across the membrane of the axon. We now know that potassium and sodium ions are unequally concentrated inside and outside cells, and that this imbalance results in a voltage difference called a membrane potential. The massive movement of potassium and sodium ions into and out of the nerve cell causes a sudden voltage change, called an action potential, and the electrical impulse enables the activity of an organism to be coordinated by the central nervous system.

In 1947, when Hodgkin and Huxley resumed their studies, they used a voltage clamp technique, which controls voltage across the axonal membrane. In a series of classic publications in 1952, they presented their highly complex mathematical model of the action potential, which predicted the movement of ions, under different conditions, through ion channels—a groundbreaking quantitative approach that superseded simple qualitative descriptions of biological events. In the 1970s and 1980s, their predictions were experimentally verified. Hodgkin and Huxley were co-recipients of the 1963 Nobel Prize for their experimental and mathematical studies of the nerve action potential. They shared this award with John Eccles, an Australian neurophysiologist, who studied transmission of nerve impulses across synapses (gaps between nerves).

SEE ALSO: Animal Electricity (1786), Neuron Doctrine (1891), Neurotransmitters (1920), Sliding Filament Theory of Muscle Contraction (1954).

This recording from an oscilloscope or CRO (cathode-ray oscilloscope) permits the researcher to readily observe changes in electrical activity (voltage changes and frequency) in a nerve over time.

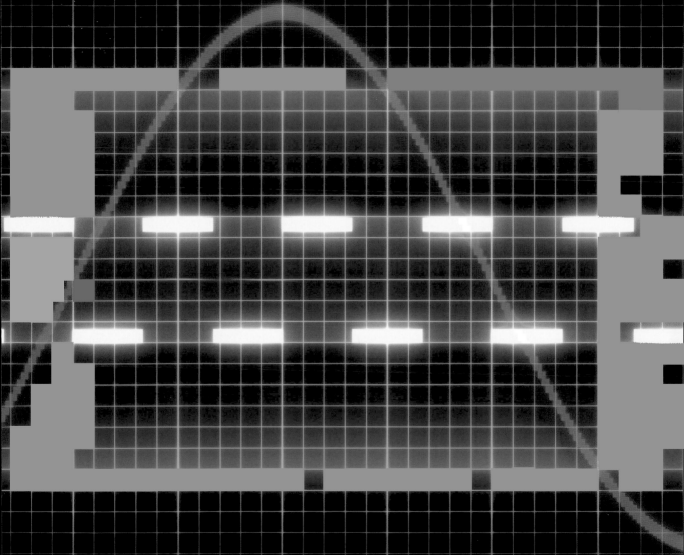

One Gene-One Enzyme Hypothesis

Archibold Garrod (1857–1936), **George W. Beadle** (1903–1989),
Edward Tatum (1909–1975)

The first clues regarding the function of genes appeared in 1902 when the English physician Archibold Garrod found that alkaptonuria, a rare disorder, was seen in families and was associated with the absence of an enzyme. In 1909, he predicted that the ability to synthesize specific **enzymes** was inherited and that the inability to produce such enzymes resulted from an **inborn error of metabolism**—a prediction that was biochemically validated in 1952.

Although the biochemical aspects of Garrod's findings were appreciated, its genetic implications were neglected into the 1930s. Geneticists of the time believed that genes were *pleiotropic*—that is, each gene had multiple primary effects. In 1941 at Stanford University, geneticist George Beadle and biochemist Edward Tatum tested the concept that gene action could be examined in discrete biochemical steps, which they evaluated in *Neurospora crassa*, a bread mold. They exposed *Neurospora* to X-rays, causing mutations and changes in their nutritional needs, which differed from the non-irradiated, wild type. For their limited nutritional requirements, molds use metabolic pathways to synthesize all the other materials they require to survive. Beadle and Tatum found that the mutant molds were unable to survive on a minimal growth media because they were unable to manufacture arginine, an essential amino acid. The researchers concluded that the multistep biochemical pathway that synthesized arginine was defective because it lacked the enzyme required for its synthesis.

Beadle and Tatum determined that the radiation-induced mutation produced a defect in a specific gene, resulting in a failure to produce a specific enzyme, and proposed the one gene-one enzyme hypothesis: the function of a gene is to dictate the production of a particular enzyme. Widely accepted at the time, it represented a unifying concept in biology, provided the first insights into the function of genes, and led to the emergence of biochemical genetics, for which Beadle and Tatum were awarded the 1958 Nobel Prize. Subsequent findings revealed that this hypothesis was an oversimplification, and that genes not only dictate enzyme synthesis but also structural proteins (such as collagen) and transfer RNA (tRNA).

SEE ALSO: Metabolism (1614), Enzymes (1878), Inborn Errors of Metabolism (1923).

Beadle and Tatum exposed the bread mold Neurospora crassa *to ultraviolet rays to induce mutations in the spores, which have a reproductive function in fungi. This photo depicts fungi-type polypores just after sporulation.*

Biological Species Concept and Reproductive Isolation

Charles Darwin (1809–1882), **Ernst Mayr** (1904–2005)

One of the most basic questions in biology was *speciation*: how one species splits into two or more new species. The question puzzled Charles Darwin in the 1830s after visiting the Galápagos Islands and seeing different species of finches. It remained a mystery until 1942, when Ernst Mayr, an evolutionary biologist, proposed the biological species concept in his book, *Systematics and the Origin of Species*. Earlier definitions of species were focused upon the physical similarities of organisms, whereas Mayr redefined the term based on their reproductive potential, arguing that members of a common species have the potential to interbreed and produce viable, fertile offspring. Reproductive isolation—the imposition of barriers that interfere with members of different species interbreeding—is the most frequent cause of speciation.

Mayr classified reproductive isolation as barriers occurring before or after fertilization and zygote formation—prezygotic or postzygotic. He noted that speciation most commonly occurs when populations of species become geographically separated, as by bodies of water (allopatric speciation), or when the two species share a common geographic area but occupy different habitats—one terrestrial, the other aquatic. In such cases, the flow of genes between these populations ceases to prevent hybrid creation. In other instances, reproductive isolation barriers may be imposed not by geography but as a result of temporal or behavioral differences in breeding, such as plants that flower at different times, or animal species with unique courtship rituals. In still other cases, an attempt to mate is thwarted by a physical incompatability, such as the shape of the genital organs.

If interspecies mating is successful, and fertilization occurs, postzygotic barriers may intervene, preventing hybrids from passing on their genes. The zygote may lack viability and not survive more than a few series of cell divisions. The hybrid may be viable, but sterile, and thus incapable of reproducing; such is the case of the mule, the hybrid offspring of a female horse and a male donkey. Finally, the initial hybrid may be fertile, but successive generations experience progressively reduced fertility, with eventual sterility.

SEE ALSO: Darwin and the Voyages of the *Beagle* (1831), Darwin's Theory of Natural Selection (1859), Biogeography (1876), Evolutionary Genetics (1937), Hybrids and Hybrid Zones (1963).

The offspring of a female horse (mare) and a male donkey (jack) is a mule, while breeding a male horse (stallion) with a female donkey (jenny) produces a hinney. In both cases, the hybrids are infertile.

Arabidopsis: A Model Plant

Friedrich Laibach (1885–1967), **George Rédei** (1921–2008)

What do *Escherichia coli* (bacterium), *Drosophila melanogaster* (fruit fly), *Caenorhabditis elegans* (roundworm), *Mus musculus* (mouse), and *Arabidopsis thaliana* (plant) all have in common? Based on the common descent of all living organisms, which share very similar metabolic pathways and common coding of hereditary information in genes using DNA, they have all been widely used as model organisms in general biological research studies. In addition, they have served as specific prototypes for studies on bacteria, insects, invertebrates, vertebrates, and plants, respectively.

In 1943, the German botanist Friedrich Laibach proposed *Arabidopsis thaliana* (thale cress, mouse-ear cress), a small flowering plant of the mustard family native to Europe and Asia—a weed with no commercial value—as a model organism. Decades after completing his doctoral research in 1907 and moving on to other research projects, he returned to *Arabidopsis* in the 1930s, and devoted the remainder of his career to its study. His research included its mutations and collections of *ecotypes*—genetically distinct varieties that have adapted their morphology and physiology to their specific and diverse environmental conditions around the world—amounting to 750 *Arabidopsis* ecotypes in total. Laibach's research on *Arabidopsis* was continued in the 1950s by, most notably, the Hungarian-born plant biologist George Rédei, who studied its mutants for decades at the University of Missouri.

A number of factors have contributed to *Arabidopsis* being embraced by biologists as a model organism for studies on plant biology, genetics, and evolution. Its small size has permitted researchers to grow thousands of plants in a very small space. In addition, it has a rapid life cycle, with a seed growing to a mature, easily cultivated plant, producing 5,000 seeds, in only six weeks. As part of his doctoral dissertation in 1907, Laibach correctly determined that the plant had only five pairs of chromosomes—among the smallest number in any plant—which facilitated pinpointing the location of specific genes. In 2000, it was the first plant whose genome had been sequenced, with 27,400 genes identified. Mutations are easily produced, and its plant cells are readily transformed with foreign DNA.

SEE ALSO: Origin of Life (c. 4 Billion BCE), Eukaryotes (c. 2 Billion BCE), Land Plants (c. 450 Million BCE), Theories of Germination (1759), Evolutionary Genetics (1937), DNA as Carrier of Genetic Information (1944), Genetically Modified Crops (1982), Human Genome Project (2003).

The Arabidopsis thaliana *(thale cress), a member of the mustard family, is widely used as a model organism in plant biology for studies in genetics and the molecular biology of flowering plants.*

DNA as Carrier of Genetic Information

Nikolai Koltsov (1872–1940), Oswald T. Avery (1877–1955), Frederick Griffith (1879–1941), Colin MacLeod (1909–1972), Maclyn McCarty (1911–2005), Francis Crick (1916–2004), James D. Watson (b. 1928)

It took years for scientists to accept the fact that DNA, and not a protein, was the critical chemical in heredity. In 1927, the Russian biologist Nikolai Koltsov first proposed that inherited traits were passed on to offspring by a "giant hereditary molecule" made up of two strands that could replicate, with each strand serving as a template. Though he never lived to see it—Koltsov had died at the hands of the secret police of the Soviet Union in 1940—this notion was confirmed one-quarter century later by Watson and Crick.

Independently, the British bacteriologist Frederick Griffith was interested in the pathology underlying pneumonia while working as a medical officer at the Ministry of Health's Pathological Laboratory during the 1920s. He injected mice with one of two forms of pneumococci—the rough non-virulent (R), or the smooth virulent (S)—with the expected fatal outcome involving the latter. But when Griffith administered a heat-killed S-form, the mice did not develop pneumonia. In the critical experiment, he injected mice with a mixture of the R-form and the heat-killed S-form, and the animals developed pneumonia and died. He concluded that the R-form was transformed to S but did not speculate about the nature of the "transforming factor."

During the 1930s and early 1940s, Oswald T. Avery, a Canadian-born physician and and foremost expert on pneumococcus, attempted to identify Griffith's "transforming factor." With his colleagues Colin MacLeod and Maclyn McCarty at Rockefeller University Hospital, in the so-called *Avery-MacLeod-McCarty experiment*, Griffith's experimental design was repeated and extended. Instead of heat killing the S-form, S-microbes were treated with chemicals that removed or destroyed various organic compounds from bacteria, including a protease enzyme that inactivated proteins. Only after the deoxyribonuclease enzyme was added, destroying DNA, was the transforming factor rendered inoperative, and in 1944 DNA was established as the critical carrier of genetic information.

SEE ALSO: Mendelian Inheritance (1866), Deoxyribonucleic Acid (DNA) (1869), Genetics Rediscovered (1900), Genes on Chromosomes (1910), Bacterial Genetics (1946), The Double Helix (1953), Human Genome Project (2003).

In the 1940s, the Avery-MacLeod-McCarty experiment *provided critical evidence that DNA—not a protein— carries genetic information.*

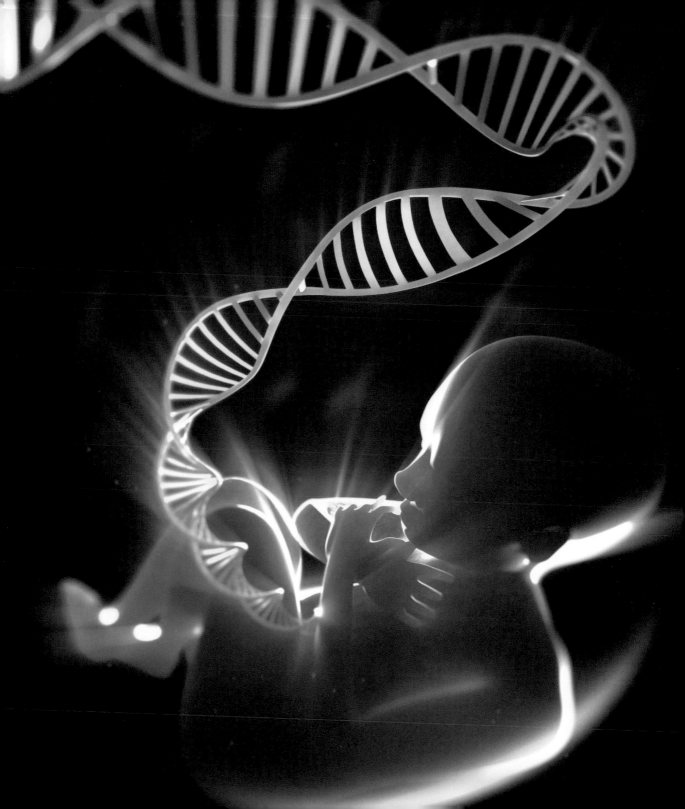

Green Revolution

Thomas Robert Malthus (1766–1834), **Norman Borlaug** (1914–2009)

The English demographer Thomas Malthus projected that the human population was increasing at a rate far greater than the production of food supplies and that, if not brought under control, mass starvation and poverty was inevitable. Happily, Malthus's predictions did not materialize in most industrialized countries. By the middle of the twentieth century, thanks to the use of modern plant breeding, improved agronomy, and the use of artificial fertilizers and pesticides, there were food surpluses. By contrast, Mexico, and developing countries in Asia and Africa, with their rapidly growing populations, experienced pervasive hunger and malnutrition.

In the early 1940s, the American agronomist Norman Borlaug, supported by grants from the Rockefeller Foundation, initiated research to increase the production of Mexican **wheat**. By 1945, he had developed varieties that were high yield and disease resistant, and he doubled the wheat-growing season. By the 1960s, Mexico was exporting one-half its wheat production. In the mid-1960s, the Indian subcontinent was immersed in war and was experiencing famine and starvation in an uncontrolled population increase. Borlaug transferred his technologically advanced approaches of modern irrigation, pesticides, high-yield crop varieties, and, perhaps most important, synthetic nitrogen fertilizer, to the **cultivation of rice** crops in India and Pakistan—once again, with remarkable success. Crop yields were increased and costs were reduced.

As might be predicted, not all of Borlaug's approaches were universally lauded. The extensive use of chemical pesticides produced human toxicity and increased cancer risks in animals. The emphasis on high-yield crop varieties decreased or eliminated the cultivation of less productive plants, thus reducing biodiversity, and in Brazil there was deforestation to increase farmlands. Unlike the small or poor farmer who lacked funds to purchase fertilizer, gain access to water for irrigation, or secure credit, large landowners were the major beneficiaries, leading to greater income inequalities.

Nevertheless, the Green Revolution has been credited with preventing widespread famine and for feeding billions of people. Borlaug was awarded the 1970 Nobel Peace Prize for increasing the world's food supply.

SEE ALSO: Amazon Rainforest (c. 55 Million BCE), Wheat: The Staff of Life (c. 11,000 BCE), Agriculture (c. 10,000 BCE), Rice Cultivation (c. 7000 BCE), Population Growth and Food Supply (1798), Factors Affecting Population Growth (1935), *Silent Spring* (1962), Genetically Modified Crops (1982).

An undated American work of art, There Were No Crops This Year, *provides a glimpse of agricultural life prior to the mid-twentieth century, when the Green Revolution eradicated the threat of starvation in developed countries.*

Bacterial Genetics

Oswald T. Avery (1877–1955), **Colin McLeod** (1909–1972),
Edward L. Tatum (1909–1975), **Maclyn McCarty** (1911–2005),
Joshua Lederberg (1925–2008)

Joshua Lederberg was impressed by Avery, McLeod, and McCarty's 1944 study showing that DNA was the critical carrier of genetic information. But many biologists questioned whether the results of genetic studies in bacteria were transferable to more complex organisms. Nevertheless, studying bacteria had a number of advantages: they were simple to grow in inexpensive culture media; they generated rapidly, reducing experiment time; they were easily handled; and they had a simple cell structure.

Animal and plant parents transfer genetic information to their offspring by the process of vertical gene transfer. Bacteria primarily reproduce by dividing into two genetically identical daughter cells (binary fission). Scientists long believed that bacteria were primitive and not suitable for genetic analysis. In 1946, Joshua Lederberg and his major advisor Edward Tatum at Yale University showed that, in bacteria, genetic material is transmitted between two organisms that are not parent and offspring by the process of gene recombination—later termed *horizontal gene transfer* (HGT). In recognition, thirty-three-year-old Lederberg and Tatum were co-recipients of the 1958 Nobel Prize. Subsequent studies have shown that HGT is common even in very distantly related bacteria and is a mechanism in bacterial evolution. It also underlies the development of drug resistance to **antibiotics**: when one bacterial cell acquires drug resistance, it can rapidly transfer the resistant genes to many other species.

There are three major modes by which HGT can spread genes between members of the same or different bacterial species: bacteria-to-bacteria transfer (*conjugation*), shown by Lederberg and Tatum (1946); virus (**bacteriophage**)-to-bacteria transfer (*transduction*, 1950), which has lead to genetic engineering work by Lederberg and his wife Esther Zimmer Lederberg, herself a prominent bacterial geneticist; and the free transfer of DNA (*transformation*). Lederberg was the leading force in microbial genetics, a founder of molecular biology, a visionary in artificial intelligence, and a spokesperson against the dangers of microbial contamination during space exploration.

SEE ALSO: Prokaryotes (c. 3.9 Billion BCE), Bacteriophage (1917), Antibiotics (1928), DNA as Carrier of Genetic Information (1944), Plasmids (1952), Bacterial Resistance to Antibiotics (1967).

Salmonella (shown) can cause severe food poisoning, and some bacterial strains are resistant to multiple antimicrobial drugs. Mechanisms leading to resistance mostly involve genes located on plasmids that are easily transferred among Salmonella and other bacteria.

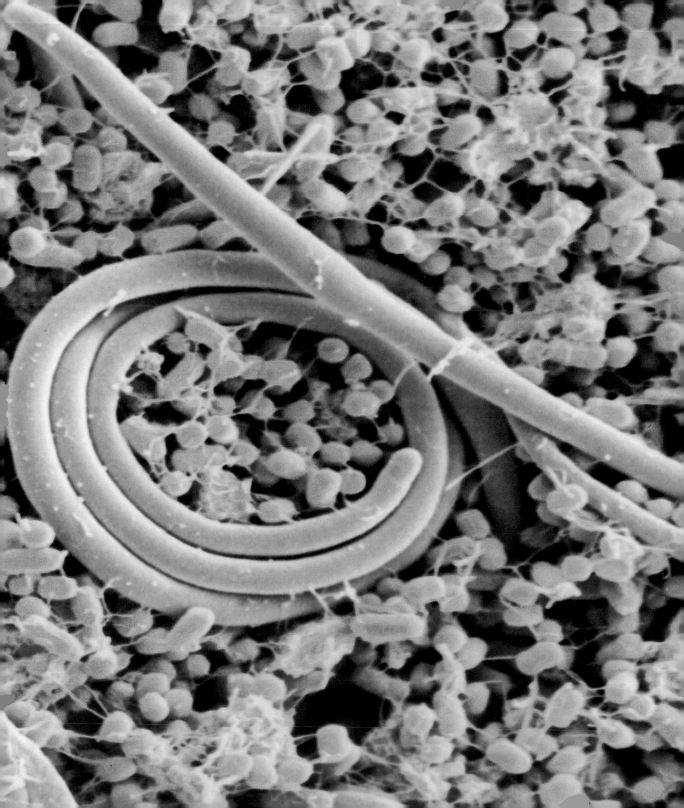

Reticular Activating System

Horace W. Magoun (1907–1991), Giuseppe Moruzzi (1910–1986)

The reticular formation (RF) consists of nerve pathways that serve as a bridge running through the central core of the brainstem and connecting to the cerebral cortex. The brainstem is an ancient area in the lower portion of the brain that controls functions that are vital for vertebrate survival, whereas the cerebral cortex is the seat of consciousness and thinking. Until the middle of the twentieth century, a state of wakefulness was believed to result from external and internal stimulation of the cerebral cortex, while inhibitory influences produced sleep. Studies conducted at Northwestern University in Chicago in 1949 by Giuseppe Moruzzi and Horace W. Magoun on the RF disproved this concept and provided new insights into sleep and wakefulness.

When Moruzzi and Magoun electrically stimulated the RF, it produced electroencephalographic (EEG) changes that simulated arousal in cats; these effects were observed even after the ascending sensory pathways leading to the cerebral cortex were destroyed. When they produced lesions to the RF, cats became comatose, even when their sensory pathways were left intact.

Thus, the reticular activating system (RAS)—of which the RF is a major component—regulates the progression from deep sleep and relaxed wakefulness to a state of heightened and selective awareness and attention. The RAS serves as a filter, capturing external stimuli that are relevant or novel, while excluding those stimuli that are familiar and repetitive (a process referred to as *habituation*). Pain signals arising from the lower body travel through the RF to the cerebral cortex, and the RF also integrates cardiovascular, respiratory, and motor responses to external stimuli.

The RAS has both cholinergic and adrenergic nerve influences. Cholinergic nerves—those having acetylcholine as their **neurotransmitter**—are believed to be the chemical mediators of arousal and wakefulness as well as **rapid eye movement (REM) sleep**; glutamate, the principal excitatory neurotransmitter in the brain, is also involved. By contrast, adrenergic nerves—with norepinephrine as their neurotransmitter—are active during deep sleep and inactive during REM sleep. Attention deficit disorder may be caused by a deficiency of norepinephrine in the RAS.

SEE ALSO: Medulla: The Vital Brain (c. 530 Million BCE), Nervous System Communication (1791), Neurotransmitters (1920), REM Sleep (1953).

Among the functions of the reticular activating system, as demonstrated in studies of cats, is mediating the shift from a state of relaxation to a period of high attention and arousal.

Phylogenetic Systematics

Carl Linnaeus (1707–1778), **Charles Darwin** (1809–1882), **Ernst Haeckel** (1834–1919), **Willi Hennig** (1913–1976)

Carl Linnaeus devised a binomial classification for plants and animals in the early eighteenth century and grouped animals and plants into increasingly inclusive hierarchical categories. However, he adopted the traditional Biblical teaching that all living beings were originally created in the form in which they now exist, so his classification was based on observable shared characteristics. But with Charles Darwin's overwhelming evidence that living organisms evolved from common ancestors, some of which may be extinct, Linnaeus's simple classification required reexamination.

In 1866, Ernst Haeckel, a biologist and early Darwin supporter, introduced the term *phylogeny* to refer to the study of the evolutionary history of the species. To construct phylogenies, the discipline of systematics seeks to understand the evolutionary interrelationships of living organisms. In his 1950 book of the same name, the German biologist Willi Hennig introduced the concept of *phylogenetic systematics*, which attempts to identify the evolutionary relationship among extant and extinct organisms.

Just as a family tree is used to trace ancestors from whom we are descended, the evolutionary history of a group of organisms can be represented by a branching diagram called a phylogenetic tree. This tree is depicted by a series of two-way branch points, with each branch point representing the divergence of two lineages from a common ancestor (for example, the most recent common ancestor of the coyote and gray wolf). The phylogenetic tree hypothesizes (but does not establish) these evolutionary linkages.

Traditional phylogenetic analysis was based on external, observable characteristics that might be misleading. Advances in molecular biology have permitted the analysis of complex sequences of genes, chromosomes and even entire genomes. Comparisons of the DNA sequences of the various genes of different organisms have shown common ancestry that would not be obvious from physical likenesses. The amount of nucleotide sequences between a pair of genomes from different organisms indicates how long ago they had this common ancestor.

SEE ALSO: Linnaean Classification of Species (1735), Darwin's Theory of Natural Selection (1859), Ontogeny Recapitulates Phylogeny (1866), Domains of Life (1990), Human Genome Project (2003), Protist Taxonomy (2005), Oldest DNA and Human Evolution (2013).

This image depicts a phylogenetic tree of life, with completely sequenced genomes divided according to the three domains of life: Archaea (green); Bacteria (blue); Eukarya (red), with the red dot denoting Homo sapiens.

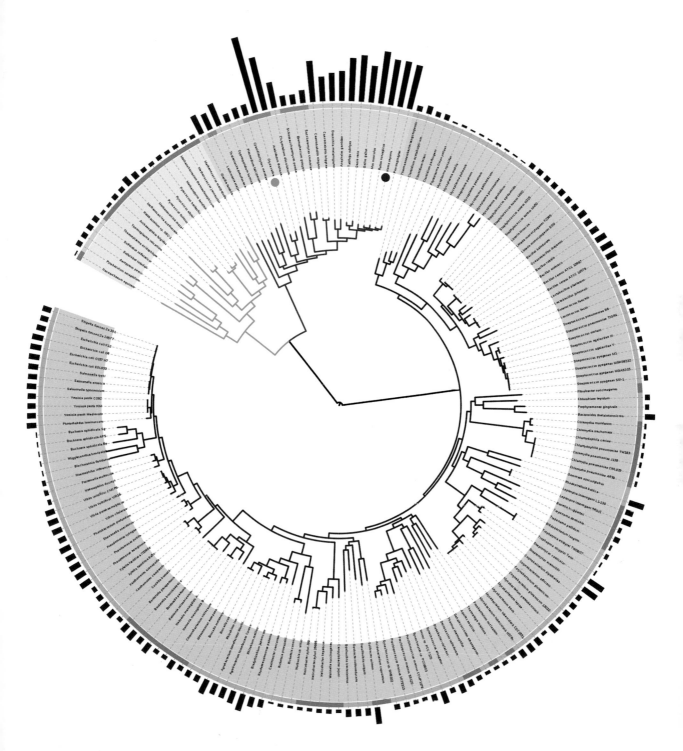

The Immortal HeLa Cells

George Otto Gey (1899–1970), **Jonas Salk** (1914–1995),
Henrietta Lacks (1920–1951), **Rebecca Skloot** (b. 1972)

In 1951, a thirty-year-old African American mother of five went to Johns Hopkins Hospital to be treated for cervical cancer. During the course of treatment, a sample of her cancerous cervix was taken and, without her permission, a piece was sent to George Gey, head of the **tissue culture** laboratory. (At that time, permission was not required to use a person's cells, and none was sought.) Eight months later, the cancer had metastasized throughout her body, and, in October, Henrietta Lacks died. On that same day, Gey appeared on television with a vial of immortal HeLa cells that he claimed held the potential to cure cancer.

When normal human cells are grown in culture, they die after dividing twenty to fifty times. The HeLa cells—the first immortal human cells—kept dividing and have continued to do so since 1951. Why they are immortal has not been determined. HeLa cells were mass-produced and distributed to laboratories throughout the world, and some researchers consider their existence to be one of the greatest medical discoveries of our time. Jonas Salk used these cells in 1954 to develop his polio vaccine, and they have served as invaluable tools to study cancer, the cell biology of tumors, anticancer drugs, and AIDS, and in genetic mapping.

Twenty-five years after her death, the Lacks family first became aware of the existence of the cells. The cells had been widely distributed and commercialized but neither Gey nor the Lacks family received any compensation, nor was Henrietta Lacks ever even acknowledged. Although occasional newspaper articles on Lacks and HeLa cells have appeared, in 2010 the story was recounted in detail in Rebecca Skloot's *The Immortal Life of Henrietta Lacks*, a book that remained on the *New York Times* Best Seller List for over two years. In March 2013, German researchers published the HeLa genome (the DNA code)—once again, without the family's permission. In August 2013, the US National Institutes of Health and the Lacks family came to an agreement that the family would have some control over who gains access to the DNA code, but still no financial compensation for the cells has been granted.

SEE ALSO: Cell Theory (1838), Tissue Culture (1902), Cell Senescence (1961), Cell Cycle Checkpoints (1970), Induced Pluripotent Stem Cells (2006).

A culture of HeLa cells, originally obtained from human cervical cancer cells. This line of cells has been dividing since 1951 and is now the most commonly used cell line in biological and drug research.

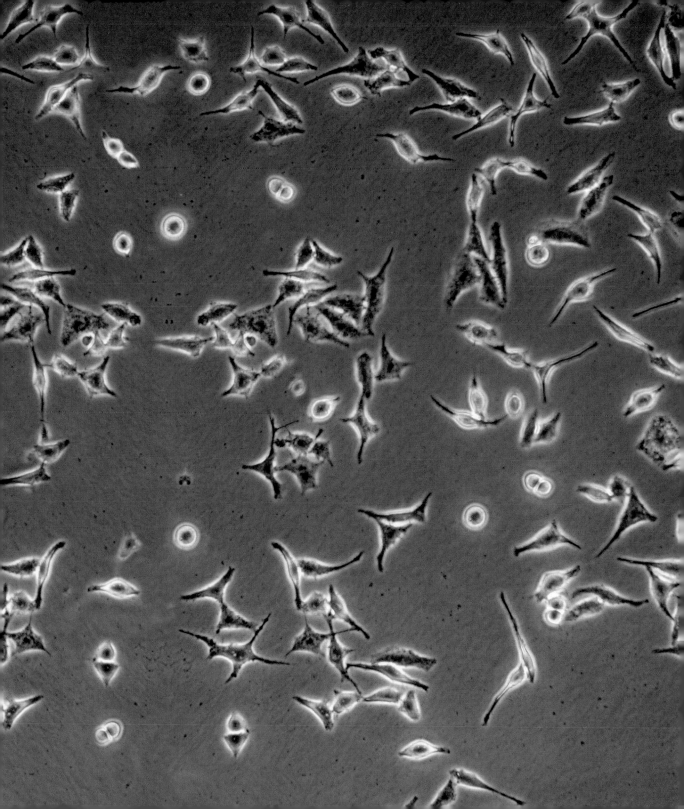

Cloning (Nuclear Transfer)

Hans Driesch (1867–1941), **Hans Spemann** (1869–1941),
Robert Briggs (1911–1983), **Thomas J. King** (1921–2000),
John Gurdon (b. 1933), **Ian Wilmut** (b. 1944)

In 1996, the world's most famous lamb, Dolly, was born at the Roslin Institute in Edinburgh, Scotland. She was the product of nuclear transfer (NT), or cloning, by Ian Wilmut. Although Dolly introduced many non-biologists to NT, it was first accomplished over 100 years earlier. In 1885, Hans Driesch separated two embryonic cells from a sea urchin, and each grew independently as a clone of the parent organism. The concept of NT and production of clones were first proposed by Hans Spemann in 1928 and involved the transfer of nuclei from differentiated (specialized) adult cells (somatic cells) or undifferentiated embryonic cells into a donor cell whose nucleus had been removed, producing a genetically identical copy.

Spemann's concept was tested and validated in 1952 at the Institute for Cancer Research in Philadelphia when Robert Briggs and Thomas J. King, using undifferentiated cells, successfully cloned northern leopard frogs. In 1962, John Gurdon, using a fully differentiated intestinal **cell nucleus**, cloned South African frogs, demonstrating that the cell's genetic potential was not reduced in specialized cells. Gurdon, whose English schoolmaster wrote a report stating that his idea of becoming a scientist was "quite ridiculous," received the 2012 Nobel Prize.

The 1993 blockbuster movie *Jurassic Park* was predicated on the notion that dinosaurs could be cloned. Although the film was criticized for some scientific inaccuracies, it popularized cloning and was immensely successful at the box office.

Dolly was the first mammal to be successfully cloned, after many failures. Her birth was perceived by some to rank among the most significant scientific breakthroughs when it was publically announced. It showed that adult cells were capable of being reprogrammed into new cells; moreover, Dolly gave birth to four lambs. These hopes were not fully realized because Dolly's cells were already old, and she was euthanized six years later in ill health. For many, their enthusiasm was tempered with apprehension and ethical concerns that human cloning might be on the horizon.

SEE ALSO: Cell Nucleus (1831), In Vitro Fertilization (IVF) (1978), Induced Pluripotent Stem Cells (2006), De-Extinction (2013).

The process of cloning results in the creation of a genetically identical copy of the original unicellular or multicellular organism.

Amino Acid Sequence of Insulin

Frederick G. Banting (1891–1941), **Charles H. Best** (1899–1978), **Frederick Sanger** (1918–2013), **Herbert Boyer** (b. 1936)

In the early 1920s, Frederick Banting and Charles Best demonstrated that a pancreatic extract was effective in treating diabetes mellitus. In 1923, this refined extract—obtained from pig and beef, with **insulin** as the active constituent—was commercialized by Eli Lilly and crystallized three years later.

In 1943, the English chemist Frederick Sanger at Cambridge University began research seeking to determine the amino acid sequence of insulin. At that time, insulin was one of the few proteins that was available in pure form, and it could be readily obtained from Boots, an English pharmacy chain. After a decade of effort, in 1951 and 1952, Sanger determined that insulin is composed of two linked chains of peptides (a string of amino acids): an A chain with twenty-one amino acids and a B chain with thirty amino acids. Insulin was the first protein to have its amino acid sequence fully determined, and Sanger concluded that all human proteins have a unique chemical sequence containing any of twenty amino acids. For his work on proteins, especially insulin, Sanger was awarded the Nobel Prize in 1958 (and in 1977 he became the only two-time recipient of a Nobel Prize in Chemistry).

Once the chemical structure of insulin was established, it was possible to synthesize this molecule in the laboratory, which was accomplished in 1963. Although the animal-derived insulin was highly effective, it was close, but not identical, to human insulin. Pig insulin differs by one amino acid and beef by three, but these seemingly subtle differences are responsible for allergic reactions in diabetic users. In 1978, researchers from the City of Hope National Medical Center, in collaboration with scientists from a then-recently founded **biotechnology** company, Genentech, led by the biochemist Herbert Boyer, synthesized the first human protein using biotechnology. In this process, a gene for human insulin was inserted into bacterial DNA; the genetically modified bacteria multiplied and served as a biological factory, producing virtually inexhaustible supplies of insulin. This human insulin, marketed as Humulin by Eli Lilly in 1982, replaced animal insulin.

SEE ALSO: X-Ray Crystallography (1912), Biotechnology (1919), Insulin (1921), Plasmids (1952).

A three-dimensional model of the insulin molecule. By convention, the following colors represent specific elements: white = hydrogen; black (shown here as dark gray) = carbon; blue = nitrogen; red = oxygen; yellow = sulfur.

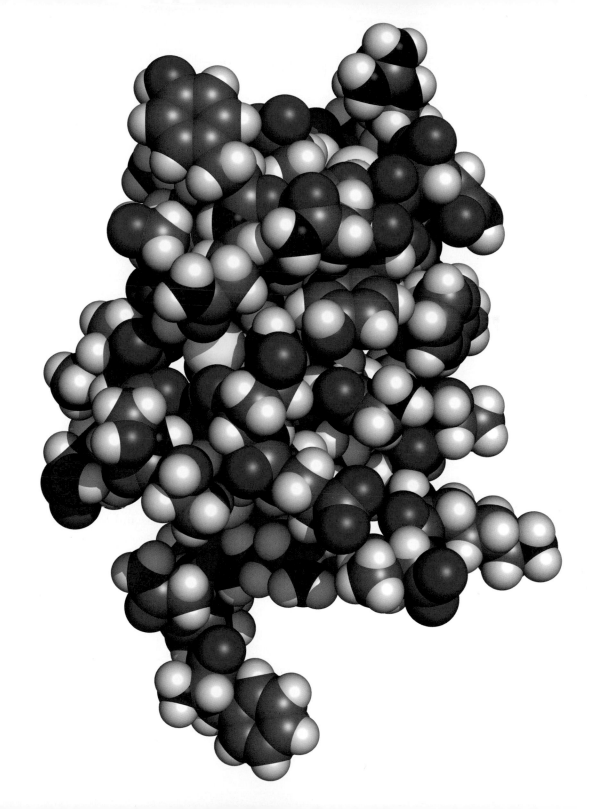

Pattern Formations in Nature

Leonardo da Vinci (1452–1519), **D'Arcy Wentworth Thompson** (1860–1948), **Alan Turing** (1912–1954)

That geometric patterns exist in nature was recognized by Leonardo da Vinci and described half a millennium later by the Scottish mathematical biologist D'Arcy Wentworth Thompson. Thompson analyzed the structure of organisms in physical and mathematical terms, and showed numerous correlations among living beings in his 1917 classic book *On Growth and Form*.

English mathematician Alan Turing examined pattern formations in nature from a highly theoretical perspective. Turing was no ordinary mathematician. During World War II, he was a leading figure at Bletchley Park, Britain's code-breaking center. His Turing machine cracked the German Enigma Machine-generated coded messages, decoding that was successfully used by the Allies in the naval Battle of the Atlantic. After the war, he was instrumental in conceiving the first computing machine and artificial intelligence. In 1952, until his death by suicide two years later, Turing turned his attention to mathematical biology and published his only biological paper, *The Chemical Basis of Morphogenesis*. (Morphogenesis refers to the "beginning of shape" and generally refers to the development of form and structure in an organism growing from embryo to adult.) In this paper, Turing proposed a mathematical model, explaining the formation of natural patterns based on physical laws that govern how certain chemicals react and spread through the skin. He then formulated a set of "reaction-diffusion" equations to produce patterns that simulated actual animal patterns.

These equations would provide the basis for explaining the formation of the diverse patterns that exist in plants and animals: the patterns on a sunflower and daisy, the stripes on a tiger and zebra fish, the spots on a jaguar, and the spacing of hair follicles on a mouse's paw. He theorized that patterns are formed by the interaction of two chemicals, called *morphogens*, which diffuse at different rates. One of these morphogens is an activator, which expresses the characteristic pattern (stripes, spots), while the other is an inhibitor, which shuts off the activator, leaving a blank space. The Turing pattern mechanism remained highly theoretical for six decades until 2012, when two chemicals were identified that behave as activator and inhibitor morphogens.

SEE ALSO: Animal Coloration (1890), Cellular Determination (1969).

Angelfish and zebras are striped, while jaguars and ladybugs are spotted. According to Alan Turing, these pattern formations are attributable to activator and inhibitor morphogens.

Plasmids

Joshua Lederberg (1925–2008), **Stanley N. Cohen** (b. 1935), **Herbert Boyer** (b. 1936)

In 1952, Joshua Lederberg, at the University of Wisconsin-Madison, introduced the general term *plasmid* to refer to DNA molecules that were separate and outside chromosomal DNA. It was Lederberg's intention to use a generic term to collectively classify a diverse group previously termed parasites, symbionts, organelles, or genes. In 1973, interest in plasmids dramatically escalated when they were found useful as tools in molecular biology and genetic engineering, primarily through the combined efforts of geneticist Herbert Boyer and biologist Stanley Cohen. They showed that it was possible to transfer a gene from one species (frog) to another (the bacterium *Escherichia coli*), and demonstrated that the transplanted gene could function normally in its new host. Plasmids also play a major role in the evolution of microbial resistance and the capacity of microbes to cause disease.

Plasmids can replicate (copy themselves) in the cell independent of chromosomes and frequently have backbone and accessory genes: The backbone genes participate in the replication and maintenance of plasmids. By contrast, the accessory genes are not essential for the survival of the host (bacterium in which the plasmid resides), but may encode functions that provide advantages to it. These include the ability to degrade environmental pollutants and use them as sources of carbon and nitrogen, or enable the host to become resistant to the toxic effects of **antibiotics** or heavy metals. In addition, plasmids can be transferred between and among bacterial species. This transfer serves as a major mechanism by which bacteria can easily and rapidly acquire a variety of traits, permitting them to adjust to a changing environment.

Plasmids have been very widely used as tools in genetic engineering, as in gene cloning, gene therapy, and recombinant protein production. In 1978, Boyer, a founder of the pharmaceutical biotechnology company Genentech, produced synthetic human insulin using this technique. A foreign DNA element, such as a gene for insulin, was spliced into the plasmid, which was introduced into the bacterial cell. Replication of the plasmid within the bacterial cell turned out large number of copies of the recombinant insulin.

SEE ALSO: Apoptosis (Programmed Cell Death) (1842), Antibiotics (1928), Bacterial Genetics (1946), Amino Acid Sequence of Insulin (1952), Bacterial Resistance to Antibiotics (1967).

Joshua Lederberg at work in his laboratory at the University of Wisconsin, 1958. In addition to his discovery that bacteria can exchange genes, he is also well known for his contributions to artificial intelligence and space exploration.

Nerve Growth Factor

Viktor Hamburger (1900–2001), **Rita Levi-Montalcini** (1909–2012), **Stanley Cohen** (b. 1922)

Two years after Rita Levi-Montalcini received her medical degree from the University of Turin in 1938, Mussolini issued a decree preventing all non-Aryan Italians from pursuing professional careers in Italy. In response, she set up a small laboratory in her home and, inspired by the work of Viktor Hamburger, studied chick embryos. In 1947, she joined Hamburger at Washington University in St. Louis, Missouri, where he was studying the growth of nerve tissues. The following year, she found that when a piece of mouse tumor was grafted on chick embryos whose wing buds had been removed, the growth of nearby nerves was stimulated.

Stanley Cohen, a biochemist, joined Levi-Montalcini in 1953 and isolated the active protein from the tumor, which they named *nerve growth factor* (NGF). NGF was found essential for the normal growth and maintenance of nerves in the peripheral nervous system (outside the brain and spinal cord) and for cholinergic (acetylcholine-containing) nerves in the brain. After going to Vanderbilt University in 1959, Cohen found and isolated another growth factor in the NGF-containing tumor. This factor, which stimulated the growth of the epidermal layer of skin and caused newborn mice to open their eyes sooner than normal, was dubbed epidermal growth factor (EGF). In 1986, Cohen and Levi-Montalcini were co-recipients of the Nobel Prize for their discoveries of growth factors.

NGF was the first of approximately fifty growth-promoting agents that have been identified and that are secreted in the blood from many different tissues. They serve as signaling molecules between cells, with each promoting the growth of specific cells. In particular, these factors stimulate cellular growth, replication, and differentiation (specialization), and have been found in a wide variety of biological species including plants, insects, and vertebrates. Growth factors are used for the medical treatment of cancers and blood and cardiovascular diseases. Among the most familiar growth factors is erythropoietin (EPO), which is produced in the kidney and stimulates the production of red blood cells. EPO has since gained notoriety as a blood doping agent in cycling and other endurance sports.

SEE ALSO: Nervous System Communication (1791).

Among the fifty-or-so growth-promoting factors that have been discovered since 1953, perhaps the most well known is erythropoietin (EPO), which stimulates red blood cell production. EPO has gained a notorious reputation as a blood-doping agent, increasing oxygen delivery to muscles, which enhances endurance performance.

Miller-Urey Experiment

Louis Pasteur (1822–1895), **J. B. S. Haldane** (1892–1964), **Harold C. Urey** (1893–1981), **Alexander Oparin** (1894–1980), **Stanley L. Miller** (1930–2007)

What was the **origin of life** on earth? For thousands of years, the prevailing scientific explanation was spontaneous generation—that life was formed from nonliving matter—a theory apparently disproven by Louis Pasteur in 1859. In the 1920s, the Soviet biochemist Alexander Oparin and the British evolutionary biologist J. B. S. Haldane independently hypothesized that conditions on Earth about four billion years ago were such that organic molecules could have been formed from simpler inorganic molecules.

During the 1950s, scientific curiosity was renewed in studying the origin of life, an interest long held by Harold Urey, a 1934 Nobel laureate. This was the subject of the Miller-Urey Experiment, performed in 1952 by Urey's graduate student Stanley Miller and reported in 1953. The experimental conditions were intended to simulate those thought to have existed in the Earth's atmosphere some four billion years ago, as theorized by Oparin in 1924. In the Miller-Urey experiment, a mixture of water and the gases ammonia, methane, and hydrogen, were continuously exposed to electric sparks—intended to simulate lightning storms, which were very common at that earlier time. After one week, organic molecules were produced and, more importantly, 2 percent of the products were amino acids, the building blocks of life. This experiment was initially interpreted as proof that life on Earth could have arisen from simple organic compounds. Moreover, evolutionary biologists generally believe that life today evolved from a common life form.

In subsequent years, this experiment and its results have been subjected to critical analysis, and a number of challenges have been made about its validity, results, and conclusions. Questions have been raised about the similarity between the compounds found in the early atmosphere and those in the experiment, and whether these chemicals were not exposed to far more electrical energy than would have occurred at that time. One of the most telling criticisms revolved about whether the amino acids on our early planet were not brought to Earth from an extraterrestrial source. In 1969, a meteorite struck earth in Murchison, Australia, and was found to contain more than ninety amino acids. The search for life outside our planet continues.

SEE ALSO: Origin of Life (c. 4 Billion BCE), Refuting Spontaneous Generation (1668).

The Miller-Urey experiment was intended to simulate conditions that existed almost four billion years ago, resulting in the production of organic compounds, including amino acids. In the experiment, simple molecules were continuously bombarded with electric sparks that were likened to lightning storms, believed to be common during Earth's early history.

The Double Helix

Linus Pauling (1901–1994), **Francis Crick** (1916–2004), **Maurice Wilkins** (1916–2004), **Rosalind Franklin** (1920–1958), **James D. Watson** (b. 1928)

Although controversy continues to cloud the assignment of credit more than six decades after the 1953 discovery of the structure of **deoxyribonucleic acid (DNA)**, no question exists regarding its cardinal significance in the transference of hereditary information, nor that it is one of the greatest scientific discoveries. In 1950, the basic elements of DNA's structure were known to consist of nitrogenous bases—adenine, cytosine, guanine, and thymine—a sugar, and a phosphate group, but the nature of the linkage among the components remained obscure. Competition for this discovery focused upon Linus Pauling at California Institute of Technology and the James Watson and Francis Crick team at the Cambridge University's Cavendish Laboratory.

Pauling, considered among the most important scientists of the twentieth century and who, in coming years, was to be the recipient of two Nobel Prizes, proposed that DNA was a triple helix—a model based upon a number of fundamental errors that led him astray. Early in 1953, Watson and Crick focused their attention on a two-chain model, with each long chain twisted about the other and traveling in opposite directions—a double helix—and with alternating sugar and phosphate groups. Their model was supported by X-ray diffraction photographic images made by Maurice Wilkins and Rosalind Franklin at King's College London. On April 25, 1953, the Watson-Crick paper appeared in *Nature* magazine, with only a terse footnote referring to Franklin's and Wilkins's "unpublished contribution."

Prior to submission of the *Nature* paper, without her permission or knowledge, copies of Franklin's outstanding photographs were shared with Watson, and one in particular was considered by many to be pivotal in the discovery of the double helix. The relative significance and importance of Franklin's contribution has not been resolved. But what appears incontrovertible is that she was never formally recognized during her lifetime and, when the Nobel Prize was awarded in 1962, she had never been nominated nor even acknowledged by Watson, Crick, or Wilkins, the recipients. (Pauling received the Nobel Peace Prize that same year.) Franklin died at the age of 37 of ovarian cancer in 1957, and deceased individuals are ineligible for the Nobel Prize.

SEE ALSO: Deoxyribonucleic Acid (DNA) (1869), DNA as Carrier of Genetic Information (1944), Cracking the Genetic Code for Protein Biosynthesis (1961).

This double-helix access ramp of a seven-level underground parking garage in Nantes, France, echoes the structure of DNA.

REM Sleep

Henri Piéron (1881–1964), **Nathaniel Kleitman** (1895–1999),
Eugene Aserinsky (1921–1998)

"To sleep, perchance to dream." Sleep was long believed to be a period of uninterrupted quiescence, when the body slowed down. In 1913, the French psychologist Henri Piéron authored the book *Le probleme physiologique du sommeil*, the first attempt to examine sleep from a physiological perspective. Piéron also sought evidence that a chemical factor ("hypnotoxin") accumulates in the brain during waking periods and eventually induces sleep.

During the 1920s, the Russian-born American physiologist Nathaniel Kleitman established the world's first sleep laboratory at the University of Chicago and exclusively dedicated his long career to sleep research; at the time, sleep research appeared on no one's scientific radar screen. Kleitman's *Sleep and Wakefulness* (1939) was the first major text on this topic and is still judged a classic; in it, he proposed that sleep consisted of a rest-activity cycle. He also frequently served as his own subject, and on one occasion, remained awake for 180 consecutive hours to study the effects of sleep deprivation.

In 1953, Kleitman's graduate student Eugene Aserinsky began studying attention in children and noted that eye closure was associated with lapses in attention; his first subject was his eight-year-old son. Aserinsky recorded children's eyelid movements electronically and monitored brainwaves using an electroencephalogram (EEG). The resulting brain tracings appeared to be associated with dreaming. Aserinsky continued his studies, recording the EEG and eye movements of sleeping adults and observed that, several times during the night, their eyes darted back and forth. These were named *rapid eye movements* or REM, and their appearance correlated with episodes of dreaming. (Ironically, Aserinsky died in 1998 when his car hit a tree, after he fell asleep.)

Sleep is not a unitary and prolonged state of quiescence. Rather, it consists of distinct phases, with REM consuming 20–25 percent of total sleep, some 90 to 120 minutes, divided over 4 to 5 periods; more than 80 percent of newborn sleep is in REM. Its biological function remains the subject of theorizing—perhaps its function is to consolidate memories or for the central nervous system development of newborns—but we do know that loss of REM results in significant physiological and behavioral abnormalities.

SEE ALSO: Circadian Rhythms (1729).

While REM represents only 20-25 percent of an adult's sleep, it represents up to 80 percent of a newborn's sleep. A dreaming infant is shown in this 1928 painting by Hermann Knopf (1870–1928).

Acquired Immunological Tolerance and Organ Transplantation

Frank Macfarlane Burnet (1889–1985), **Peter B. Medawar** (1915–1987)

In 1940, the British biologist Peter Medawar was called upon to consult on a severely burned airman whose plane crashed near Medawar's Oxford home during the Battle of Britain. This led to a series of studies in which he and a colleague experimented with skin grafts and their sustainability. They observed that when the burn victim received a graft of his own skin (an autograft), it was successfully retained. By contrast, skin grafts from an unrelated donor failed to graft permanently and were rejected within two weeks; subsequent grafts were rejected even more rapidly. Medawar suspected that an underlying immunological reaction was responsible and later found that by suppressing this reaction with cortisone-like drugs, he could delay the rejection of the graft.

Working independently during the 1940s, Frank Macfarlane Burnet, an Australian virologist, was intrigued by immune tolerance in pregnancy, where the fetus and **placenta**—both foreign tissues—are not rejected by the maternal immune system. He introduced the concepts of *self* and *non-self* to immunology, which helped to explain autoimmunity, where the body generates antibodies against its own tissues, viewing them as *non-self* and attempting to destroy them.

Burnet had established the theoretical basis for acquired immunological tolerance. In 1953, Medawar provided experimental supporting evidence for this theory, which lead to successful transplantation of solid organs, and for which they were co-recipient awardees of the 1960 Nobel Prize. Medawar established that during embryonic development, and shortly after birth, immune cells develop that can destroy foreign (non-self) cells. In his key experiment in 1953, Medawar injected tissue cells from adult mice (donors) into developing mouse embryos (recipients). After birth, the recipient mice were able to tolerate skin grafts from their donors but rejected grafts from other unrelated mice. These results established acquired immunological tolerance and provided the basis for later work that developed approaches for suppressing the rejection of organ and tissue transplants.

SEE ALSO: Placenta (1651), Innate Immunity (1882), Adaptive Immunity (1897), Ehrlich's Side-Chain Theory (1897).

This c. 1998 USA stamp promoted organ and tissue donation. Organs that can be transplanted include kidneys, heart, lungs, liver, pancreas, and intestines, while transplantable tissues include corneas, heart valves, bone, cartilage, and ligaments.

Organ & Tissue Donation
Share your life...

1998

USA
32

Sliding Filament Theory of Muscle Contraction

Thomas Henry Huxley (1825–1895), **Aldous Huxley** (1894–1963),
Alan L. Hodgkin (1914–1998), **Andrew F. Huxley** (1917–2012),
Hugh E. Huxley (1924–2013)

The basic mechanical events associated with muscle contractions are common in all animals, whether it be an octopus grasping prey with its tentacles or a track star competing in the 100-meter dash. In 1954, two unrelated English biologists named Huxley independently discovered the mechanism by which skeletal (voluntary) muscles contract and published their findings in back-to-back articles in the journal *Nature*.

The elder of these two Huxleys, Andrew, came from a distinguished family whose members included the biologist Thomas Henry Huxley (grandfather), as well as the writer Aldous (half-brother). Hugh Huxley, by contrast, grew up in a middle-class home. Both Huxleys attended Cambridge University and had their studies interrupted by service during World War II. After the war, Andrew resumed work with Alan Hodgkin, studying the nerve **action potential**—work for which they were co-recipients of the 1963 Nobel Prize. In 1952, he determined how muscle contracts using a microscope of his own design. Hugh Huxley resumed his doctoral studies in 1948, focusing on the molecular structure and function of skeletal muscle using X-ray diffraction and electron microscopy. He continued this work at Massachusetts Institute of Technology in 1952, and in 1954, published his sliding filament theory of muscle contraction, using different methods from Andrew Huxley, but reaching the same basic conclusions.

Skeletal muscle consists of fibers that run parallel to the length of the muscle. Within each fiber (muscle cell) is a myofibril that has a striped appearance formed by a repeating series of thousands of sarcomeres, the contractile unit of the muscle. Within each sarcomere is a series of actin (thin) and myosin (thick) filaments, which lie in parallel to each other. During contraction, the thin actin filaments change their length, while the length of the myosin filaments remains unchanged. Hugh Huxley proposed that when actin slides past myosin, it creates muscle tension.

SEE ALSO: Animal Locomotion (1899), X-Ray Crystallography (1912), Electron Microscope (1931), Action Potential (1939).

Atlas flexes his muscles in this statue at Pathos, Cyprus. Atlas is commonly depicted carrying the Earth on his shoulders, but in the original myth, as a punishment, he was made to bear the weight of the heavens.

Ribosomes

Albert Claude (1898–1983), **George Palade** (1912–2008)

The combination of cell fractionation and electron microscopy opened a new frontier in biology, making possible the visualization of the cell's interior contents and determining their biological functions. In 1930, the Belgian biologist Albert Claude, at Rockefeller University, devised the process of cell fractionation, in which a cell is ground up to release its contents and centrifuged at different speeds to separate the contents according to weight. Claude's cell fractionation process was refined in 1955 by his student George Palade, a Romanian-born American, who used the **electron microscope** to study these cell fractions. Palade was first to identify and describe "small granules," which were given the name *ribosome* in 1958, and found to be the site of protein synthesis within the cell. Claude and Palade (the latter often called the father of modern cell biology and the most influential cell biologist ever) were co-recipients of the 1974 Nobel Prize.

THE PROTEIN FACTORY. All living organisms have ribosomes within each of their cells that are directed by the genetic code to function as factories, carrying out the synthesis of proteins. Cells having high rates of protein synthesis, such as the pancreas, have millions of ribosomes. DNA carries instructions to messenger RNA (mRNA) for building specific proteins. Transfer RNA (tRNA) then brings the amino acids to the ribosome where they are sequentially added to a growing protein chain.

Ribosomes found in eukaryotic cells (animals, plants, fungi) and prokaryotic (bacterial) cells have a similar structure and function. In the former, they are attached to rough-endoplasmic reticular membranes and, in the latter, suspended in the cytosol, the fluid component of cytoplasm. That ribosomes are found across all kingdoms of life suggests that the ribosome evolved early in the evolutionary process. Palade determined that ribosomes are made up of large and small subunits and that there are subtle differences in density (mass per unit volume) between the ribosomes in prokaryotic and eukaryotic cells. This is of practical significance in the treatment of bacterial infections. Certain **antibiotics**, such as erythromycin and the tetracyclines, selectively inhibit the protein synthesis in bacteria without having such effects in the patient's cells.

SEE ALSO: Prokaryotes (c. 3.9 Billion BCE), Eukaryotes (c. 2 Billion BCE), Cell Nucleus (1831), Cell Theory (1838), Antibiotics (1928), Electron Microscope (1931), Lysosomes (1955), Cracking the Genetic Code for Protein Biosynthesis (1961).

The primary function of ribosomes is the manufacture of proteins. The image depicts a model of a eukaryotic ribosome, which differs in structure from a prokaryotic ribosome.

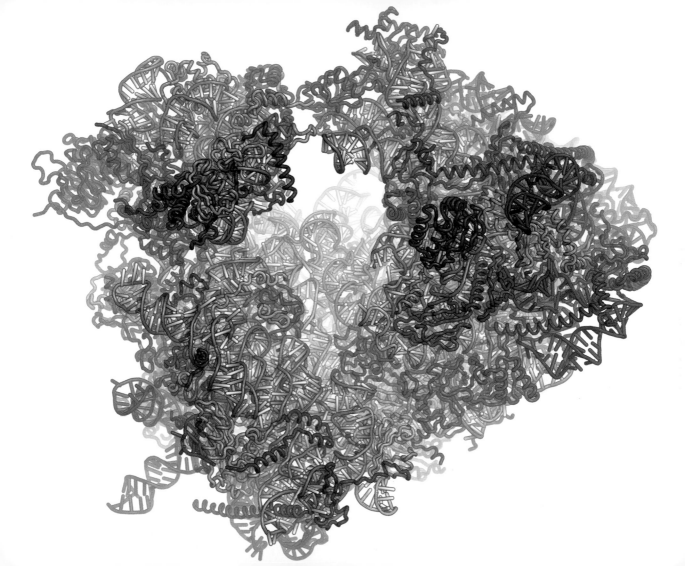

Lysosomes

Alex B. Novikoff (1913–1987), Christian de Duve (1917–2013)

Christian de Duve, a cytologist-biochemist at the University of Louvain in Belgium, was using ultracentrifugation to separate and examine the contents of cells. In 1949, while exploring the action of **insulin** on liver cells, his focus was diverted by an unexpected observation. Prior to placing cells in the ultracentrifuge, he homogenized them with either a pestle or an electric blender and then added the **enzyme** acid phosphatase. To his surprise, only the cell fraction that was homogenized with the electric blender lost most of its enzyme activity. Further study in 1955 revealed the presence of a previously undiscovered intracellular organelle—one that had a sac-like structure surrounded by a membrane.

The contents of this organelle had lytic properties (capable of breaking down tissues), and de Duve called the organelle *lysosome*. In collaboration with the electron microscopist Alex Novikoff, the presence of lysosomes was visually confirmed. De Duve never retuned to investigating insulin and liver cells but was a co-recipient of the 1974 Nobel Prize for his discovery of lysosomes.

A CELLULAR DIGESTIVE SYSTEM. Lysosomes play an important role in health and disease. When functioning normally, their contents contain about fifty acid hydrolase enzymes that are capable of breaking down proteins, nucleic acids, carbohydrates, and fats. Whereas there are conflicting reports as to whether lysosomes are present in plants, they are in all animal cells, with greatest numbers in disease-fighting cells such as white blood cells. They serve as the digestive system of the cell, breaking down materials taken from outside the cell, such as viruses and bacteria, as well as playing a cellular housekeeping role within the cell, ridding it of excess and worn-out organelles. Lysosomes also play a role in protecting cells during periods of prolonged starvation. By the process of autophagy ("self-eating"), lysosomes can digest intracellular components, with their metabolites recycled to synthesize molecules that are essential for the cell's survival.

When lysosomes fail to degrade substances that are normally broken down, these substances may accumulate, causing cellular malfunctions and organ damage. There are approximately fifty of these rare genetic lysosomal storage diseases, which include Gaucher's disease and Tay-Sachs disease.

SEE ALSO: Metabolism (1614), The Liver and Glucose Metabolism (1856), Enzymes (1878), Insulin (1921), Inborn Errors of Metabolism (1924), Ribosomes (1955).

In this image of the interior structure of a plant cell, lysosomes are depicted as small orange spheres. Unlike animal cells, plant cells have a cell wall.

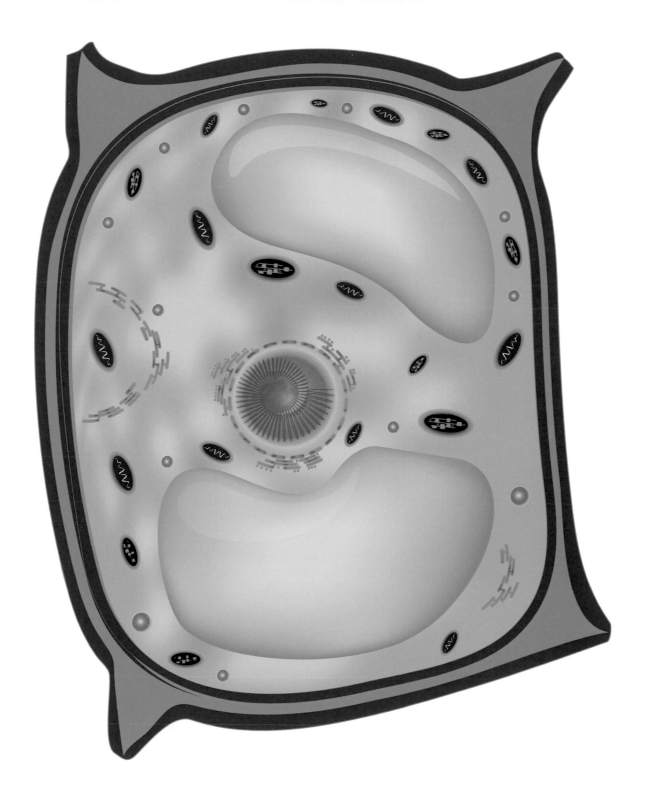

Prenatal Genetic Testing

John H. Edwards (1928–2007), **Giuseppe Simoni** (b. 1944)

Chromosomal abnormalities occur in about 1 in 200 live births, although most fetuses with such abnormalities die before birth. Factors increasing this risk include pregnancy after the age of thirty-five, previously having a child or fetus with a birth defect, and a family history of chromosomal abnormalities. A number of common tests are available to screen for or diagnose genetic abnormalities in the fetus. After the third month of pregnancy, ultrasound can detect whether the fetus has any obvious structural defects.

Removal of amniotic fluid for medical examination goes back to the late 1870s. In 1956, John H. Edwards discussed the use of amniotic fluid obtained from amniocentesis for the "antenatal detection of hereditary disorders." This fluid surrounds the fetus and contains cells used to prepare a karyotype, which displays chromosomes arranged in pairs. This procedure is commonly performed between the fifteenth and twentieth weeks of pregnancy to detect such conditions as Down syndrome (trisomy 21)—in which there is an extra chromosome 21–spina bifida, cystic fibrosis, and Tay-Sachs disease.

An alternative diagnostic procedure to amniocentesis is chorionic villus sampling (CVS), which is usually performed during the tenth and twelfth weeks of pregnancy, thus the results are determined much earlier. CVS, first performed in 1983 by the Italian biologist Giuseppe Simoni at the Biocell Center, can detect more than 200 genetic abnormalities. The chorion is a portion of the fetal membrane that forms on the fetal side of the placenta; the chorionic villi are small, finger-like projections in the chorion that are removed for study during the procedure. Since the villi are of fetal origin, they can provide a sample of the genetic makeup of the fetus.

In 2011, cell-free fetal DNA tests became available. Unlike amniocentesis and CVS, the cell-free tests are noninvasive and only involve a blood test about the tenth week of pregnancy. Since they only evaluate DNA fragments in the bloodstream, unlike the older tests, cell-free is only a screening test (as for Down syndrome) and not a diagnostic test of a genetic defect.

SEE ALSO: Placenta (1651), Eugenics (1883), Inborn Errors of Metabolism (1923).

There are two types of tests that can be conducted during pregnancy for Down syndrome: a screening test using ultrasound (sonogram), which gives an indication that there is a higher risk of a disorder; and a diagnostic test, such as amniocentesis or chorionic villus sampling, which provides a firm diagnosis.

DNA Polymerase

Francis Crick (1916–2004), **Arthur Kornberg** (1918–2007),
James D. Watson (b. 1928)

Watson and Crick's classic paper describing the chemical structure of **DNA** appeared in 1953, causing initial skepticism by some scientists about its significance. Watson-Crick suggested in their paper that a mechanism for copying DNA remained to be determined. American biochemist Arthur Kornberg, then in the microbiology department at Washington University in St. Louis, Missouri, recognized the paper's significance. Consequently, he became interested in how the body synthesizes nucleic acids—in particular, DNA. During these studies, working with the relatively simple bacterium *Escherichia coli*, in 1956 he discovered the **enzyme** that assembles the building blocks of DNA. This enzyme, called DNA polymerase I, is present with some variation in every living organism. Kornberg's papers describing these findings were initially rejected but later accepted and published in 1957 in the prestigious *Journal of Biological Chemistry*. In 1959, he was a co-recipient of the Nobel Prize for determining "mechanisms in the biological synthesis of DNA."

BIOLOGICAL COPY MACHINE. The discovery of DNA polymerase I, commonly designated pol I, is highly significant in biology because it plays a central role in the process of life by contributing to our understanding of how DNA is replicated and repaired. Prior to cell division, pol I duplicates the entire contents of a cell's DNA. This is followed by the parent cell passing one copy of its DNA to each daughter cell; thus, genetic information is transferred from one generation to the next. Kornberg found that pol I reads an intact DNA strand and uses it as a template to synthesize a new strand, which is identical to the original strand—a process not unlike a copying machine generating duplicate documents.

However, unlike a copy machine that blindly copies the document regardless of its contents, some members of the seven subclasses of DNA polymerases—such as pol I—have the ability to proofread the original DNA template, detecting, removing, and correcting errors, thereby producing a new error-free DNA strand. Other DNA polymerases merely replicate but do not repair, thus perpetuating mutations in the genome or leading to the possible death of the cell.

SEE ALSO: Deoxyribonucleic Acid (DNA) (1869), Enzymes (1878), Bacterial Genetics (1946), The Double Helix (1953), Central Dogma of Molecular Biology (1958), Polymerase Chain Reaction (1983).

There are seven subclasses of DNA polymerase (model shown). Some, such as pol I, engage in quality control—reading, detecting, and correcting errors in DNA prior to making a copy.

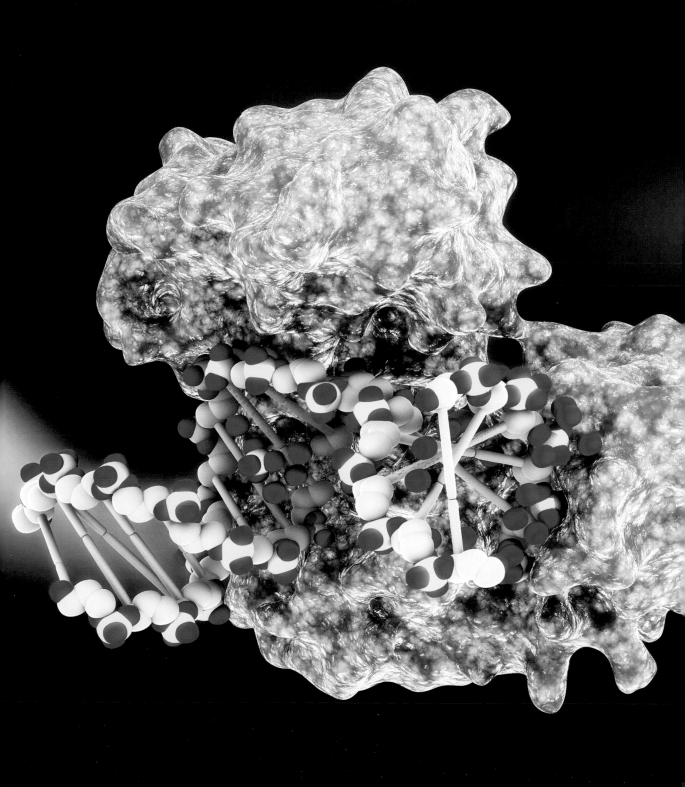

Second Messengers

Earl W. Sutherland, Jr. (1915–1974)

When faced with a predator in the wild, the potential prey has one of two options: fight or flight. To prepare for action, the body responds by increasing heart rate, breathing faster, activating voluntary muscles, and increasing blood glucose (sugar). These bodily responses to stress are mediated by releasing the hormone epinephrine (adrenaline) from the adrenal glands. The glucose obtained from carbohydrate sources can be immediately utilized for energy generation or stored in the liver and muscle as glycogen for later use. When epinephrine is released, it binds to a receptor protein on the surface of the liver or muscle, and this serves as a signal, setting into motion a series of biochemical reactions, culminating in glucose release. This is a three-stage process, the first involving hormone-receptor binding (reception), and concluding with glucose formation (response) in the third stage. But what occurred in the second stage was a mystery.

The American pharmacologist Earl Sutherland had studied these reactions during the 1940s and 1950s and knew that the **enzyme** glycogen phosphorylase was directly involved. But, when he added this enzyme and epinephrine to liver slices in a test tube, no glucose was formed. Sutherland sought to determine the nature of the missing second stage—transduction—and identify the linking chemical responsible for converting the signal hormone (or first messenger) on the liver cell surface into a response within the cell.

That linking chemical—the second messenger—was cyclic adenosine monophosphate or cAMP. In a series of 1956–1957 papers, Sutherland described the sequence of events: The epinephrine-receptor union activates the enzyme adenylyl cyclase located on the liver cell surface which, in turn, promotes the conversion of adenosine triphosphate (ATP) to cAMP. Through a series of subsequent enzyme-catalyzed reactions, glycogen phosphorylase is activated and glycogen is broken down to glucose. Sutherland was awarded the 1971 Nobel Prize for demonstrating the biological role of cAMP.

As a second messenger, cAMP plays a role in such diverse cellular activities as energy **metabolism**, division and differentiation, ion movement, and muscle contractions, and has been shown to be involved in signal transduction in animals, plants, fungi, and bacteria.

SEE ALSO: Metabolism (1614), The Liver and Glucose Metabolism (1856), Enzymes (1878), Negative Feedback (1885), Secretin: The First Hormone (1902).

When confronted by an adversary, the choice is "fight or flight." In either case, the body prepares itself by releasing epinephrine, which activates an increase in blood glucose. The second messenger (cAMP) provides the link between the activation of a receptor on the surface of a liver cell and the release of energy-providing glucose.

Protein Structures and Folding

Christian B. Anfinsen (1916–1995)

Proteins commonly produce their actions by recognizing and binding to other molecules, and for these interactions to occur, the shape of the protein must match the shape of the other molecule. This may be exemplified by the interaction of an antibody protein and an antigen, and between the opioid receptor protein and morphine or heroin.

All proteins have three and sometimes four levels of structures: the primary structure consists of a simple chain of amino acids arranged in a linear fashion; the secondary structure has folding or coiling within the protein structure; the tertiary structure is the three-dimensional shape of a folded protein; a quaternary level is when two or more peptides are joined together to form a single large protein. Proteins are only able to function biologically when their chains are folded into three-dimensional shapes.

From the mid-1950s, Christian Anfinsen, an American biochemist at the National Institutes of Health, studied the relationship between protein structure and its function. For this purpose, he selected the protein ribonuclease, an **enzyme** that breaks down ribonucleic acid (RNA). Ribonuclease is stable, of small size, well studied, and readily available in purified form from commercial sources. In 1957, Anfinsen determined that after the three-dimensional structure of ribonuclease was disrupted and lost its biological activity, it spontaneously refolded and returned to its native (normal), fully functional shape, with its enzyme activity restored. Many other proteins respond in the same manner as ribonuclease.

From these experiments, Anfinsen concluded that the information required by a protein to assume its final three-dimensional configuration is encoded in its primary structure—namely, its amino acid sequence. Moreover, according to Anfinsen's "thermodynamic hypothesis," ribonuclease assumes this three-dimensional structure because this structure is most stable. In 1972, Anfinsen was awarded the Nobel Prize in chemistry for establishing a connection between the amino acid sequence of a protein and its biologically active shape.

A number of diseases—as Alzheimer's, Parkinson's, and Huntington's—have been associated with an accumulation of misfolded proteins, all of which are thought to have an amyloid protein origin, increase with age, and may have a genetic basis.

SEE ALSO: Enzymes (1878), Adaptive Immunity (1897), Ehrlich's Side-Chain Theory (1897), Inborn Errors of Metabolism (1923).

Immunoglobulin M (shown), by far the largest antibody in the human circulatory system, is the first antibody to appear on the scene during an infection, and its detection is often used in the diagnosis of infectious diseases.

Bioenergetics

Rudolf Clausius (1822–1888), **William Thomson, aka Lord Kelvin** (1824–1907), **Hans Krebs** (1900–1981), **Hans Kornberg** (b. 1928)

Bioenergetics describes how living organisms extract energy from their environment to fuel basic energy-consuming activities, including using adenosine triphosphate (ATP) as a source of chemical energy. Living organisms can provide for their energy needs as autotrophs or heterotrophs. Autotrophs, which include plants and algae, use the highly efficient process of **photosynthesis**, which converts energy from sunlight to ATP. Heterotrophs, by contrast, ingest and break down complex organic molecules contained in externally obtained nutrients to generate energy.

Given the wide diversity of living organisms, you might reasonably expect that many mechanisms evolved to generate energy. Not so. Glucose is broken down by the same chemical pathways in bacteria as it is in higher organisms. All organisms use ATP as an intermediate in energy **metabolism**. Metabolism refers collectively to chemical reactions that break down complex chemicals to generate energy and make ATP (catabolism), and those chemical reactions that consume energy and ATP to form complex molecules from simpler ones (anabolism).

In 1957, Hans Krebs and Hans Kornberg, both German-born English biochemists, authored the 85-page booklet *Energy Transformations in Living Matter*, the first publication on thermodynamics in living beings that linked biology and biochemistry. There are two laws of thermodynamics (transformations of energy), which were developed over decades during the nineteenth century by many scientists, including William Thomson (Lord Kelvin) in 1848 and Rudolf Clausius in 1850. The first law states that all the energy in the universe is constant and that it cannot be created or destroyed, only converted to other forms of energy. The chemical energy extracted from nutrients is converted into energy used to process food for use in anabolic reactions and to support living processes. The second law states that energy transformations are inefficient because some energy is lost and unavailable to do work. Loss of heat from the body, such as during exercise, would represent such an energy loss. **Energy balance** is that aspect of bioenergetics that examines the body's harmony between the intake of energy and its expenditure.

SEE ALSO: Metabolism (1614), Photosynthesis (1845), Homeostasis (1854), Enzymes (1878), Mitochondria and Cellular Respiration (1925), Energy Balance (1960).

Glucose is a secondary source of energy and an intermediate in ATP production, which is broken down by the same biochemical reactions in humans and bacteria. The ball-and-stick representation of alpha-D-glucopyranose depicts its three-dimensional structure, with white = hydrogen atoms, black = carbon, and red = oxygen.

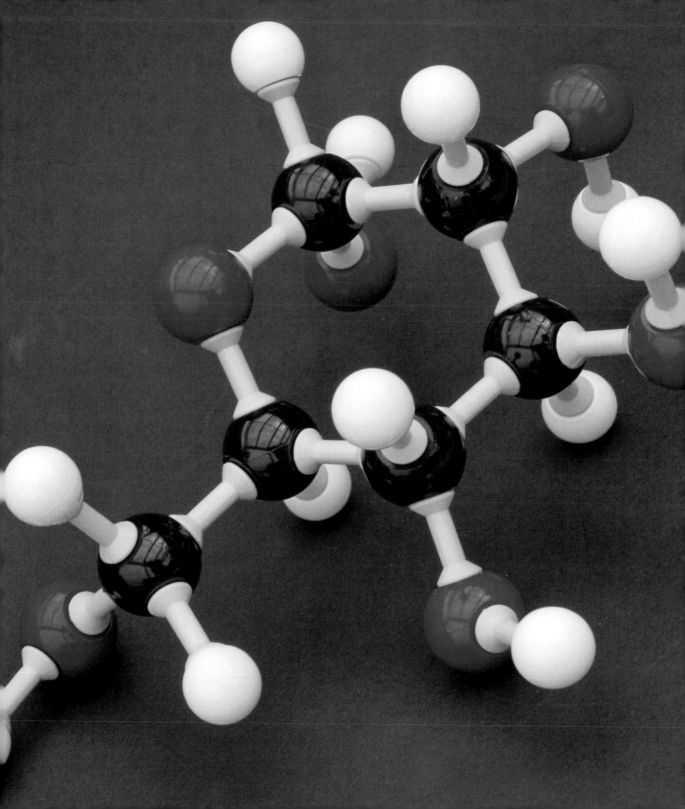

Central Dogma of Molecular Biology

Francis Crick (1916–2004), **James D. Watson** (b. 1928), **Howard Temin** (1934–1994), **David Baltimore** (b. 1938)

In 1958, five years after Watson and Crick discovered the molecular structure of **deoxyribonucleic acid (DNA)—the double helix**—Crick proposed the central dogma of molecular biology, and this he popularized in a paper in *Nature* in 1970. In its basic terms, the central dogma states that genetic information flows in only one direction from DNA ("transcription") to RNA ("translation") to proteins.

Information is "transcribed" from a section of DNA to a newly assembled piece of messenger RNA (mRNA); mRNA makes a copy of one of the two strands of DNA, which serves as a template. The mRNA then travels from the nucleus to the cytoplasm where it binds to a **ribosome**. The ribosome translates the instructions as a codon, a three nucleotide sequence that spells out the order in which amino acids are to be added to the growing peptide chain. The final step involves the faithful replication of DNA to a daughter cell, carried out by the process of **mitosis**.

As originally formulated, the sequence was never translated backwards from DNA to RNA. When the enzyme *reverse transcriptase* was independently discovered in 1970 by Howard Temin at the University of Wisconsin-Madison and David Baltimore at MIT, this upset the premise of the central dogma. For this work, Temin and Baltimore were co-recipients of the 1975 Nobel Prize. Subsequently, it was found that reverse transcriptase is present in retroviruses, such as the human immunodeficiency virus (HIV), and converts DNA from RNA. In addition, and as another exception to the central dogma, not all DNA is involved in programming the synthesis of proteins. Some 98 percent of human DNA is noncoding DNA (dubbed "junk DNA"); its biological function has not yet been determined.

Semantic issues were also raised. In his 1988 autobiography, *What Mad Pursuit: A Personal View of Scientific Discovery*, Crick commented that the term "dogma" was ill-advised. He chose not to use the word "hypothesis," which, in retrospect, would have been far more appropriate. Dogma is a belief that cannot be doubted—certainly not the case when used here.

SEE ALSO: Deoxyribonucleic Acid (DNA) (1869), Mitosis (1882), The Double Helix (1953).

This image depicts the flow of genetic instructions from DNA, to RNA, to the production of amino acids, which link together to form proteins.

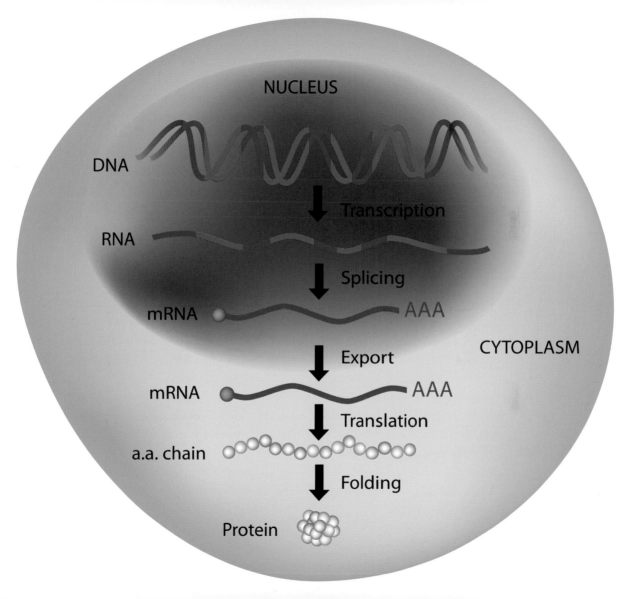

Bionics and Cyborgs

Nathan S. Kline (1916–1883), **Jack Steele** (1924–2009),
Manfred E. Clynes (b. 1925), **Martin Caidin** (1927–1997)

Bionic men and women and cyborgs, blending biology and technology, have been familiar action characters in novels, television, and the movies since the 1970s. The term *bionic* was coined in 1958 by Jack Steele, a US Air Force physician, and has been variously described as referring to "like life" or a contraction of biology plus electronics. Two years later, Manfred Clynes, a scientist-investor, and Nathan Kline, a clinical psychopharmacology pioneer, introduced *cyborg* ("cybernetic organism"), an enhanced human who could survive in extraterrestrial environments. *Cyborg* was the title of Martin Caidin's 1972 novel, which served as the inspiration for the television series *The Six Million Dollar Man* (1974–1978) and the spin-off *The Bionic Woman* (1976–1978). The most famous cyborg is the Terminator, in a multipart movie series of the same name. A cyborg is typically an organism that has been modified to possess greatly enhanced mental and physical human capabilities; the designations bionic and cyborg are commonly used interchangeably.

Bionics has different meanings when used in technology and in biomedicine. In the real world of science and technology, bionics (also called biomimicry, biomimetics) refers to the application of biological methods and systems found in nature to the design of engineering systems—that is, adapting the use of a function rather than attempting to imitate its structure. Products developed using this approach include: Velcro (1948), based on how the hooks and loops of burrs of burdock attach to clothing and animal fur; fabrics and paints that repel dirt and water based on the surface of a lotus flower leaf—the "lotus effect" (1990s); and sonar and ultrasound imaging, which simulate the function echolocation by bats.

In a biomedical context, the focus of bionics is on the replacement of organs or a body part with a mechanical or synthetic version that functions like the missing or defective human part. (By contrast, a prosthesis replaces a body part but cannot function independently.) Cochlear implants have been used with success to aid the profoundly deaf since the 1970s, and a fully functional artificial heart has been available since 2004. Research is underway to develop bionic hands and limbs.

Will the nuclear family of the future be redefined to include bionic parents and their children?

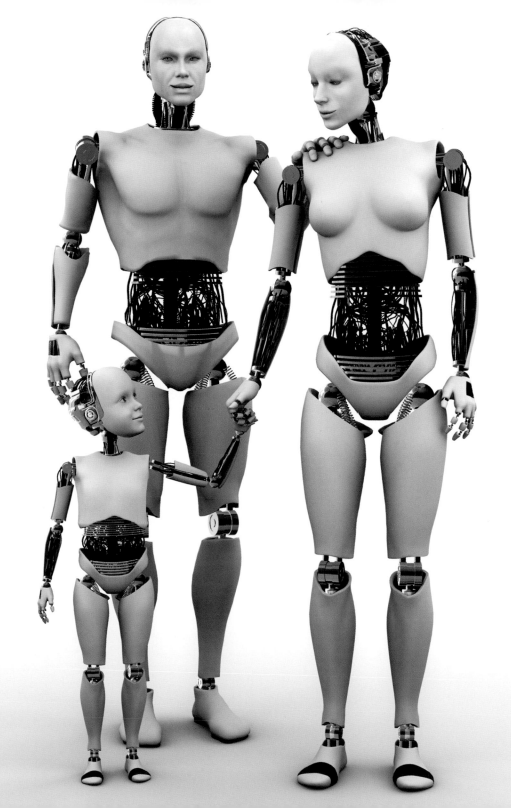

Pheromones

Adolf Butenandt (1903–1995), Martha McClintock (b. 1947)

Courtship in moths can begin as a long-distance relationship. When female moths are available for breeding, they emit a signal that can be detected by a potential male mate 6 miles (10 kilometers) away. What is the nature of this signal? Working over a period of twenty years, in 1959 Adolf Butenandt at the Max Planck Institute in Munich removed glands from 500,000 female oriental silkworm moths (*Bombyx mori*), and from these glands, located at the tip of the abdomen, isolated and characterized a chemical he named *bombykol*. When male moths were exposed to bombykol, they wildly beat their wings in a "flutter dance." Butenandt, a 1939 recipient of the Nobel Prize in Chemistry, designated such chemical factors that trigger a social response in the same species as *pheromones*.

Chemical senses are the oldest of the senses and found in all organisms, including bacteria. In addition to reproductive signals, chemical senses also signal alarm and the presence of predators and food. Once a forager ant finds food, it lays down a pheromone trail that other ants follow to the food. Pheromones produced by the queen bee attract males during her departures from the hive, when she is seeking a mate. (A common pheromone is present in both the Asian elephant and 140 species of moths.) In **insects**, such as moths and butterflies, males detect pheromones via hairlike olfactory receptors on their antennae.

In mammals, reptiles, and amphibians, pheromones are detected by the vomeronasal organ (VNO), located at the base of the nasal septum, and their message is carried to the brain. In 1971, while still an undergraduate student at Wellesley College, Martha McClintock reported that when female college students lived in close proximity, their menstrual cycle became synchronized; this was dubbed the McClintock effect. She also collected compounds from the underarms of women during different phases of their menstrual cycle, which resulted in alterations of the time of their cycles. The validity of the McClintock effect has since been challenged based on methodological and data analyses considerations. Moreover, the actual existence of pheromone molecules and the presence of the VNO in adult humans have been questioned.

SEE ALSO: Insects (c. 400 Million BCE), Insect Dance Language (1927), Sense of Smell (1991).

Pheromones play an important role in the mating of insects, including moths, where males have been known to travel miles following a trail of a female-emitted pheromone in the air. Pheromone traps have been developed that are used for insect control. This image depicts the oriental silkworm moths (Bombyx mori) used by Adolf Butenandt to isolate the original pheromone.

Energy Balance

Nicholas Clément (1779–1842), Claude Bernard (1813–1878)

In its attempt to maintain well-being, the body strives to achieve **homeostasis**, a concept Claude Bernard introduced in 1854 to describe a stable and constant internal environment. Such stability commonly includes body temperature and pH but also energy. **Bioenergetics** is the study of the flow of energy in living organisms. To achieve such balance, our energy intake must equal energy expenditure. The energy intake is determined by diet and includes the food energy (calories) and the amount of food consumed. Energy expenditure is based on physical or external work being performed and the internal heat produced. The internal heat includes: the basal metabolic rate (BMR), which is the amount of energy expended when at rest that is sufficient to enable vital organs and systems to continue to function normally; and the thermic effect of food—that is, the energy cost associated with biologically processing food for use and its storage for later expenditure.

As is all too obvious, gaining imbalances may occur when energy intake exceeds its expenditures, and it commonly results from overeating and a sedentary lifestyle. Such excess energy is primarily stored as fat, leading to weight gain. Conversely, a losing imbalance happens when the energy intake is less than the energy expenditure; this is the consequence of under-eating, digestive disorders, or other disease states.

In 1960, the International System of Units established a set of standards used in commerce and science, and these have been adopted worldwide by virtually all nations, with the notable exception of the United States. In a food-related context, the joule (J) or kilojoule (kJ) is the unit of energy. Food packages in the European Union use both the kJ and the metric system unit of energy, the calorie (c) or kilocalorie (kcal); in the United States, labels only designate the Cal (1 Cal = 1 EU kcal or 4.2 kJ). One Cal is defined as the amount of energy required to raise the temperature of one kilogram (2.2 pounds) of water by $1°$ Celsius ($1.8°$ Fahrenheit). There are many claimants for the individual first to use the calorie in nutrition, but one prominent contender is Nicholas Clément in 1824.

SEE ALSO: Metabolism (1614), Homeostasis (1854), Bioenergetics (1957), Optimal Foraging Theory (1966).

The international unit of energy, the joule, is named after the English physicist James Prescott Joule (1818–1869), who in 1845 devised a "heat apparatus" (shown) that was able to estimate the "mechanical equivalent of heat"—that is, the work required to raise the temperature of a fixed volume of water by $1.8°F$ ($1°C$).

Chimpanzee Use of Tools

Louis Leakey (1903–1972), **Jane Goodall** (b. 1934)

The ability to make and use tools dates back to our earliest human forebears, and it was long believed that "man, the toolmaker" was the unique toolmaker. This claim for uniqueness was dashed in 1960 by the firsthand observations of Jane Goodall, a 26-year-old woman without a college degree.

Goodall was born in England in 1934 and, as a child, developed a passionate interest in and love of animals and Africa. In 1958, Louis Leakey, a noted paleontologist, hired Goodall as a secretary in Kenya. Leakey was so impressed with her ability to organize his research notes into presentations, he sent her to observe chimpanzees in the Gombe Stream Game Preserve in Tanganyika (now the Gombe Stream National Park in Tanzania). Within three months after her arrival in 1960, Goodall made two startling discoveries: Chimps were believed to be herbivores, but on occasion they ate small insects. Moreover, she observed groups of monkeys hunt for and eat meat, such as young pigs and smaller monkeys.

More dramatic was their use of tools. On another occasion, Goodall was observing a chimp feeding on termites, a favorite food. Thick grass blades were used to dig holes in a termite mound. The chimp repeatedly placed the grass stalk inside a hole, pulled out the stalk covered with insects, and then proceeded to use the lips to remove and eat the termites.

Other scientists have seen chimps scrape out food with a stick, just as humans use a spoon to scoop out food. Some chimps learn to use leaves to help them reach water perched in hollows high in trees. They first take a handful of leaves, chew them, and then dip the resulting "sponge" into the pool and suck out the water. Chimps also possess rudimentary tool-making skills. After removing a twig from a tree, they strip off the leaves and use the stem as a tool to catch insects. Elsewhere in Africa, chimps have been observed to crack nuts with rocks. As Leakey noted, "We must now redefine 'man' redefine 'tool' or accept chimpanzees as humans."

SEE ALSO: Primates (c. 65 Million BCE), Anatomically Modern Humans (c. 200,000 BCE).

Before Jane Goodall's observation that chimps use tools for food and water acquisition, it was generally believed that humans were the only species that made and used tools. Depending upon how one defines a tool, its use has also been reported in other mammals, birds, fish, cephalopods, and insects.

Cellular Senescence

Alexis Carrel (1873–1944), Leonard Hayflick (b. 1928)

It was common knowledge during the first half of the twentieth century that animal cells grew indefinitely. In 1912, Alexis Carrel, a French-born Nobel Prize-winning surgeon at Rockefeller Institute, started an experiment in which chick heart cells growing in culture continued to be viable for thirty-four years. The dream of immortal cells ended in 1961 when Leonard Hayflick, an American cell biologist then at the Wistar Institute in Philadelphia, showed that most human cells have a natural limit of reproducing forty to sixty times before senescence occurs and they die; this has been called the Hayflick limit. (The immortality of Carrel's cells has been attributed to the possible inadvertent addition of fresh cells.) Some cells—such as human ova and sperm or cells of perennial plants, sponges, lobsters, hydra, and cancer—are immortal and, barring being killed, can divide indefinitely. What accounts for these differences?

Chromosomes containing DNA are located in the nucleus of each of our cells. At the tip of the spindle-shaped chromosome is a cap or telomere that keeps the ends of chromosomes from sticking to one another and prevents individual strands of DNA from linking. But telomeres have another function: cellular aging. Telomeres have been likened to cellular clocks that set the rate at which cells age and die. Each time a normal cell undergoes **mitosis** (cell division), its telomere shortens a bit and, when it become sufficiently short, the cell dies. Limiting the number of cell divisions may be beneficial by protecting the cell from developing cancer.

By contrast, cancer cells grow telomeres after each division, and this has been attributed to the activity of the enzyme telomerase; normal human cells also have telomerase, but the gene responsible for its activity is suppressed. There are several fascinating implications but, to date, none have been actualized. Potential anticancer drugs are being tested that act by preventing cancer cells from producing telomerase. Conversely, activation of telomerase might be used as an anti-aging treatment—a modern-day fountain of youth—or given to treat conditions associated with premature aging, but this benefit may carry the increased risk of tumor development.

SEE ALSO: Mitosis (1882), The Immortal HeLa Cells (1951), Cell Cycle Checkpoints (1970)

Lobsters have been estimated to live up to 60 years of age, continuing to grow throughout their lifetime without weakening or becoming less fertile. Rather than imbibing the "Fountain of Youth," their longevity has been attributed to their ability to produce telomerase throughout adulthood.

Cracking the Genetic Code for Protein Biosynthesis

George Gamow (1904–1968), **Francis Crick** (1916–2004),
Rosalind Franklin (1920–1958), **Robert W. Holley** (1922–1993),
Har Gobind Khorana (1922–2011), **Marshall Warren Nirenberg** (1927–2010),
James D. Watson (b. 1928), **J. Heinrich Matthaei** (b. 1929)

The structure of DNA was determined in 1953 by Watson, Crick, and Franklin, with strands of the double helix consisting of four nucleotides: adenine (A), thymine (T), cytosine (C), and guanine (G); in RNA, uracil (U) replaces T. But how was the genetic information contained in the DNA molecule translated to the biosynthesis of a protein?

The Russian physicist George Gamow postulated that a three-letter nucleotide (codon) could define up to sixty-four amino acids, more than sufficient to code for all twenty amino acids used to build proteins. In 1961, Marshall Nirenberg, with J. Heinrich Matthaei at the National Institutes of Health, sought to determine what amino acid would be formed after a single nucleotide was added to a reaction mixture. UUU produced the amino acid phenylalanine, cracking the first letter in the genetic code. Shortly thereafter, the addition CCC was found to yield proline. Har Gobind Khorana at the University of Wisconsin-Madison produced more complex sequences composed of repeated two-nucleotide sequences, the first of which was UCUCUC, read as serine-leucine-serine-leucine … ; subsequently, the remainder of the codons were determined.

In 1964, Robert Holley, at Cornell University, discovered and established the chemical structure of transfer RNA (tRNA), thus providing the link between the role of messenger RNA (mRNA) and **ribosomes**. The information needed to make a protein is first attached to tRNA and then translated to messenger mRNA in a ribosome. Each tRNA only recognizes one set of three nucleotides in mRNA, and tRNA binds to only one of the twenty amino acids. A protein is formed by the addition of one amino acid at a time. Nirenberg, Khorana, and Holley were jointly awarded the 1968 Nobel Prize.

Apart from variations, the genetic codes used by all forms of life are very similar. Based on the theory of evolution, the genetic code was established very early in the history of life.

SEE ALSO: Deoxyribonucleic Acid (DNA) (1869), DNA as Carrier of Genetic Information (1944), The Double Helix (1953), Ribosomes (1955), Central Dogma of Molecular Biology (1958), Bioinformatics (1977), Genomics (1986), Human Genome Project (2003).

This image depicts the relationship between the codon (the three-letter nucleotide consisting of adenine, thymine, cytosine, and guanine or uracil) and the encoding of amino acids.

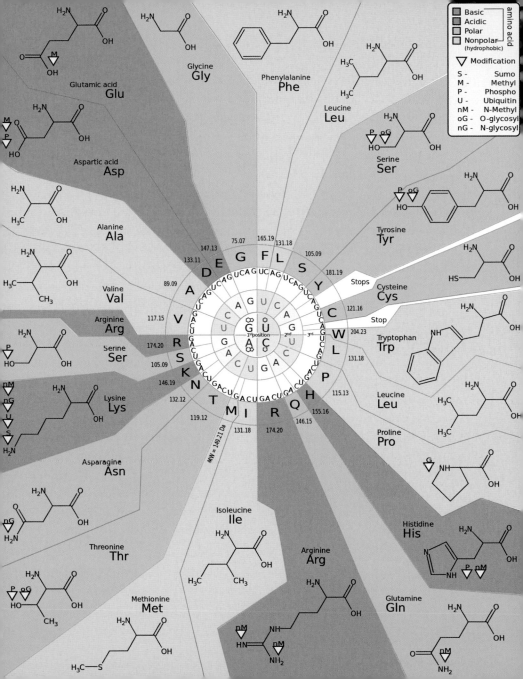

Operon Model of Gene Regulation

Jacques Monod (1910–1976), François Jacob (1920–2013)

Cells abhor squandering energy, and it takes a considerable amount of energy to synthesize proteins. Thus, it would be highly inefficient and wasteful for cells to manufacture proteins if they were not needed. François Jacob and Jacques Monod, French biologists at the Pasteur Institute in Paris, determined how this process is regulated in eukaryotic cells, using as a model *Escherichia coli*, a bacterium that inhabits the gut of animals, including humans.

Glucose is a highly efficient source of energy and is preferred by *E. coli*, if available. To utilize lactose (milk sugar) as an alternative energy source, it must first be broken down into two simpler sugars, glucose and galactose, by the enzyme ß-galactosidase. When Jacob and Monod cultured *E. coli* in glucose, only three units of ß-galactosidase were produced. However, when lactose was substituted for glucose, the production of ß-galactosidase increased by 1,000 times in fifteen minutes. In their classic 1961 study, these scientists showed that this enzyme was inducible—that is, its manufacture was "turned on" when needed—by the *lac* operon.

The *lac* operon consists of three genes, inducing **enzymes** that control the breakdown of lactose and its utilization. Also present is a repressor protein that can shut down the *lac* operon, which is the default situation. The *lac* operon is normally in the "off" position when lactose is absent and the ß-galactosidase enzyme is neither needed nor produced. When lactose is present, the action of the repressor is nullified, and the *lac* operon is transcribed into messenger RNA for the production of the lactose-utilizing enzymes. As lactose is broken down and its levels within the cell fall, the repressor protein again becomes available to shut down the *lac* operon, since the additional synthesis of ß-galactosidase is no longer needed. Jacob and Monod, who were both decorated at the highest level for their service to France during World War II, were co-recipients of the 1965 Nobel Prize for demonstrating the process by which the operon control mechanism occurs at the genetic level.

SEE ALSO: Prokaryotes (c. 3.9 Billion BCE), Metabolism (1614), Enzymes (1878), One Gene-One Enzyme Hypothesis (1941), The Double Helix (1953), Central Dogma of Molecular Biology (1958).

An illustration of Escherichia coli, *a common resident of the intestines of animals, which Jacob and Monod used to formulate their model on the genetic control of the manufacture of enzymes.*

Thrifty Gene Hypothesis

James V. Neel (1915–2000)

Obesity has assumed epidemic proportions in many countries in the world. In the United States, two-thirds of adults are overweight, one-half of whom are obese. Obesity is the leading preventable cause of death worldwide and has been strongly linked to such diseases as type 2 diabetes and heart disease. In 1962, James Neel, a prominent research medical geneticist at the University of Michigan Medical School, proposed the thrifty gene hypothesis, which sought to explain the tendency of certain ethnic groups (such as Native Americans) to have obesity and diabetes. With additional evidence about mechanisms underlying diabetes, the differentiation between type 1 (insulin-dependent) and type 2 (non-insulin dependent) diabetes, and alternate theories about factors leading to obesity, Neel revised his original thrifty gene hypothesis in 1998 to be more general, and not diabetes-specific.

Fat was not always viewed negatively; in fact, it was perceived to confer certain advantages, including serving as a depot for the long-term storage of energy. Throughout much of human history, famine, unfavorable climatic conditions, or the absence of prey made the supply of food for the hunter-gatherer problematic. Humans evolved complex physiological and genetic systems (such as Neel's thrifty gene) to protect against starvation and preserve body fat. In addition, body fat keeps individuals warmer in the cold, both by serving as insulation and by emitting heat when burned. Such a situation would have been beneficial when early humans left the warm climes of Africa and migrated to colder climates—in particular, northern Europe. Fat also provides physical protection, as is seen in pregnant women, who add layers of fat to keep their fetus warm and protected.

The thrifty gene theory has been subject to challenge. In particular, scientists have been unable to find proof or even evidence for its existence. In its stead, simple, commonsense explanations for obesity have been offered: over the past century, energy-saving devices have largely reduced the need for heavy manual labor. Virtually all experts agree that the contemporary prevalence of obesity is the result of our relative lack of physical activity and the abundance and consumption of attractive, unhealthy food choices.

SEE ALSO: Leptin: The Thinness Hormone (1994).

During the eighteenth and nineteenth centuries, waist cincher corsets, as shown in this 1791 cartoon by English caricaturist Thomas Rowlandson (1756–1827), helped women attain the then-desired waist measurement of 19 inches (48 centimeters).

A LITTLE TIGHTER

Silent Spring

Paul Müller (1899–1965), Rachel Carson (1907–1964)

In 1962, *Silent Spring* appeared and was instrumental in launching the environmental movement in the United States. Rachel Carson, a marine biologist and former science editor for the US Fish and Wildlife Service, previously authored a number of natural history books, including *The Sea Around Us* (1951), a *New York Times* best seller for eighty-six weeks.

Four years in preparation, *Silent Spring* documented evidence that pesticides had an adverse effect on the environment that spread far beyond their intended insect targets and extended to fish, birds, and even humans; Carson thought that these chemicals should be called *biocides*. The book's title alludes to a spring in which bird songs are absent as all birds had vanished because of pesticides. She called not for a ban but for more responsible use and careful management of pesticides and greater awareness of their impact on the ecosystem.

Of these pesticides, she focused particular attention on DDT, invented by Paul Müller in 1939, and used highly effectively during World War II in eradicating mosquito-carriers of the malaria parasite in the Pacific and controlling lice responsible for typhus in Europe. When a single application was applied to crops, DDT killed insect pests for weeks and even months. However, runoffs containing DDT were often deposited in nearby waterways and ingested by fish that were the prey of bald eagles—the national symbol since 1782. DDT interfered with the eagle's calcium metabolism and impaired its ability to produce strong eggshells; shells were so thin they broke during incubation. The population of bald eagles, peregrine falcons, and brown pelicans fell precipitously, and these birds were classified as endangered species.

Notwithstanding a firestorm of criticism by the chemical industry, *Silent Spring* was critically acclaimed by both the scientific community and the public. In 1970, the Environmental Protection Agency was created, and, in 1972, the use of DDT was banned in the US and, thereafter, throughout most of the world. The bald eagle has since returned to healthy numbers. Critics of the ban continue to assert that DDT's removal from the market is responsible for the millions of deaths caused by malaria.

SEE ALSO: Food Webs (1927), Factors Affecting Population Growth (1935), Green Revolution (1945), Biological Magnification (1979), Depletion of the Ozone Layer (1987).

It is generally accepted among scientists that DDE, a metabolic breakdown product of DDT, causes eggshell thinning in many bird species, including bald eagles, with eggshells unable to support the weight of the incubating bird.

Hybrids and Hybrid Zones

Ernst Mayr (1904–2005)

In 1942, the evolutionary biologist Ernst Mayr defined a species on the basis of their potential to interbreed and produce viable, fertile offspring. He described that speciation—how one species splits into two or more species—can result when populations of the same species become geographically separated over time, a physical separation that serves as a reproductive barrier. In his 1963 book, *Animal Species and Evolution*, Mayr described the outcome when different but closely related species come into contact, breed, and produce hybrid offspring. Despite interbreeding, the hybrids and their two populations of parents are identifiably different. Since hybrids are often sterile, this prevents the movement of genes from one species to the other, thereby keeping each species distinct.

Hybrid zones are overlapping geographic regions, varying in width from hundreds of feet to thousands of miles, which exist between the populations of two closely related but genetically different species, including hybrids. Evolutionary biologists have been keenly interested in hybrid zones because they provide three possible examples of how speciation might occur in nature. If the reproductive barriers responsible for speciation become strengthened, interbreeding will end, and fewer hybrids will be formed. Conversely, if the reproductive barriers break down or weaken, the two parent species will be able to freely breed. Unlike the sterile hybrids, the gene pools of the two parent strains can mix, become more alike, and eventually fuse, becoming a single species. The third scenario involves maintenance of the status quo and the hybrid zone: the reproductive barriers remain intact, and hybrid organisms continue to be produced.

Examples of hybrid zones and hybrids are seen in both plants and animals, with plants hybridizing more readily than animals, both naturally and through horticultural intervention. Plant hybrids are commonly fertile and capable of reproducing. Among animals, the liger is a hybrid of a lion and tiger, and mussels (genus *Mytilus*) actively hybridize worldwide. Not all efforts of hybridization are successful. The European honeybee was bred with the African bee in hopes of developing a tamer, more manageable hybrid. Instead, the killer bee was produced.

SEE ALSO: Darwin's Theory of Natural Selection (1859), Evolutionary Genetics (1937), Biological Species Concept and Reproductive Isolation (1942), Punctuated Equilibrium (1972).

Hybrids are the product of mating two members of different species but the same genus, such as this liger, the offspring of a lion and a tiger. The hybrid shows the traits and characteristics of both parents and is usually sterile, preventing the movement of genes from one species to the other, thereby keeping both species distinct.

Brain Lateralization

Wilder Penfield (1891–1976), **Herbert Jasper** (1906–1999),
Roger Wolcott Sperry (1913–1994), **Michael Gazzaniga** (b. 1939)

In the 1940s, at McGill University's Montreal Neurological Institute, the famed Canadian neurosurgeon Wilder Penfield was treating severely epileptic patients by surgically destroying specific brain areas from which the seizure was thought to originate. Prior to operating, he applied very slight electrical stimulation to discrete regions of the motor and sensory cortex and, with his colleague, the neurologist Herbert Jasper, mapped the body part that responded to stimulation. Together, they constructed a homunculi ("little man") map representing specific parts of the body affected by motor and sensory brain sites.

Studies conducted at the California Institute of Technology during the 1960s provided greater insight into brain lateralization (functional specialization). The left and right cerebral hemispheres (sides) of the brain are almost identical in appearance and yet are very different in carrying out functions. The two hemispheres normally communicate with each other through a thick band of nerve fibers called the *corpus callosum*. Since the 1940s, large portions of this band had been severed to treat severe epilepsy, resulting in split-brain patients; these operations are now rare. The psychobiologist Roger Sperry and his graduate student Michael Gazzaniga tested the functioning of each hemisphere independent of the other in split-brain humans and monkeys. In about 1964, they found that while each hemisphere was able to learn, one hemisphere had no perception of what the other hemisphere learned or experienced.

The results of these studies led to the conclusion that the left and right hemispheres are specialized in performing different functions. The left brain is primarily concerned with analytical, verbal, and language-processing tasks, while the right side handles the senses, creativity, feelings and facial recognition. Sperry was awarded the 1981 Nobel Prize for his split-brain discoveries.

Individuals are often characterized as being left-brain or right-brain thinkers. Left-brain persons are said to be more logical, fact-oriented, linear thinkers, and concerned with structure and reasoning, while those labeled as being right-brained are said to be feelings-oriented, intuitive, creative, and musical. Although this makes for interesting conversation at cocktail parties, there is no compelling anatomical or physiological evidence to support these labels, and most scientists regard this characterization as a myth.

SEE ALSO: Localization of Cerebral Function (1861).

The left brain is said to control analytical, structured thinking, while the right brain is believed to influence creativity. This left-brain, right-brain distinction is popular but has generally been discounted by neuroscientists.

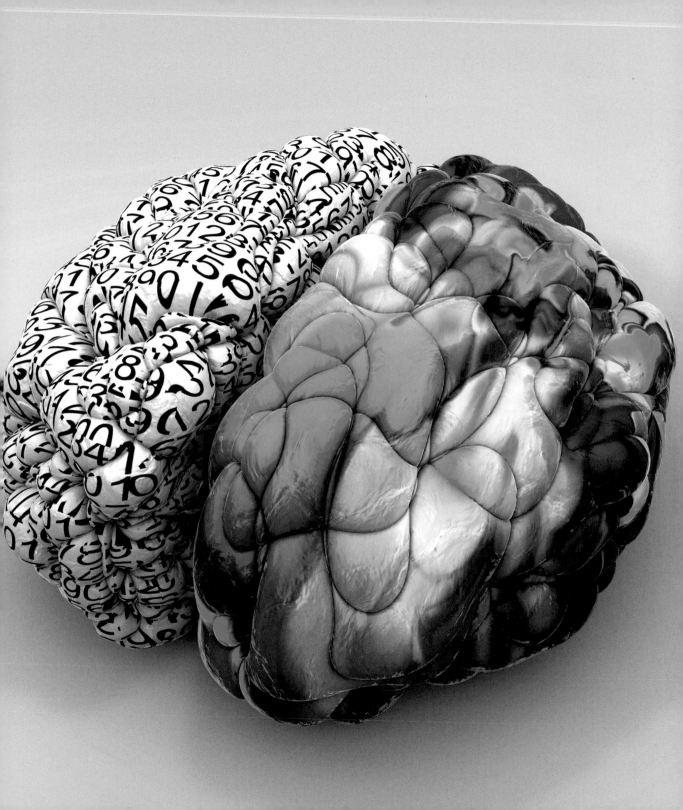

Animal Altruism

Charles Darwin (1809–1882), William D. Hamilton (1936–2000)

The notion of altruism—the selfless concern for others—is a traditional virtue in many cultures, and "the Golden Rule" is a core belief in many religions. Humans are said to be altruistic if the act is performed with the conscious intention of helping another. But acts that would be considered altruistic are performed by animals not perceived to be capable of conscious thought. When analyzing altruism, animal biologists look at the consequences of the act and not its conscious intention.

Moreover, the behavior performed may benefit other organisms at a great cost to the altruist. From an evolutionary perspective, some of these acts seem at odds with **Darwin's theory of natural selection**, which anticipates that animals will act to enhance their own survival and successful reproduction, thus providing themselves with a competitive advantage. Evolutionary success for an organism depends upon leaving behind as many of its genes as possible. Consider then that worker bees evolve without the ability to reproduce and only exist to benefit the hive and guarantee the propagation of the sole queen bee, defending her to the death from attack.

In 1964, W. D. Hamilton, one of the great evolutionary theorists of the twentieth century, proposed the inclusive fitness or kin selection hypothesis to explain this altruistic behavior. Hamilton postulated that the abnormally close genetic relationship of the sterile female bee altruists in the same colony encourages them to ensure one another's survival—and, by extension, the survival of the non-sterile queen—as the probability of spreading their genes is more dependent on the survival of the queen than on their own individual survival. Similarly, vervet monkeys, squirrels, and American robins sound a vocal alarm when detecting the presence of a potential predator, at the cost of revealing their own location and exposing themselves to attack. Vampire bats share their blood with less fortunate colony members who have not fed. Altruistic behavior is also seen among nonrelatives of the same species, who exchange favors, which is called reciprocal altruism: "you scratch my back and I'll scratch yours," both literally and figuratively.

But from an evolutionary perspective, it is difficult to explain why dogs adopt orphan cats and squirrels, or why dolphins have saved humans from attacking sharks. Perhaps a good deed is its own reward.

SEE ALSO: Darwin's Theory of Natural Selection (1859).

Kin selection theory predicts that animals will behave more altruistically toward their relatives than toward unrelated members of their own species. Moreover, research studies have shown that the closer the relationship, the greater degree of altruism.

Optimal Foraging Theory

Robert MacArthur (1930–1972), **Eric Pianka** (b. 1939)

The ability to successfully forage for food occurs in different ways by different animals. Social insects learn to forage—that is, their behavior is modified based on past experience—while nonhuman primates learn by emulating their peers or elders. By contrast, this behavior is genetically influenced in the fruit fly (*Drosophila melanogaster*).

Cost-Benefit Analysis in Nature. In 1966, Robert MacArthur and Eric Pianka, then at Princeton University, developed the optimal foraging theory, based on the familiar economic principle of cost-benefit analysis. Animals find food sources that provide them with maximum caloric benefit while requiring the least expenditure of energy. Foraging costs include "handling," such as searching for the prey and then catching, eating, and digesting it. The ease of obtaining food must also be weighed against the risk of predators. The mule deer *(Odocoileus hemionus)*, found in the Zion Canyon in southwestern Utah, forages for plants in open areas, although food is less plentiful than in forested areas and more energy must be expended to locate it. This animal prefers the open areas because it is less vulnerable to attack by mountain lions (*Puma concolor*), which can stalk their prey while remaining concealed in the woods.

The optimal foraging theory describes optimal behavior, yet feeding in the wild does not always present ideal conditions, and the forager may be faced with a number of constraints and trade-offs. If the forager is too specialized or selective in its food choices, excessive energy will be expended in the search. Conversely, animals whose tastes are generalized or nondiscriminating will pursue unprofitable foods that may provide less benefit.

The cost-benefit of foraging behavior is also influenced by the population of prey in a given geographic area. If the region has low prey density, the forager will spend most of its time searching for food and will eat almost any prey it encounters. But, where there is high prey density enabling new prey to be caught almost immediately, and the bulk of energy expenditures are devoted to catching, eating, and digesting, the forager can select food with the most favorable cost-benefit ratio.

SEE ALSO: Metabolism (1614), Bioenergetics (1957), Energy Balance (1960).

The benefits of an animal successfully foraging for food must be weighed against its costs—namely, exposing itself to predators. The mule deer (shown) forages for plants in open areas rather than in forested areas, where food is more plentiful. Feeding in the forest makes the deer more vulnerable to attack by mountain lions concealed in the woods.

Bacterial Resistance to Antibiotics

When penicillin was first introduced in the 1940s, it ushered in a new era in the treatment of infectious diseases that were previously untreatable and often fatal. Penicillin was the first **antibiotic**, a substance derived from bacteria or **fungi** that kills or controls the growth of other microbes. More antibiotics were developed that were chemical modifications of natural antibiotics and drugs made in the laboratory.

DASHED HOPES. Early on, many experts believed that antibiotics would eradicate infectious diseases that had long plagued humans and animals, relegating such diseases to medical history books. Unfortunately, that initial enthusiasm was tempered when many infectious microbes were found resistant to these drugs. For example, in 1967, the first penicillin-resistant strain of the staphylococcus that caused pneumonia appeared in Australia. More frightening is a recent report that 70 percent of bacteria causing hospital-acquired infections are now resistant to at least one of the antibiotics used to treat them.

Bacterial resistance results from two general mechanisms: mutations and horizontal gene transfer. Normally, an antibiotic binds to a critical microbial protein, preventing the protein from functioning normally. If that function involves the synthesis of DNA that codes for the manufacture of an essential protein or the bacterial cell wall, the bacteria will be killed. If, however, bacteria have a mutation in their DNA that interferes with the antibiotic's attachment to that protein, the bacteria will survive. Based on the process of natural selection, the surviving mutant bacteria will better compete for resources and survive. Resistance can also result from horizontal gene transfer (swap DNA), where one microbe receives a resistant (R) gene or DNA from another antibiotic-resistant microbe. This mechanism does not involve evolution, since no new DNA is being formed.

Resistant bacteria can chemically inactivate the antibiotic, prevent it from attaching to the bacteria, or prevent it from gaining access or building up inside the bacterial cell. Many consequences of bacterial resistance may occur: higher and more dangerous doses of the antibiotic must be used; more expensive drugs may be required; or the patient may fail to recover.

SEE ALSO: Prokaryotes (c. 3.9 Billion BCE), Fungi (c. 1.4 Billion BCE), Darwin's Theory of Natural Selection (1859), Genes on Chromosomes (1910), Antibiotics (1928), Bacterial Genetics (1946), Plasmids (1952), Human Microbiome Project (2012).

Several years after the introduction of the antibiotic methicillin in 1959, reports surfaced of methicillin-resistant Staphlococcus aureus *(MRSA, shown). The development of resistance has been attributed to horizontal gene transfer via plasmids.*

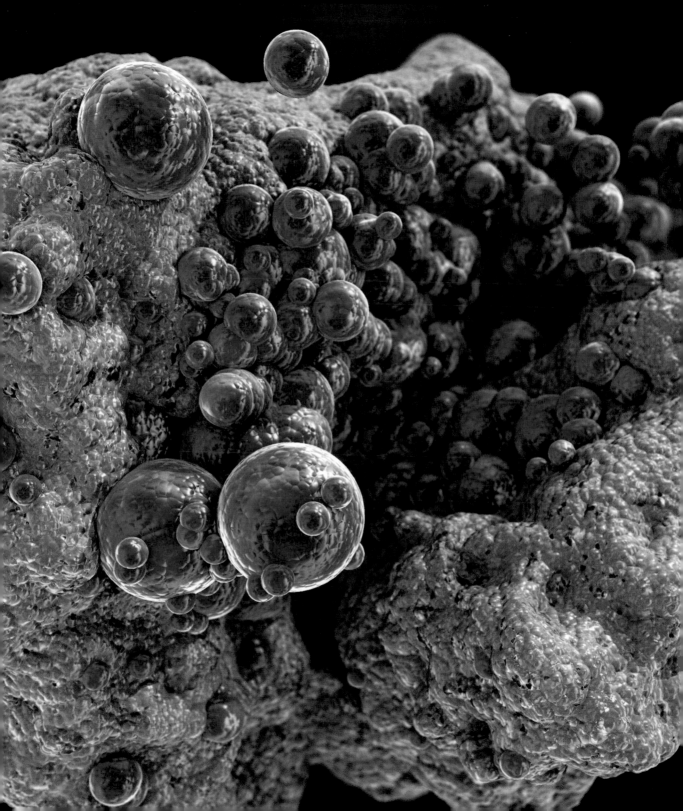

Endosymbiont Theory

Konstantin Mereschkowski (1855–1921), Lynn Margulis (1938–2011)

The endosymbiont theory helps us understand evolution because it explains the origin of organelles in eukaryotic cells—those in plants, animals, fungi, and protists. Symbiosis, which occurs at all levels of biological organization, involves two organisms that cooperate for their mutual benefit to gain a competitive advantage—for example, insect pollination of flowers or the digestion of food by gut bacteria. In eukaryotic cells, **mitochondria** and chloroplasts are organelles involved in the generation of energy required to carry out cell functions. Mitochondria, the site of cellular respiration, use oxygen to break down organic molecules to form ATP (adenosine triphosphate), while chloroplasts in plants—the sites of **photosynthesis**—use energy derived from the sun to synthesize glucose from carbon dioxide and water.

ADDING ONE ORGANELLE AT A TIME. According to the endosymbiont theory, small bacteria (alpha proteobacteria) containing mitochondria were engulfed by primitive eukaryotic cells (protists). In the ensuing symbiotic relationship, the bacterium (now called the *symbiont*) provided its evolving mitochondria, the generator of energy, while the eukaryotic cell offered protection and nutrients. By an analogous process, a eukaryotic cell engulfed a photosynthetic cyanobacterium that, in time, evolved into a chloroplast. In this description of primary endosymbiosis, one living organism has been engulfed by another. When the product of this primary endosymbiosis is engulfed by another **eukaryote**, secondary endosymbiosis is said to have occurred. This provides the basis for incorporating additional organelles and expands the number of environments in which eukaryotes can survive.

The endosymbiotic theory was first proposed in 1905 for chloroplasts by the Russian botanist Konstantin Mereschkowski (who rejected Darwin's theory of evolution and actively promoted eugenics), the idea was expanded to include mitochondria in 1920. Endosymbiotic theory gained no scientific traction until 1967, when it was reintroduced by Lynn Margulis, a biology professor at the University of Massachusetts, Amherst (and former wife of the late astronomer Carl Sagan). Her paper was rejected by fifteen journals before being accepted, and is now considered a milestone in endosymbiont theory.

SEE ALSO: Prokaryotes (c. 3.9 Billion BCE), Eukaryotes (c. 2 Billion BCE), Photosynthesis (1845), Darwin's Theory of Natural Selection (1859), Ecological Interactions (1859), Mitochondria and Cellular Respiration (1925), Protist Taxonomy (2005).

This image depicts the symbiosis between a fly agaric mushroom (Amanita muscaria) and a birch tree. The mushroom receives sugar ($C_6H_{12}O_6$) and oxygen from the tree in exchange for minerals and carbon dioxide.

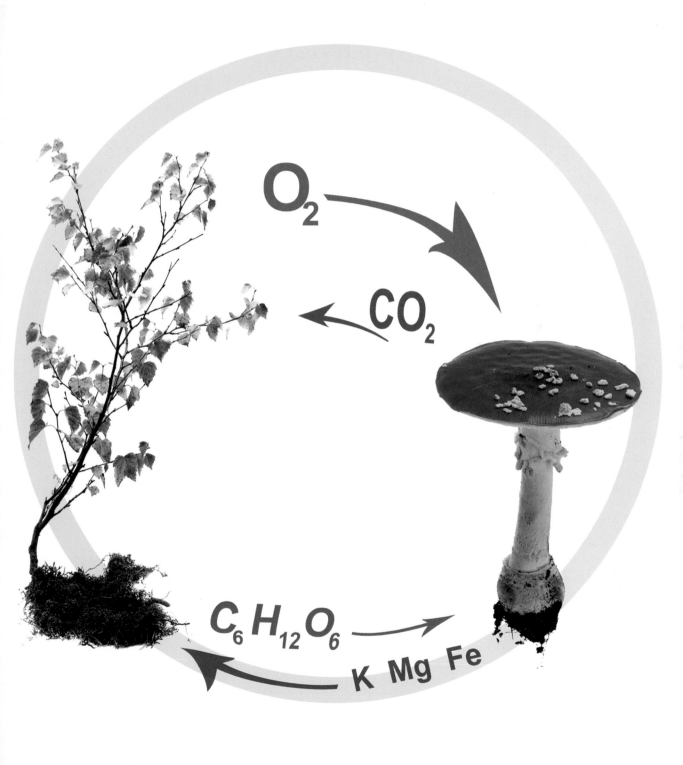

Multi-Store Model of Memory

Aristotle (384–322 BCE), **William James** (1842–1910),
Richard Atkinson (b. 1929)

The study of memory has intrigued scientists and philosophers for thousands of years. Aristotle envisioned memories to be imprinted on the mind, like carving on a wax tablet. He drew a distinction between the memory of animals and recollection by humans. Animals can remember where to find their food, whereas humans *recollect*, deliberately searching their memories that help one reflect on the present, past, and future.

In 1890, the American psychologist-philosopher William James was the first to propose two systems of memory (dichotomous memory). Primary memory, now referred to as short-term memory (STM), is the initial repository for information and is continually accessible for conscious inspection. It is fleeting in duration, with information consciously retained for seconds to minutes. Secondary memory, or long-term memory (LTM), persists for an unlimited period and can be brought to consciousness, if desired.

The concept of a multi-store model was proposed in 1968 by Richard Atkinson and Richard Shiffrin at Stanford University, and it provided the first comprehensive framework for information processing in memory. In their Atkinson-Shiffrin model, information flowed from sensory memory (SM), passing through STM into LTM. SM is information acquired from the environmen—usually visual and auditory in nature—that persists for milliseconds to several seconds. Since we are constantly bombarded with sensory information, happily only a fraction of it passes to the next stage: STM, or working memory. SM and STM have a limited capacity. Information stored in STM is retained for twenty to thirty seconds—long enough to satisfy an immediate need, such as looking up a telephone number—and then it is rapidly forgotten. We retain LTM for days to years, outside our consciousness, but it can be retrieved back into working memory when needed.

Neuroscientists believe that the information contained in both STM and LTM is stored in the cerebral cortex. From an evolutionary perspective, organizing information into STM and LTM, and the delay in transmitting between them, permits LTM to be gradually incorporated into our existing store of knowledge and experience, enabling the establishment of more meaningful associations that may aid survival.

SEE ALSO: Localization of Cerebral Function (1861), Neuron Doctrine (1891), Associative Learning (1897).

Dolphins have been found to have extremely long memories—at least twenty years—which is longer than that of elephants. A dolphin's social memory serves them well because they leave one group and join others multiple times during their lifetimes.

Hypothalamic-Pituitary Axis

Geoffrey W. Harris (1913–1971), **Rosalyn Yalow** (1921–2011),
Roger Guillemin (b. 1924), **Andrew V. Schally** (b. 1926)

The pituitary gland, a grape-size gland located at the base of the brain, consists of two primary lobes: the anterior pituitary produces and secretes six hormones, while the posterior pituitary secretes two hormones. These anterior pituitary hormones stimulate endocrine glands, regulating their hormonal secretions. In the 1930s, the English anatomist Goeffrey Harris hypothesized that the hypothalamus, which lies directly above the pituitary, controls it by secreting its own hormones, but he was unable to identify such hypothalamic hormones and to prove his hypothesis. Although only the size of an almond, the pituitary controls a wide range of basic bodily functions, as well as emotions. During the late 1950s and into the 1960s, Roger Guillemin and Andrew V. Schally—first as collaborators at Baylor University in Houston, Texas, and later as rivals—successfully identified a number of hypothalamic hormones. These hormones are secreted at the base of the hypothalamus and travel through a number of blood vessels to the anterior pituitary, where they either stimulate or inhibit the release of specific hormones.

In 1968, the first such hypothalamic hormone was isolated and chemically characterized: Thyrotropin-releasing hormone (TRH), which stimulates the release of thyroid-stimulating hormone (TSH) from the anterior pituitary. TSH travels in the blood to the thyroid gland, promoting the secretion of the thyroid hormones. The hypothalamus and anterior pituitary do not function in isolation; rather, they receive messages or **negative feedback** from nerves throughout the body, which modulate or turn off additional TRH and TSH secretion. Other hypothalamic hormones include luteinizing hormone-releasing factor, adrenocorticotropin-releasing hormone, and somatotropin. Guillemin and Schally, the founders of neuroendocrinology—the interaction between the central nervous system and endocrine glands—were co-recipients of the 1977 Nobel Prize, which they shared with Rosalyn Yalow for her discovery of the radioimmunoassay (RIA) of these hormones.

SEE ALSO: Nervous System Communication (1791), Homeostasis (1854), Negative Feedback (1885), Secretin: The First Hormone (1902), Thyroid Gland and Metamorphosis (1912).

This illustration depicts the hypothalamic-pituitary axis. Hormones from the hypothalamus stimulate or inhibit the release of hormones from the anterior pituitary gland. These pituitary hormones, in turn, travel in the blood to different endocrine glands, activating the release of specific hormones.

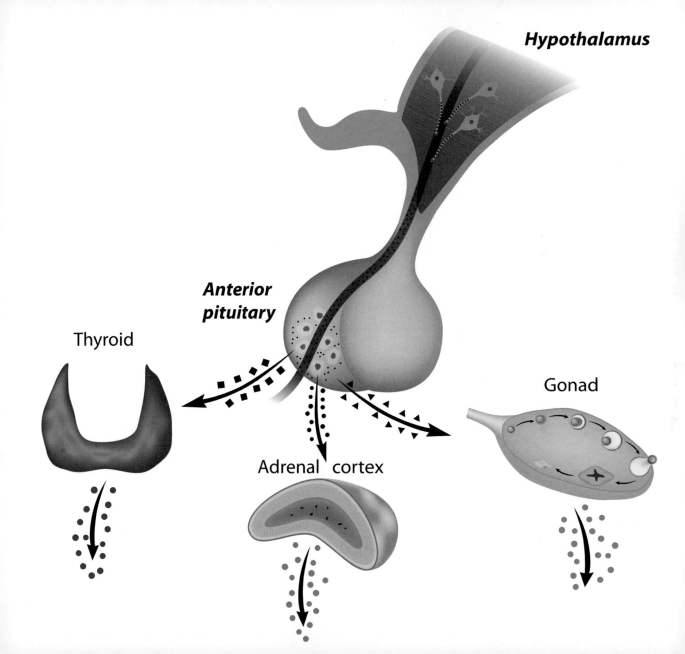

Hypothalamus

Anterior pituitary

Thyroid

Gonad

Adrenal cortex

Systems Biology

Ludwig von Bertalanffy (1901–1972)

The term *system* refers to a group of parts or components that interact and that are dependent upon one another to form a more complex whole. Researchers have adopted *micro* or *macro* approaches to investigating systems. Biological scientists, when looking at living organisms, have traditionally focused upon the individual components of the organism. Not infrequently, they have intensively studied and gathered as much information as possible, sometimes over a lifetime, on a specific enzyme system, brain part, or photosynthetic pigment. This has been referred to as a reductionist approach to scientific research: all the information gathered on individual components is used to formulate a complete description of the system, referred to as the "bottom-up" approach.

In 1968, the Austrian-born biologist Ludwig von Bertalanffy proposed turning over the reductionist model 180 degrees and using a "top-down" approach, which he referred to as a *general system theory*. The basic elements of this theory are applicable to problems in many disciplines, including engineering, the social sciences, and biology. Rather than adopting the reductionist approach of studying individual components in isolation, in systems biology researchers look at organisms as an integrated network of genes, proteins, biochemical reactions, and physiological responses that give rise to life. In systems biology, researchers approach all components and interactions among those components as parts of a single system, with the interactions responsible for the form and function of the entire system. Thus, the whole is viewed to be greater than the sum of its parts.

Few individual biologists can fully understand a complex biological system by exclusively using their specialized and relatively circumscribed backgrounds to study individual components parts. Bertalanffy envisioned systems biology as an integrated multidisciplinary study that called upon the expertise of biologists, physicists, computer scientists, mathematicians, and engineers. Such a systems approach might, for example, be applied to construct mathematical models that predict the consequences of climatic changes resulting from a reduction in rainfall on plant life, which affects crop supply and, in turn, food supply for human consumption.

SEE ALSO: Homeostasis (1854), Ecological Interactions (1859), Coevolution (1873), Biosphere (1875), Global Warming (1896), Endosymbiont Theory (1967), Depletion of the Ozone Layer (1987).

The study of the universe utilizes a systems approach, allowing astrophysicists to predict cosmic events billions of years before their actual occurrence. The Cat's Paw Nebula, shown in this photograph made from a combination of exposures, lies in the constellation of Scorpius (The Scorpion), close to the center of the Milky Way—relatively near to Earth at a distance of 5,500 light years.

Cellular Determination

Thomas Hunt Morgan (1866–1945), **Alan Turing** (1912–1954),
Lewis Wolpert (b. 1929), **Christiane Nüsslein-Volhard** (b. 1942)

Scientists have long wondered how a simple fertilized egg transforms into a highly differentiated multicellular organism. How do cells know where to migrate? Why do some cells become neurons and others bone? *Morphogenesis* is the process that determines the spatial distribution of cells during embryonic development that result in the overall body plan. In 1901, the American evolutionary biologist-geneticist Thomas Hunt Morgan observed that **regeneration** of worms occurs at different positions and at different rates on the worm's body. Morgan suggested that morphogenesis results from signals that are released from localized groups of cells ("organizing centers") to cause differentiation of the cells around them. One-half century later, in his paper "The Chemical Basis for Morphogenesis" (1952), Alan Turing proposed that chemicals, which he called *morphogens* ("form producers"), start from a homogeneous distribution and organize into spatial patterns based on their concentration.

"FRENCH FLAG MODEL." In 1969, Lewis Wolpert, a South African–born British professor of developmental biology at the University College London, envisioned morphogens to be secreted from a group of source cells that travel at different concentrations to serve as signaling mechanisms, acting directly on target cells to produce a response. The intensity of this response is based on the morphogen concentration at the target cell. By way of illustration, he used the "French flag model," which has three broad vertical strips of blue, white, and red. Cells closest to the source (blue strip) would receive the highest concentration of morphogen, which would activate high-threshold target genes; cells farther from the source (white) would receive lower levels of morphogen activitating low-activity genes; those cells farthest (red) away would not be activated. Different combinations of target cells would be activated based on their distance from the source. During the 1980s, the German biologist Christiane Nüsslein-Volhard, based on the French flag model and morphogens, determined the genetic basis for the head-to-tail body plan of the fruit fly—work for which she was a co-recipient of the 1995 Nobel Prize.

SEE ALSO: Regeneration (1744), Genes on Chromosomes (1910), Embryonic Induction (1924), Pattern Formations in Nature (1952).

The "French flag model" has been used to illustrate how the relative concentration of morphogen determines the distribution of cells during embryonic development. This poster depicts small children bearing toy muskets and the French flag while saluting wounded World War I soldiers on Bastille Day, 1916.

14 JUILLET 1916
JOURNÉE DE PARIS

AU PROFIT DES ŒUVRES DE GUERRE
DE L'HÔTEL DE VILLE

EDITEUR IMP. H.CHACHOIN, 108, rue Folie-Méricourt.

Cell Cycle Checkpoints

Leland Hartwell (b. 1939), **R. Timothy Hunt** (b. 1943), **Paul Nurse** (b. 1949)

Cell division plays a critical role in the life of the organism, including such essential functions as reproduction, growth, and development, as well as the renewal and repair of worn out or damaged cells. Many anticancer drugs act by disrupting specific phases of the cell cycle.

The cell cycle is a continuous process that results in the formation of two daughter cells that arise from a single dividing parent cell. There are two major parts of the cell cycle (aka cell division cycle): interphase and mitotic phase. While in interphase, the cell grows and chromosomes are replicated, which alternates with the mitotic phase, during which **mitosis** (nuclear division) and cytokinesis (cell division) occur. The interphase, which consumes all but one to two hours of a twenty-four-hour cell cycle, has three phases: G_1, S, and G_2. G_1 and G_2 are gaps between the end of cell division in the mitotic phase, during which time the environment is assessed for errors prior to moving to the next step.

Starting in about 1970, checkpoint mechanisms at the gaps were identified that review the environment to ensure that the preceding step in the cell cycle has been completed or corrected. If so, the "go" signal is given, and the DNA is replicated in the S phase. The "go" signal involves the proteins cyclin and a cyclin-dependent kinase (Cdks). If conditions are not right, corrections are made or the cell is destroyed; an incorrectly divided cell may result in cancer. In 1991, the Nobel Prize was awarded to Paul Nurse, Leland Hartwell, and R. Timothy Hunt for their discovery of these protein molecule checkpoints that regulate cell division in the cell cycle.

At the conclusion of the cell cycle, the parent cell has doubled in size, the chromosomes have doubled in number, and the cell has divided in half, forming two genetically identical daughter cells to reinitiate the cycle. Cell cycles vary from ten to twenty-four hours in rapidly growing intestinal cells, to once annually in liver cells, to never for mature nerve or muscle cells.

SEE ALSO: Cell Theory (1838), Meiosis (1876), Mitosis (1882), DNA as Carrier of Genetic Information (1944).

This illustration depicts the cell division cycle. During the interphase (I), which consists of three phases (G_1, S, G_2), the cell grows and chromosomes are replicated. At the mitotic phase (M), representing 1–2 hours of the total 24-hour cycle, nuclear division (mitosis) and cell division occur.

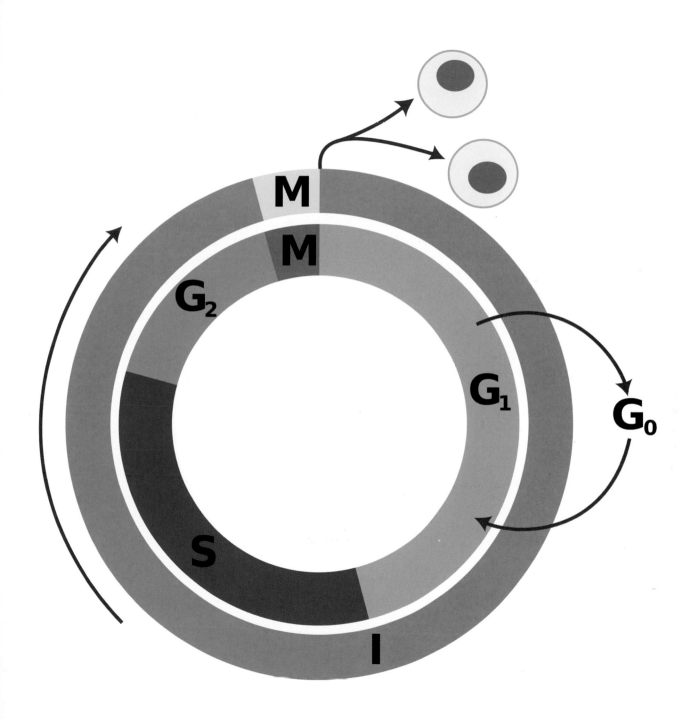

Punctuated Equilibrium

Charles Darwin (1809–1882), **Ernst Mayr** (1904–2005),
Stephen Jay Gould (1941–2002), **Niles Eldredge** (b. 1943)

How did species evolve—gradually by "creeps" or dramatically by "jerks"? Darwin's *Origin of Species* (1859) explains evolution as being smooth and occurring by steady and gradual changes in the species. This explanation of evolution by natural selection is widely accepted by evolutionary biologists, but it does not account for the sudden and unaccounted for appearance of numerous new species in the fossil record, many of whose antecedents remain to be discovered. Darwin acknowledged these gaps and partially explained them as resulting from imperfect preservation of the fossil record. But he also noted that not all species changed at the same rate and to the same extent.

In 1972, the evolutionary biologists-paleontologists Niles Eldredge, at the American Museum of Natural History, and Stephen Jay Gould, at Harvard, proposed an alternative explanation of evolution—one that occurred by "jerks" that would account for the sudden debut of new fossilized species—which they called *punctuated equilibrium*. According to this hypothesis, most new species originated after dramatically splitting from a parent species, rather than gradually changing from the parent. During their relatively early independent existence, the branched population briefly (in geological terms) underwent its major changes in appearance. Thereafter, the new independent species, small in number, remained in an extended state of stasis (equilibrium), with only modest changes in appearance for the rest of their existence—a period that might be millions of years in duration.

The foundation of punctuated equilibrium was built upon Ernst Mayr's well-accepted theory of geographic (allopatric) speciation, popularized in 1963. In Mayr's theory, speciation—the formation of distinct species from a parent species—occurred when a small group became physically separated and branched from the mass of the parent population over a relatively short period, a time insufficient to leave a definitive fossil record. Punctuated equilibrium is considered to be an important model of evolution, but it is highly controversial and often misunderstood in many aspects, including the mistaken notion that it is a refutation of Darwin's theory of evolution by natural selection.

SEE ALSO: Paleontology (1796), Fossil Record and Evolution (1836), Darwin's Theory of Natural Selection (1859), Genes on Chromosomes (1910), Evolutionary Genetics (1937), Biological Species Concept and Reproductive Isolation (1942), Hybrids and Hybrid Zones (1963).

A 1981 British postage stamp depicts Charles Darwin and Galápagos Island finches with beaks of different sizes and shapes, which were a building block in his developing theory of natural selection.

26p

C. Darwin

Sustainable Development

The notion that natural resources are not inexhaustible goes back hundreds of years to forestry management models seeking to balance the consumption of trees with their replacement. The goal of sustainable development is to responsibly use natural resources to meet present needs without compromising the ability of future generations to meet their needs. In recent decades, the boundaries of sustainable development have expanded beyond environmental protection—the focus of the "green" movement—to also include economic growth, social equality, and cultural protection.

In 1972, the United Nations Conference on the Human Environment was convened in Stockholm, the first major international meeting dealing with how human activity was affecting the environment; it highlighted the problems of pollution, destruction of natural resources, and damage to species. The 1992 Rio Earth Summit, attended by more than 100 countries in Rio de Janeiro, dealt with climate change by advocating limiting emissions of the greenhouse gases carbon dioxide and methane. In addition, they argued for maintaining biological diversity and using biological resources in a sustainable manner, such as by reducing deforestation.

International conferences have faced the challenge of balancing the desires of developed countries with the needs of those that are developing. The developed countries have become increasingly concerned about environmental issues and have sought to reduce the environmental impact of industry's continued growth. Nevertheless, 80 percent of the world's natural resources are being consumed by 20 percent of the world's population. There is ever-increasing emphasis on investment in financially viable green technologies, energy efficiencies, and the use of environmentally friendly renewable resources, such as wind and solar energy power.

Developing nations aspire to reach the higher levels of economic growth that industrialized countries have achieved. Driven by economic constraints, they have resorted to resource extraction and using the least expensive methods to achieve such goals as industrialization—methods that impose a high environmental cost. The challenge is to harmonize prosperity with ecology, to maintain continued economic growth without undue environmental harm.

SEE ALSO: Population Growth and Food Supply (1798), Global Warming (1896), Factors Affecting Population Growth (1935), Green Revolution (1945), Energy Balance (1960), Depletion of the Ozone Layer (1987).

Renewable energy is derived from resources that are being continually replenished and include sunlight (solar energy), wind, rains, tides, waves, and geothermal heat.

Parental Investment and Sexual Selection

Robert L. Trivers (b. 1943)

The well-being of offspring, indeed their survival and future reproductive success, depends upon the effort their parents make for their benefit from the time of copulation. In 1972, Robert Trivers, an American evolutionary biologist and sociobiologist then at Harvard University, proposed a parental investment theory. The time, energy, resources, and risk that parents expend for the benefit of their offspring constitutes their investment, with the nature of that investment varying among taxonomic groups and between sexes.

Trivers noted that prior to the offspring's birth, the male only invests a small amount of time and effort to achieve reproductive success—just sufficient to copulate. His evolutionary return is great, spreading his genes, after which he can move on and seek another mate(s). By contrast, female members of the species invest in the gestation of their offspring, undergo the extended mental and physical costs that accompany pregnancy, and during pregnancy cannot reproduce. The postnatal parental investment varies among taxonomic groups. With few exceptions, the offspring of aquatic invertebrates, fish, and amphibians receive little or no postnatal care by either parent. Prenatal and postnatal parental investment by one and commonly both bird parents involves preparing the nest, guarding the eggs, and caring for the brood. With mammals, and humans in particular, after a nine-month-long pregnancy and period of nursing, the investment by both, but sometimes by only one parent, is extensive and may persist for decades.

Such a relative difference in the investment by each parent, Trivers argues, plays a profound influence on selection of a mate, with the female far more discriminating in her choice. Males compete with one another for the opportunity to mate, with success determined by such factors as size, strength, and bright coloration, an indicator of health and vitality. Females prefer males to be physically fit, with superior physical traits (i.e., good genes to pass to their offspring), high status (alpha males), and resources. For species in which both parents participate in the care of the offspring, females will select a male perceived to be interested in assisting.

SEE ALSO: Sexual Selection (1871), Animal Coloration (1890).

While female lions hunt in groups of three to eight for food, males are the protectors of the cubs. The cubs are vulnerable to attacks by hyenas and leopards; however, the greatest threat comes from other male lions.

Lucy

Mary Leakey (1913–1996), **Yves Coppens** (b. 1934),
Maurice Taieb (b. 1935), **Donald Johanson** (b. 1943)

In 1974, Lucy, a 3.2-million-year-old *Australopithecus afarensis*, made her debut—or at least her skeletal remains did, with an age determined by **radiometric dating**. *A. afarensis* may represent one of the earliest members of the hominid species, the human branch of the evolutionary tree. Unlike other anthropological findings, where fossils or a few bone fragments have been discovered, this one finding represented 40 percent of an entire skeleton. Based upon the size of the pelvic opening, it was inferred that these were the remains of a female; she was 43 inches (3.1 meters) tall and weighed ~ 66 lbs (30 kilograms). For many years, Lucy's skeleton was exhibited in well-publicized tours in the United States, and in 2013 the remains were returned home. A plastic replica and related artifacts are now displayed in the Ethiopian capital, Addis Ababa, at the National Museum of Ethiopia.

First signs of fossil remains near Hadat, in northeastern Ethiopia, were made in 1972 by the French geologist Maurice Taieb. To explore the site, he arranged for a tri-national team of scientists that included an American anthropologist, Donald Johanson, the British archeologist Mary Leakey, and the French paleontologist Yves Coppens. During the second season of their field study, in 1974, Lucy was found and was named after a Beatles song played in their field camp, "Lucy in the Sky with Diamonds."

Examination of the pelvic and leg bones led to the conclusion that Lucy was bipedal, while the capacity of the cranial cavity was comparable to that of an ape, around one-third the size of modern humans. This has led scientists to conclude that bipedalism preceded an increase in brain size during human evolution, a view that is contrary to what was previously believed. Some researchers have questioned whether Lucy and *A. afarensis* are, in fact, ancestors to modern humans. Among other remains found, most in the same area of Africa, there is no evidence that *A. afarensis* used shaped tools or fire.

SEE ALSO: Primates (c. 65 Million BCE), Neanderthals (c. 350,000 BCE), Anatomically Modern Humans (c. 200,000 BCE), Radiometric Dating (1907), Oldest DNA and Human Evolution (2013).

A reproduction of Australopithecus afarensis, *one of the longest-lived and (thanks to Lucy) most well-known early human species. Remains found in Eastern Africa date back to 3.85–2.95 million years ago, suggesting that this species survived more than 900,000 years—almost four times as long as* **Anatomically Modern Humans** *thus far.*

Cholesterol Metabolism

Adolf Windaus (1876–1959), **Feodor Lynen** (1911–1979),
Konrad Emil Block (1912–2000), **Robert B. Woodward** (1917–1979),
Joseph L. Goldstein (b. 1940), **Michael S. Brown** (b. 1941)

When cholesterol comes to mind, it is reflexly associated with atherosclerosis, heart attack, and stroke. However, this solid steroid alcohol is essential for building and maintaining animal cell membranes as well as their permeability and fluidity, which allows proteins and other compounds to move within the membrane's two layers. It is the starting molecule for the biosynthesis of such steroid compounds as bile, which plays a role in the digestion and absorption of fats, and for the synthesis of vitamins A, D, E, and K, the adrenal gland hormones cortisol and aldosterone, and the male and female sex hormones. Cholesterol is also a key component in the myelin sheath that surrounds and insulates the axon, which facilitates conduction of nerve impulses.

Cholesterol was first found in the bile and gallstones in 1769 and in the blood in 1833. Later research focused on its chemistry and **metabolism**, and the health risks associated with elevated levels. In 1903, Adolf Windaus determined its chemical structure. In 1951, the preeminent organic chemist Robert Woodward synthesized it.

During the 1950s, Konrad Emil Block and Feodor Lynen, working independently, determined the biosynthesis of cholesterol. Block traced its synthesis from a 2-carbon acetate to the 27-carbon, four-ring structure of cholesterol, involving twenty-six enzymes. Biosynthesis of cholesterol is regulated by the existing body levels of cholesterol via a **negative feedback** system; a higher dietary intake leads to decreased biosynthesis and vice versa. In 1974, Michael S. Brown and Joseph L. Goldstein at the University of Texas Southwestern Medical School identified a series of molecules that regulate cholesterol metabolism. Statins, inhibitors of cholesterol synthesis at its rate-limiting step (slowest step in reaction), are among the world's most widely used drugs.

To date, thirteen Nobel Prizes have been awarded to researchers who have studied cholesterol, and it has been said, perhaps without hyperbole, that it is "the most highly decorated small molecule in history." These Nobel laureates include Windaus (1928), Woodward (1951), Block and Lynen (1964), and Brown and Goldstein (1976).

SEE ALSO: Metabolism (1614), Negative Feedback (1885), Progesterone (1929), Stress (1936).

Blocked arteries caused by a buildup of cholesterol—a disease called atherosclerosis—are a major cause of death in Western countries. As the blood flow becomes blocked, blood clots develop within the artery. When the clots break off, they can block blood flow to the arteries in the heart and brain, resulting in a heart attack and stroke, respectively.

Sense of Taste

Kikunae Keda (1864–1936)

During our early school years, we learned there were four primary tastes—sweet, salty, sour, and bitter—and each was selectively sensed on our tongues. The tongue map was memorized by generations of students since 1901, when it was developed by the German scientist, D. P. Hanig. Today, we know there are five primary tastes. The fifth taste, umami (Japanese "good taste" or "good flavor"), common in foods containing monosodium glutamate (MSG), was discovered in 1907 by the Japanese chemistry professor Kikunae Keda. In 1974, Virginia Collings found that only very small variations in the sensitivity of different parts of the tongue to tastes and taste sensations are distributed throughout the tongue. In short, the tongue map is a myth.

The four primary tastes provided early humans with clues about the nature of the food they planned to ingest: sweet tastes were rich in calories; salty provided nutrient value; sour signaled spoiled or unripened food; and bitter warned that the food was potentially toxic. Tastes are chemical senses recognized by specialized receptor cells present in the taste buds that are contained in goblet-shaped papillae—the bumps on the tongue. Fifty such receptors may be found on a single taste bud, and each primary taste triggers a receptor. Each receptor cell has a protrusion, the gustatory hair, which reaches to the tongue's outer surface through a taste pore. After a tasty molecule mixes with saliva, it enters the taste pore, interacts with the gustatory hair receptor, and stimulates a taste message transmitted to the gustatory areas in the cerebral cortex.

Studies going back to the 1930s and continuing to recent times provide a basis for our impression that some individuals have greater sensitivity to taste than others. Using propylthiouracil (a drug used in thyroid disorders) as a test substance, 50 percent of subjects perceived that it had a bitter taste, 25 percent could not taste it ("nontasters"), while 25 percent reported it to be intensely bitter ("supertasters"). Supertasters are more common among females and individuals from Asia, Africa, and South America, and their greater sensitivity has been attributed to having a greater number of taste receptor cells.

SEE ALSO: Nervous System Communication (1791), Neuron Doctrine (1891), Action Potential (1939), Sense of Smell (1991).

Some individuals find the taste of cruciferous vegetables (such as Brussels sprouts and broccoli) to be extremely bitter—a taste that has been attributed to the same chemical found in propylthiouracil.

Monoclonal Antibodies

Kitasato Shibasaburō (1853–1931), **Paul Ehrlich** (1854–1915), **Emil von Behring** (1854–1917), **Michael Potter** (1924–2013), **Cesar Milstein** (1927–2002), **Georges Köhler** (1946–1995)

At the turn of the twentieth century, the German physician-scientist Paul Ehrlich proposed the concept of a *magic bullet*: a compound that could selectively target and kill a disease-causing organism without causing harm to the patient. Ehrlich had conceived of a magic bullet in 1890, early in his scientific career, when Emil von Behring and Kitasato Shibasaburō introduced an antiserum for treating diphtheria and tetanus. The immunity conferred to these diseases resulted from the production of specific antibodies in response to the bacterial toxins, which served as antigens. There was and continues to be hope that monoclonal antibodies would be "magic bullets."

During the 1950s, Michael Potter, at the National Cancer Institute of the National Institutes of Health, perfected a technique for growing plasma cell tumors (plasmacytomas) in mice that produced highly specific antibody molecules in response to specific antigens. Potter freely shared these mouse plasma cells with scientists around the world, including Cesar Milstein and Georges Köhler at the Laboratory of Molecular Biology in Cambridge, England. In 1975, Milstein, a biochemist born in Argentina and a naturalized British citizen, and Köhler, a German postdoctoral fellow, fused mouse spleen cells, a rich source of beta-lymphocytes from the plasmacytomas, and mouse myeloma cells to produce hybridomas.

These hybridomas produce *monoclonal antibodies*, or antibodies that are identical to one another because they are produced by one type of immune cell type; moreover, they can be produced in an unlimited supply. The development of monoclonal antibodies is considered one of the most significant advances in biomedical research in the twentieth century, earning Nobel Prizes for Milstein and Köhler in 1984. Milstein did not seek patent protection for his technique, causing major reverberations at the highest levels of the British government. It was predicted that monoclonal antibodies would have a wide range of therapeutic uses, be safe and highly selective in their effects, and be easy to manufacture. As of June 2014, thirty monoclonal antibody-derived products had been approved by the US Food and Drug Administration for the treatment of cancers, autoimmune diseases, inflammatory disorders, and as diagnostic agents.

SEE ALSO: Adaptive Immunity (1897), Ehrlich's Side-Chain Theory (1897).

The depicted monoclonal antibody (mAb) is an immunoglobulin G (IgG) molecule, the most abundant class of antibodies found in the blood and lymph.

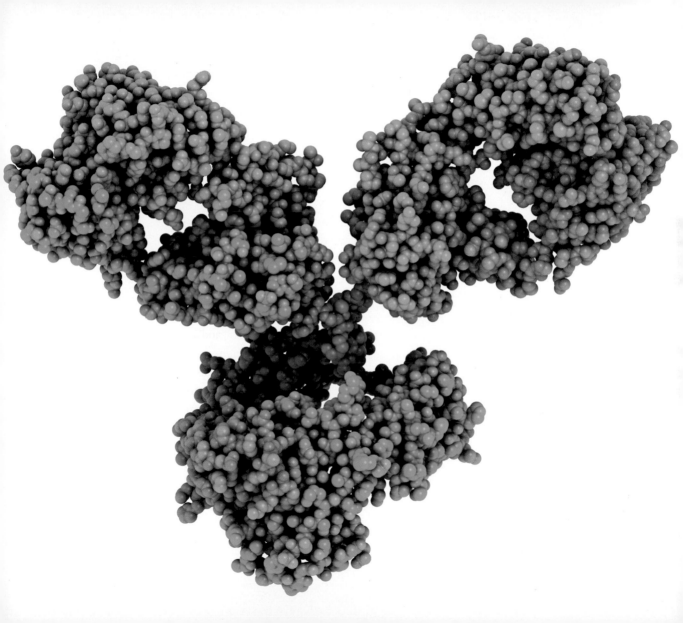

Sociobiology

Edward O. Wilson (b. 1929), Stephen Jay Gould (1941–2002)

Based on the tenets of natural selection, those individuals with variations in genetic makeup that promote their chances of survival and offer the opportunity to procreate will multiply at the expense of those with less advantageous genes. In 1975, the American biologist and entomologist E. O. Wilson authored *Sociobiology: The New Synthesis*, in which he sought to explain animal and human behavior on the basis of evolutionary theory and natural selection, suggesting that species behave in a manner that maximizes the chances of transmitting their genes to successive generations. In many species of mammals, for example, the mother's innate protective behavior helps her offspring to survive and reproduce.

Sociobiologists primarily focus on instinctive behavior and on group rather than individual behavior. Inherited adaptive behaviors in nonhuman animals are well accepted by evolutionary biologists, and they represent an active area of research. But Wilson has argued that *human* behavior is shaped as much, if not more, by genetic influences than by culture. Following his line of reasoning, social and environmental factors can have only limited influence in changing human behavior. Sociobiology's application to humans has generated intense controversy and criticism.

Among those in the vanguard of criticism was Stephen Jay Gould, a paleontologist and popular science writer. Gould and other evolutionary biologists have rejected biologic determinism in humans and argued that human behavior may be influenced by genetic makeup but is not determined by it. Accepting the belief that a human's genetic makeup is intractable and that it controls one's destiny and the need to maintain the status quo, it can be used to justify entrenching the ruling elite, legitimizing authoritarian political policies, and a host of other social injustices, including racism and sexism.

Prior to the 1980s, sociobiology and behavioral ecology were more or less synonymous; behavioral ecology studies the ecological and evolutionary basis for animal behavior. To avoid the controversy of attempting to apply theories of the evolution of animal behavior to humans, researchers in this field limit their studies to animals and much prefer the designation "behavioral ecologist."

SEE ALSO: Insects (c. 400 Million BCE), Darwin's Theory of Natural Selection (1859), Eugenics (1883), Associative Learning (1897), Parental Investment and Sexual Selection (1972).

Sociobiology explains this snow monkey's innate protective behavior toward her baby as based upon her efforts to enable the baby's survival and reproduction, which guarantees the spread of her genes.

Cancer-Causing Genes

Francis Peyton Rous (1879–1970), **J. Michael Bishop** (b. 1936),
Harold Varmus (b. 1939)

In 1911, Peyton Rous discovered that the (Rous) sarcoma virus (RSV), an oncogenic virus, could cause cancer in chickens. This finding, the first to show that a virus could cause cancer in any species, was finally recognized by the Nobel Committee over fifty years later, in 1963, when Rous was eight-four years old. RSV was later found to be a retrovirus that contains RNA instead of DNA and can be transcribed into DNA by the enzyme reverse transcriptase, which is present in the virus. The abnormal DNA can enter the chromosomes of normal cells and redirect their activity, resulting in cancer.

In 1976, Michael Bishop and Harold Varmus at the University of California, San Francisco, used RSV to demonstrate how malignant tumors are formed from normal cell genes. Oncogenes refer to specific parts of the genetic material of **viruses** that can direct the transformation of a normal cell into a cancer cell, when influenced by other parts of the virus or by radiation or certain chemicals. They found that the oncogene in RSV was not a true viral gene but rather a normal cell gene—a proto-oncogene—that the virus had acquired during replication in the host cell and carried along. Proto-oncogenes are genes that code for (send instructions for synthesis of) a kinase enzyme, which triggers signals that stimulate normal cell growth and division. The Bishop-Varmus discovery has led to the isolation of many cellular genes that normally control growth and development but that can become mutated, causing cancer.

During the cell cycle, damaged DNA in genes is normally repaired or destroyed. If such repair-destruction mechanisms are inadequate, mutations build up and genetic damage is passed on to the daughter cells. Cancers arise when normal cells in the body suffer an irreversible change in their genetic material. Two classes of genes regulate the growth of cells: proto-oncogenes and tumor suppressor genes (TSG), mutations in either of which might potentially result in cancer. Mutations in proto-oncogenes may result in overstimulation and uncontrolled cell growth while mutations in TSG may remove the brakes that control cell growth.

SEE ALSO: Viruses (1898), Cancer-Causing Viruses (1911), Bacterial Genetics (1946), Central Dogma of Molecular Biology (1958), HIV and AIDS (1983).

In this photograph, National Cancer Institute director Harold Varmus—co-recipient of the 1989 Nobel Prize for demonstrating how oncogenes in viruses can cause cancers—gives a presentation in 2010.

Bioinformatics

Frederick Sanger (1918–2013), Paulien Hogeweg (b. 1943)

Biological data is being generated from multi-disciplinary laboratories throughout the world at a mind-boggling rate that is sufficient to overwhelm even the most sophisticated research teams. Nowhere is this more evident than in molecular biology, where progress has been propelled by advances in genomic technologies. **Genomics** is the sequencing, assembling, and analysis of the structure and function of the complete set of DNA within the cell of an organism. In 1975, Frederick Sanger, who elucidated the **amino acid sequence of insulin** two decades earlier, developed the first DNA sequencing technique, and in 1977 he determined the 5,386 nucleotides in the first fully sequenced DNA-based genome of a **bacteriophage** (a virus that infects bacteria). Since this time, progress in genomics has expanded 100 million-fold! The **human genome project**, completed in 2003, sequenced 20,500 genes. The challenge is no longer the acquisition of information but rather the ability of researchers to utilize it to advance their studies.

MAKING SENSE OF DATA. Bioinformatics, a term coined in 1970 by Paulien Hogeweg, a Dutch theoretical biologist, is a science that merges biology, computer science, and information technology into a single discipline. It involves the use of information technology to acquire, store, manage, and analyze information in biological databases. These databases are designed so that researchers can access and retrieve existing information and add new information as it is generated. At the next level, it seeks to develop mathematical algorithms, data mining techniques, and other resources that aid in the analysis of existing data and permit its comparison with existing information. It ultimately seeks to uncover new biological insights and obtain a global perspective from which fundamental concepts in biology can be determined. Gaining an all-inclusive picture of the normal activities of the cell will provide a foundation for an understanding of their deviation in disease.

In addition to DNA and amino acid sequencing, as well as predicting the amino acid sequence of proteins, bioinformatics has made it possible to trace the evolution of organisms by measuring changes in their DNA, analyze highly complex regulatory systems that lead to a change in the activity of proteins, and seek mutations present in cancer cells.

SEE ALSO: Bacteriophages (1917), Amino Acid Sequence of Insulin (1952), Cancer-Causing Genes (1976), Genomics (1986), Human Genome Project (2003), Human Microbiome Project (2012), Oldest DNA and Human Evolution (2013).

Researchers are potentially being buried under an onslaught of new and increasingly detailed data and information, made possible by technological advances involving circuit boards like this one.

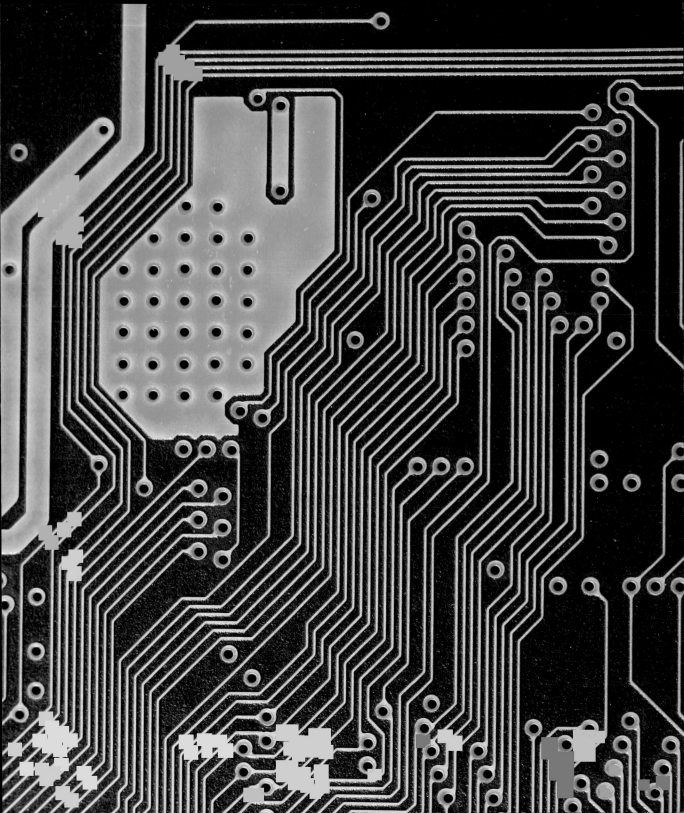

In Vitro Fertilization (IVF)

Walter Heape (1855–1929), **Gregory G. Pincus** (1903–1967),
Min Chueh Chang (1908–1991), **Patrick C. Steptoe** (1913–1988),
Robert G. Edwards (1925–2013)

In 1978, an almost century-old dream became a reality. Louise Joy Brown, the first successful in vitro fertilization (IVF) baby, was born in Oldham, England, thanks to the decade-long efforts of obstetrician-gynecologist Patrick Steptoe and physiologist Robert Edwards. The news was greeted with cheers by childless parents and derision by some church leaders accusing them of "playing God." At the time Edwards was awarded the Nobel Prize in 2010 (Steptoe was deceased), an estimated four million "test tube babies" had been born. Louise Brown conceived and delivered a child naturally in 1999.

The genesis for IVF started at Cambridge University in 1891 when Walter Heape successfully transplanted embryos into a rabbit, which gave birth to a litter of six. In 1934, reproductive biologist Gregory Pincus (co-discoverer of the oral contraceptive decades later) and E. V. Enzmann at Harvard first proposed that mammalian eggs could undergo normal development in vitro (outside the body). The notion of creating embryos in laboratories appeared two years earlier in Aldous Huxley's *Brave New World*. Fiction was realized in 1959, when M. C. Chang, a reproductive biologist at the Worcester Foundation for Experimental Biology, was the first to achieve IVF in rabbits after fertilizing newly ovulated eggs with sperm in a flask.

IVF seems relatively simple in principle, but optimizing conditions for successful fertilization took Steptoe and Edwards ten years to perfect. Normally, women produce a single egg each month. In IVF, fertility drugs are used to induce "super ovulation," causing release of multiple eggs. The eggs are retrieved from the woman's ovaries by the process of follicular aspiration. If no eggs are produced, donated eggs can be used. Sperm and the egg(s) are mixed (insemination) in vitro, with the egg usually fertilized after several hours; sperm may also be injected into the egg (intracytoplasmic sperm injection or ICSI). The fertilized egg divides, becoming an embryo, and is permitted to incubate in a flask where, after 3–5 days, it is implanted into the uterus. The success rate of giving birth to a live baby after IVF decreases with age: 41–43 percent for women under thirty-five and 13–18 percent for those over forty-one.

SEE ALSO: Placenta (1651), Spermatozoa (1677), Theories of Germination (1759), Ovaries and Female Reproduction (1900), Timing Fertility (1924).

In this digital illustration, a glass needle injects sperm into an egg extracted from a woman's ovary during IVF.

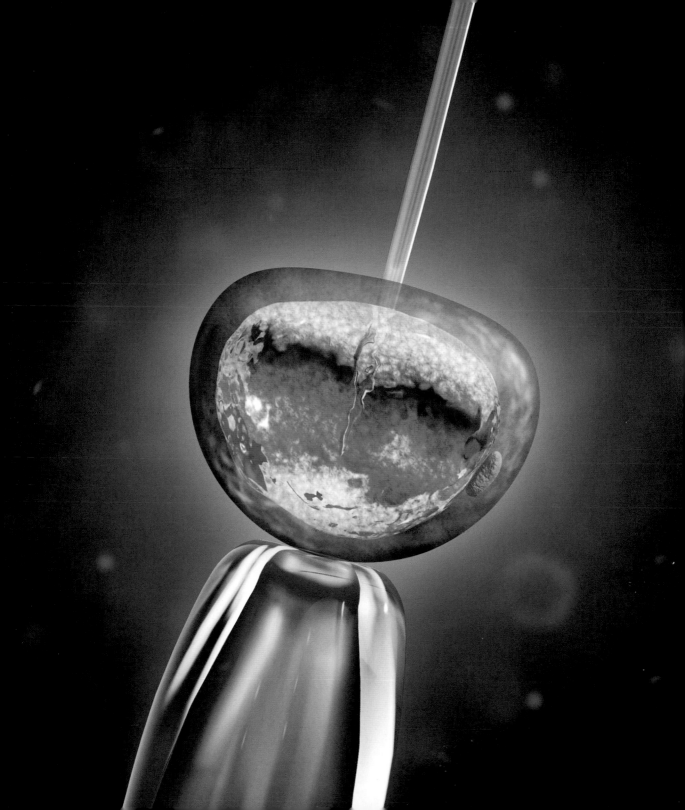

Biological Magnification

Rachel Carson (1907–1964)

Polychlorinated biphenyls (PCBs) had been used from 1929 until 1979 for hundreds of beneficial industrial and commercial applications. But leaky equipment and the illegal or improper dumping of PCB wastes have led to environmental exposure, which continues today with disastrous consequences. With growing evidence that PCBs caused cancer as well as adverse effects on the immune, reproductive, endocrine, and nervous systems, in 1979 their production was banned in the US in accordance with the Toxic Substances Control Act of 1976.

PCBs are very stable chemicals that do not readily break down in the environment. They are insoluble in water but very soluble in fat tissue and remain in the body for extended periods of time. In short, PCBs are perfect chemicals for biological magnification. Other similar common toxic substances include pesticides (such as DDT) and heavy metals (arsenic, mercury, lead). Once fat-soluble pollutants are ingested by birds or mammals, they become stored in tissues and internal organs. Unlike chemicals that dissolve in water, they are poorly excreted in the urine and, therefore, become concentrated in the body, increasing their toxic potential.

Biological magnification (biomagnification or bioaccumulation) is the process whereby a harmful substance becomes increasingly concentrated as it moves up the food chain. This is illustrated by PCB and its increasing concentration in the body (in parts per million) as it ascended the food chain in the Great Lakes: phytoplankton, the base of the food chain (0.025) → zooplankton (0.123) → smelt, a small fish (1.04) → trout, a larger fish (4.83) → herring gull eggs (124)—biological magnification of almost 5,000 times!

The indiscriminate use of DDT resulted in a catastrophic decline in the population of bald eagles, peregrine falcons, and brown pelicans. This was highlighted in Rachel Carson's book *Silent Spring* (1962), which led to the banned use of DDT in the US in 1972. From 1932 until 1968, the Chisso Corporation dumped methylmercury into wastewater that emptied into Minamata Bay in Japan, accumulating in fish and shellfish, which were eaten by the local residents and their animals. Extremely severe nerve toxicity (Minamata disease) resulted, with almost 1,800 human deaths documented.

SEE ALSO: *Silent Spring* (1962).

History is replete with examples of heavy metals, chemicals, and pesticides entering the food chain and, through the process of biological magnification, causing the widespread death of wildlife.

Can Living Organisms Be Patented?

Louis Pasteur (1822–1895), **Ananda Chakrabarty** (b. 1938)

Scientists in the petrochemical industry were long aware of bacteria that were capable of metabolizing hydrocarbons into simpler, harmless substances. But since no single strain of bacteria could metabolize all hydrocarbons present in crude oil, multiple strains were used in oil spills. Not all these strains could survive under different environmental conditions, and these strains sometimes competed with one another, thereby reducing their effectiveness.

In 1971, the Indian American microbiologist Ananda Chakrabarty, at the General Electric Company, discovered **plasmids** capable of degrading crude oil. These plasmids could be transferred to the bacterium *Pseudomonas* to create a genetically engineered species that does not exist in nature. This "oil-eating" bacterium could consume oil several magnitudes faster than the earlier four strains combined and break down two-thirds of the hydrocarbons found in a typical oil spill. But effectiveness aside, could a living organism be patented?

The right to grant patents was enshrined in Article I, Section 8 of the US Constitution to "promote the Progress of Science and useful Arts…" It grants a monopoly of fixed term to the inventor in exchange for publicly sharing knowledge of the invention. In 1873, a United States patent was issued to Louis Pasteur for a purified yeast cell. The Plant Patent Act of 1930, intended to foster agricultural innovation, carved out plants, indicating that they were an exception and could be patented. In 1980, Sidney Diamond, Commissioner of the Patent and Trademark Office, challenged the patentability of "oil-eating" *Pseudomonas* on the basis that, as bacteria, they were the products of nature.

In their 1980 findings in the case of Diamond v. Chakrabarty, the US Supreme Court, in a 5-4 decision, held that "the fact that micro-organisms are alive is without legal significance for the purposes of patent law" and that "anything under the sun made by man" is patentable. An avalanche of **biotechnology** patent applications and approvals followed this landmark decision, including the "Harvard Mouse" (1988), the first transgenic animal, and genetically engineered crops (1990); only Canada prohibits patents on higher life forms, such as mice. However, in June 2013, the US Supreme Court held that naturally occurring DNA sequences were ineligible for patents.

SEE ALSO: Artificial Selection (Selective Breeding) (1760), Microbial Fermentation (1857), Biotechnology (1919), Plasmids (1952), Genetically Modified Crops (1982), Deepwater Horizon (BP) Oil Spill (2010).

A 2013 Supreme Court decision disallowing the patenting of human genes involved a breast cancer test based on the presence of a defective $BRCA_1$ gene. This decision could also affect other naturally occurring compounds, including those isolated from plants, proteins from human or animal sources, and microorganisms from the soil or sea.

Genetically Modified Crops

The use of genetically modified crops (GMC) is highly controversial, emotional, political, and economic, and pits scientific advances in **biotechnology** against what some believe to be major societal and health risks. The US is the largest producer of GMC. Many of its most respected scientific and health groups have concluded that GMC are "substantially equivalent" to non-GMC (conventional) foods and do not require specific "GM" (genetically modified) labeling. By contrast, some of the twenty-eight European Union member states are not convinced that these foods are safe and have resisted their importation and opposed growing GMC to meet the nutritional needs of their own people.

GM is based on the 1947 observation of genetic recombination—that is, the natural transfer of DNA between organisms. GMC involves altering the genetic makeup of plants by inserting one or more genes into the plant's genetic pool to enhance traits deemed desirable. This is most commonly accomplished by using a biolistic device (gene gun) that shoots DNA into the plant under high pressure, or via agrobacteria, which are bacterial plant parasites that naturally transfer genes. The most common of these agrobacteria is *Bacillus thuringiensis* or Bt toxin, a naturally occurring pesticide that reduces the need for chemicals. In 1982, **tobacco** was the first GMC to have been conferred resistance to herbicides, and it continues to be the most widely used "model plant" to study plant genetics. The first commercially available GMC in the US (1994) was the Flavr Savr tomato, which had a longer shelf life. The most common GMCs are corn, papaya, and soy.

Critics seeking to halt or limit the use of GMC have raised issues about their safety: their ability to cause allergic reactions, to potentially contaminate non-GM crops accidentally (with the creation of "superweeds"), and to disrupt ecodiversity. In addition, critics have voiced concern about perceived undue corporate control over the farmers' use of their products—90 percent of the world's GM seed patents are held by Monsanto Company. Countering these arguments is the production of safe crops that are resistant to viral diseases, tolerant to drought and frost, and increase the food supply and nutritional content in resource-poor developing countries.

SEE ALSO: Tobacco (1611), Population Growth and Food Supply (1798), Mendelian Inheritance (1866), Biotechnology (1919), Green Revolution (1945), Bacterial Genetics (1946).

A conceptual image of a genetically modified organism (GMO): corn growing in a pea pod.

HIV and AIDS

Luc Montagnier (b. 1932), **Robert Gallo** (b. 1937),
Françoise Barré-Sinoussi (b. 1947)

In 1981, an increasing number of gays and intravenous drug users had a marked deficiency of white blood cells, an essential component for the immune system—a condition later called *acquired immunodeficiency syndrome* or AIDS. As cases rapidly spread globally, laboratories sought to determine its cause. Some fiercely competitive scientists eagerly sought priority for this discovery, and nowhere was this competition more intense than between Robert Gallo and Luc Montagnier.

In 1976, Gallo and his colleagues at the National Cancer Institute of the National Institutes of Health were the first to successfully grow T-cells (a type of white blood cell), and discover HTLV, the first retrovirus identified in humans in 1981. In May 1984, Gallo published a series of papers in the prestigious journal *Science*, reporting that he had isolated a related retrovirus, HTLV-III, and that it caused AIDS. In that same issue of *Science*, Montagnier, at the Pasteur Institute in Paris, described LAV, a virus he isolated from an AIDS patient and described that its role in AIDS "remains to be determined."

Priority for discovering the viral cause of AIDS was not only the subject of an acrimonious dispute between scientists but also international contention between the governments of the United States and France that required their respective presidents—Ronald Reagan and François Mitterrand—to resolve. At issue was which government would be awarded a patent for testing and detecting the virus. It was decided, in a Solomonic tradition, that each individual would be given equal billing for the discovery, patent royalties would be divided equally, and that the virus be given a neutral name: *human immunodeficiency virus* or HIV.

In 2008, the Nobel Prize was jointly awarded to Montagnier and his colleague Françoise Barré-Sinoussi, but not Gallo, a decision that "surprised" Montagnier. It is now generally (but not universally) agreed that although Montagnier's laboratory was first to isolate HIV, Gallo was first to attribute the cause of AIDS to HIV and generated the background science, making the discovery of HIV possible. There were an estimated 34 million living individuals with AIDS in 2013.

SEE ALSO: Adaptive Immunity (1897), Viruses (1898), DNA as Carrier of Genetic Information (1944), Viral Mutations and Pandemics (2009).

The discovery that human immunodeficiency virus (HIV, shown) is the cause of AIDS led to the transformation of AIDS from a death sentence into a chronic, treatable disease.

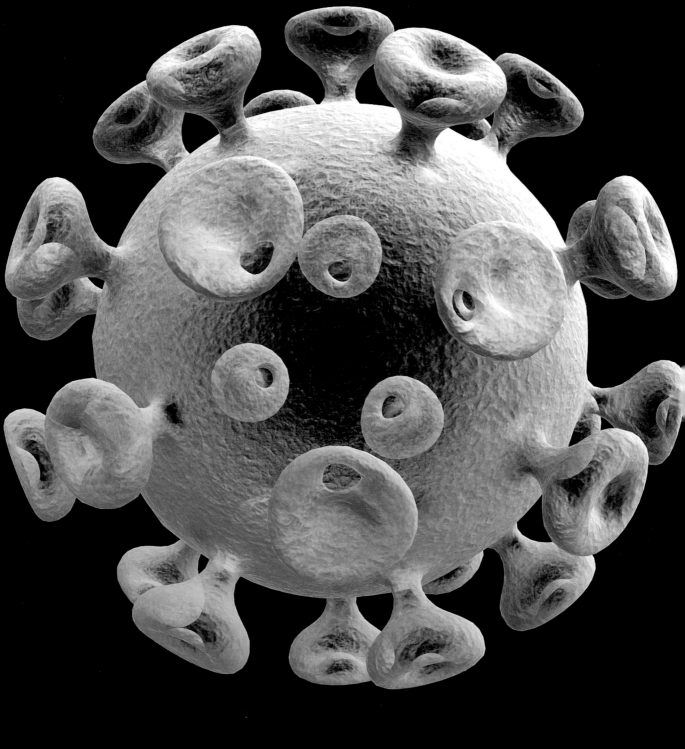

Polymerase Chain Reaction

Kary B. Mullis (b. 1944)

MASS-PRODUCING DNA. Starting with a very limited quantity of DNA, which may be contained in an impure sample, and using a test tube, some basic reagents, and a source of heat, the polymerase chain reaction (PCR) makes possible the production of millions of purified copies of DNA in only several hours. Prior to the introduction of this procedure, reproduction of DNA was difficult to perform, involved its cloning in bacterial cells, and consumed weeks to complete. PCR was developed in 1983 by Kary Mullis, an American biochemist working at Cetus, a California biotechnology company. In 1991, the PCR patent was sold for $300 million, and two years later Mullis was a co-recipient of the 1993 Nobel Prize.

The PCR procedure involves three primary steps, conducted at different temperatures: In the first step, the double-stranded DNA sample is subjected to a high temperature that causes it to split into two pieces of single-stranded DNA. Each strand serves as a template of the sequence of the DNA to be copied. After the addition of a primer, the polymerase enzyme (Taq) moves along the template, reading it and assembling a copy of the double-stranded DNA molecule. This is repeated thirty to forty times in an automated cycler, with the number of copies of the template increasing in an exponential manner with each cycle.

The applications of PCR range from molecular biological research to such applied forensic applications as crime-scene fingerprint analysis. More specifically, PCR has been used to create transgenic animals as models of human disease, to diagnose genetic defects, to detect the AIDS virus in human cells, to establish paternal relationships, and in criminal investigations to link persons of interest to samples of blood and hair. Evolutionary biologists have been able to generate large quantities of DNA from trace amounts found in fossil remains and from a 40,000-year-old frozen woolly mammoth. PCR analysis, for example, has revealed that red pandas are more closely related to raccoons than to great pandas.

SEE ALSO: DNA Polymerase (1956), HIV and AIDS (1983), DNA Fingerprinting (1984), De-Extinction (2013).

The Southern blot method, a common laboratory procedure, is used for the detection of a specific DNA sequence in a DNA-containing sample. Applications include showing genetic relationships, such as to establish paternity, or **DNA** *fingerprinting. The method was named after its inventor, the British biologist Edwin Southern (b. 1938).*

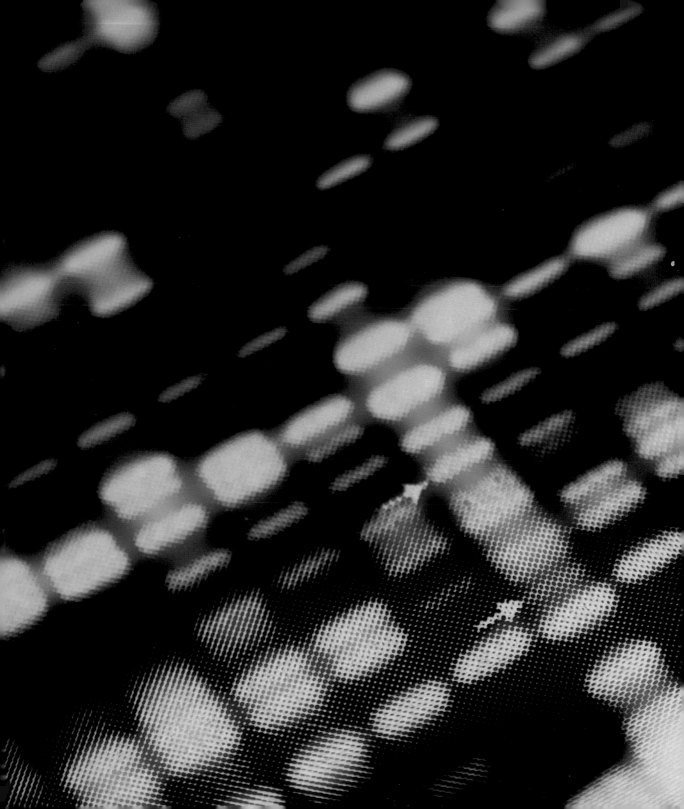

DNA Fingerprinting

Alec Jeffreys (b. 1950)

Since the early years of the 1900s, fingerprint evidence has served as one of the most common methods of crime-scene investigation and has been used to solve more crimes than any other procedure. Although there appears to be little doubt about the uniqueness of fingerprints (apart from identical twins), questions have been raised about the ability of examiners to accurately interpret them, particularly those unintentionally left at a crime scene.

In 1984, Alec Jeffreys, a genetics professor at the University of Leicester, unexpectedly found similarities and differences between the DNA of a technician's family members while examining X-ray films of DNA. Three years later, his DNA fingerprinting methods were commercialized and are now not only used in criminal investigations but also for paternity determination, for identification of victims of catastrophic events (such as 9/11), to match organ donors, and to determine livestock pedigree. In 1992, DNA evidence was used to establish that Nazi doctor Josef Mengele was buried in Brazil under a fictitious name.

Forensic laboratories can use blood (the most accurate), semen, saliva, hair, or skin, since identical DNA is present in every cell in the body. Of the total DNA, 99.9 percent is identical among us and only 0.1 percent unique. There is an estimated one in 30 billion chance that two persons, who are not monozygotic twins, will have identical DNA fingerprints. DNA fingerprinting—also called genetic fingerprinting, DNA profiling, and DNA typing—is based on analyzing minisatellites, which are certain DNA sequences that do not contribute to gene function and are recurrently repeated within the gene. DNA is extracted from cell samples, purified, and placed on a gel subject to an electric current (electrophoresis).

United States courts have generally accepted the reliability of DNA analysis, and the results can be included as evidence. Questions have been raised about its accuracy, the cost of testing, and the ability of technicians to avoid sample contamination or to accurately analyze or interpret the findings. Ethical issues have been raised as to whether samples taken without consent violate US Constitutional protections of privacy; in 1985, the US Supreme Court decided it did not.

SEE ALSO: DNA Polymerase (1956), Polymerase Chain Reaction (1983), Human Genome Project (2003), Human Microbiome Project (2012).

An example of DNA fingerprinting, in which ten different individuals are tested for six loci (locations of a DNA sequence on a chromosome).

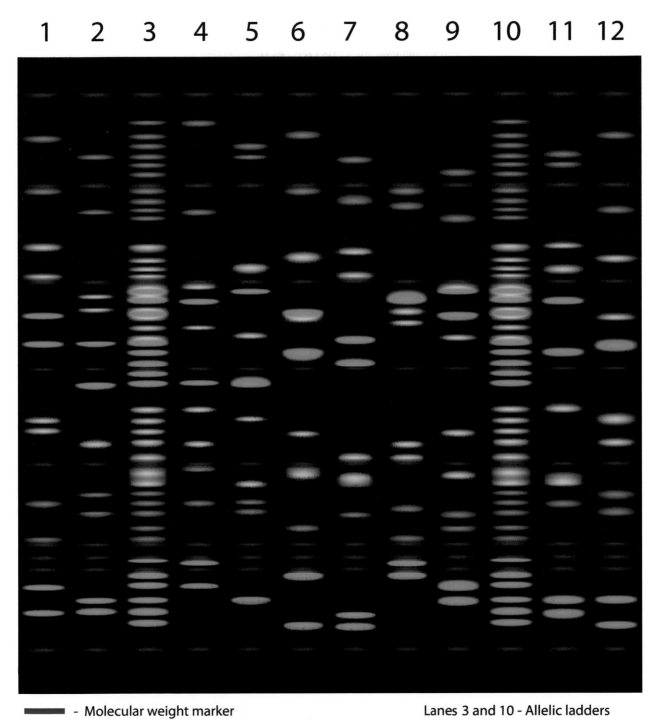

1 2 3 4 5 6 7 8 9 10 11 12

▬▬▬ - Molecular weight marker

Lanes 3 and 10 - Allelic ladders

6 loci are tested : 2 ▬▬▬ ; 2 ▬▬▬ ; 2 ▬▬▬

Genomics

Francis Crick (1916–2004), **Frederick Sanger** (1918–2013), **Rosalind Franklin** (1920–1958), **James D. Watson** (b. 1928), **Thomas Roderick** (1930–2013), **Walter Gilbert** (b. 1932), **Craig Venter** (b. 1946)

THE BIG PICTURE. Genomics is a subset of genetics. Whereas genetics looks at individual genes, genomics views an entire system, including the mapping, sequencing, and functional analysis of the genome—the entire genetic material of an organism. The foundations for genomics were laid in the 1980s with advances in technology that enabled the sequencing of DNA, and the collection and analysis of the mountains of data generated. The field matured in the 1990s, with progress continuing to this day. The name *genomics* was coined in 1986 by Thomas Roderick, a geneticist at Jackson Laboratory in Bar Harbor, Maine.

Mapping and sequencing the complete genome of an organism required advances in analytical tools. In 1953, the DNA structure was determined by James Watson, Francis Crick, and Rosalind Franklin and found to consist of four nucleotide bases: adenine, guanine, cytosine, and thymine. The laborious task of sequencing DNA—determining the precise order of these nucleotides within the DNA molecule—was expedited by more rapid methods introduced by Frederick Sanger and Walter Gilbert in the 1970s (for which they received the 1980 Nobel Prize) and by semi-automated and fully automated DNA sequencing machines, in 1986 and 1987, respectively. In 1995, the genome of the first free-living organism, *Haemophilus influenzae*, with 1.8 billion base pairs, was published by Craig Venter's laboratory at the Institute for Genomic Research, followed in 2003 by the mapping the human genome and its 3.3 billion base pairs.

A parallel challenge was the storage of this tremendous volume of data in an accessible and readily retrievable format, and the capability to analyze and interpret these data looking for functional correlations. This need has been satisfied by the field of **bioinformatics**, where information processing systems have been developed to scan DNA sequences and search information libraries for genes that correlate with certain functions, including disease states. Comparative genomics directly compares the DNA sequences between and among organisms. Fully 99.9 percent of nucleotide bases are exactly the same in all humans, and a high degree of similarity exists between the DNA of humans and insects, supporting the genetic code's early origin in the history of life.

SEE ALSO: The Double Helix (1953), Bioinformatics (1977), Human Genome Project (2003), Human Microbiome Project (2012).

This eukaryotic phylogenetic tree, showing the genetic relationship between plants, fungi, and animals, is based on genomic analysis.

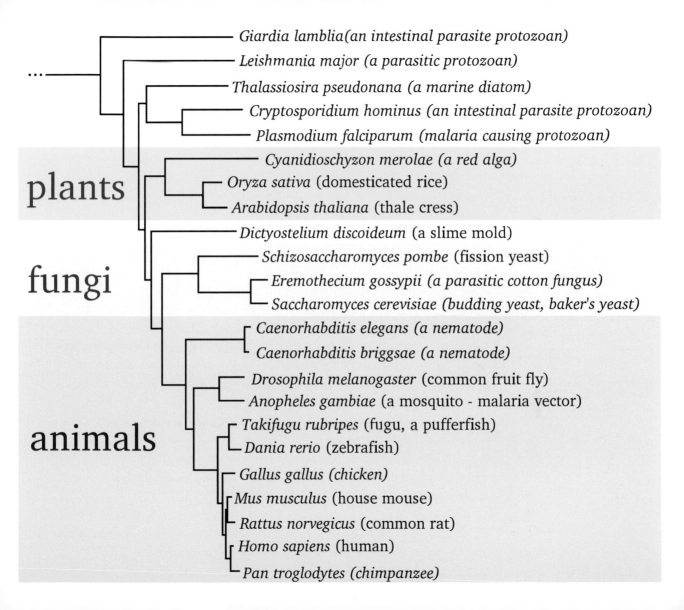

Mitochondrial Eve

Allan Wilson (1934–1991), **Rebecca L. Cann** (b. 1951), **Mark Stoneking** (b. 1956)

A 1987 paper in the prestigious journal *Nature* reported that "all mitochondrial DNA stems from one woman" and that she had lived in Africa some 200,000 years ago. The paper, authored by Rebecca L. Cann, Mark Stoneking, and their doctoral advisor, Allan Wilson, at the University of California, Berkeley, aroused intense interest and controversy for many reasons, and it continues to do so.

The authors referred to the samples they analyzed as "mitochondrial DNA," whereas the press dubbed them "mitochondrial Eve"—far more memorable, but also subject to misinterpretation. This Eve was not the single and only woman living at the time, as was said of the Eve of Genesis. In addition, the literal Biblical interpretation computed the age of humans to a time measured in thousands of years, not 200,000 years. Moreover, many evolutionists believed that humans evolved in separate parts of the world at about the same time, rather than the "Out of Africa" theory, in which **Anatomically Modern Humans** originated in Africa and then migrated worldwide.

Cann and her colleagues analyzed mitochondrial DNA (mtDNA) and not nuclear DNA (nDNA), the latter responsible for transmitting the color of our eyes, racial characteristics, and susceptibility of certain diseases; mtDNA only codes for manufacturing proteins and performing other mitochondrial functions. Present in all cells of our body, nDNA is a merger of our mother's and father's DNA (recombination), whereas mtDNA is derived virtually exclusively from the maternal side with few if any mtDNA contributed from the sperm. Closely related individuals have almost identical mtDNA, with occasional mutations arising over thousands of years. It is assumed that the fewer the number of mutations, the shorter the period of time since common ancestors diverged.

Proponents of mitochondrial Eve do not suggest that this Eve was the first woman or only woman living at the time. Rather they estimate that some catastrophic event occurred, dramatically reducing Earth's population to some 10,000–20,000, and that only this Eve had an unbroken line of female descendants. Eve was said to be the most recent common ancestor from whom all living humans descended.

SEE ALSO: Anatomically Modern Humans (c. 200,000 BCE), Mitochondria and Cellular Respiration (1925), Domains of Life (1990).

Adam and Eve, completed after 1536 by the German Renaissance painter Lucas Cranach the Younger (1515–1586).

Depletion of the Ozone Layer

Since the 1970s, there has been a decline of 4 percent each decade in the total volume of ozone in the ozone layer. Such depletion increases exposure to UV-B radiation, significantly affecting all inhabitants of the biological world. Overexposure to UV-B has been associated with an increased risk of the skin cancer malignant melanoma, cataracts, and a weakened immune system. A reduction in crop yield is associated with changes in the distribution and metabolism of nutrients in plants, along with changes in developmental phases. Even marine life is affected, as evidenced by a reduction in phytoplankton production, the lowest level in the aquatic food chain. Damage also occurs during the earliest developmental stages of fish, shrimp, crabs, and amphibians.

Ozone is a rare, naturally occurring gas, about 90 percent of which is found in a layer of atmosphere that begins 6–10 miles (10–17 kilometers) above the Earth's surface and extends up some 30 miles (50 kilometers). This atmospheric layer is the stratosphere, and the ozone found there is called the ozone layer. Sunlight is the most common form of ultraviolet (UV) radiation, and ozone absorbs UV-B radiation so that only a small fraction reaches the Earth's surface.

Associated with ozone depletion has been a one-third decrease in the ozone layer above the polar caps, called the ozone hole, during the Antarctic springtime (September through early December). The primary causes of this depletion have been attributed to chlorofluorocarbons (CFCs) and hydrofluorocarbons (HCFCs), formerly used as propellants in spray cans, refrigerants, in foam and insulating products, and as electronic solvents. These volatile substances are carried to the stratosphere where they are broken down by UV radiation, releasing chlorine atoms, which react with ozone (O_3), causing its breakdown and depletion.

International awareness of CFC's role led to the 1987 adoption of the Montreal Protocol on Substances that Deplete the Ozone Layer, an international treaty calling for a reduction in the use of CFCs and other ozone-depleting compounds; as of 2010, there were 190 nation signatories. It has been estimated that if CFCs cease to be used, normal levels of ozone will be restored by 2050.

SEE ALSO: Global Warming (1896), Food Webs (1927), *Silent Spring* (1962), Sustainable Development (1972), Skin Color (2000).

In this NASA satellite image from October 1, 1998, the hole in the ozone layer over Antarctica is shown in purple.

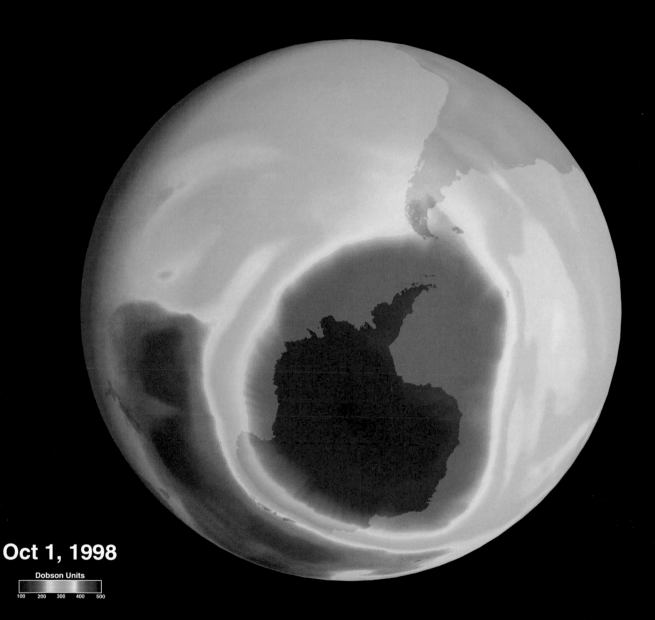

Oct 1, 1998

Dobson Units

100 200 300 400 500

Domains of Life

Carl Linnaeus (1707–1778), **Ernst Haeckel** (1834–1919),
C. B. van Niel (1897–1985), **Roger Y. Stanier** (1916–1982),
Carl Woese (1928–2012), **George E. Fox** (b. 1945)

An impetus for classification came during the seventeenth century when new plants and animals were arriving in Europe. In 1735, Carl Linnaeus, a pioneer in the science of taxonomy (also called *systematics*), developed a hierarchical system of biological nomenclature in which the highest rank, inclusive of all lower levels, was the *kingdom*, and these were two: animal and vegetable (plant). With the growing realization that unicellular organisms were unaccounted for, in 1866 Ernst Haeckel proposed the addition of a third kingdom, *Protista*.

In the 1960s, Roger Y. Stanier and C. B. van Niel devised a four-kingdom classification system based on the distinction between prokaryotic and eukaryotic cells, with the latter having a cell membrane enclosing its nucleus. Furthermore, they proposed a higher and more inclusive rank termed *superdomain* or *empire*. The Empire Prokarya encompassed the Kingdom Monera (bacteria), and the Empire Eukarya included the Kingdoms Plantae, Animalia, and Protista.

Until the mid-1970s, all classifications were based on the outward appearance of cells, namely their anatomy, morphology, embryology, and cell structure. In 1977, Carl Woese and George E. Fox at the University of Illinois at Urbana-Champaign classified organisms based on a comparison of their genes at a molecular level. In particular, they compared the nucleotide sequences in a subunit of ribosomal rRNA, the molecules that undergo evolutionary changes. In 1990, they introduced the concept of three domains of cellular life: the Domain Archaea, a disparate collection of prokaryotic organisms, among the most ancient found on Earth and capable of adapting to extreme environments (*extremophiles*); the Domain Bacteria; and the Domain Eukarya, which was subdivided into Kingdoms **Fungi** (yeasts, molds), Plantae (flowering plants, ferns), and Animalia (vertebrates, invertebrates). More recently, their Protista Kingdom has been subdivided into more discrete kingdoms. The final chapter on classification has not been written, with systems proposed that contain two to eight kingdoms.

SEE ALSO: Prokaryotes (c. 3.9 Billion BCE), Eukaryotes (c. 2 Billion BCE), Fungi (c.1.4 Billion BCE), Linnaean Classification of Species (1735), Endosymbiont Theory (1967), Protist Taxonomy (2005).

The rainbow colors in the Grand Prismatic Spring in Yellowstone National Park, Wyoming—the world's third largest hot spring—result from resident thermophilic microbes (extremophiles of the Kingdom Archaea) that favor temperatures ranging from 1,880°F (870°C) at the center to 1,470°F (640°C) at the rim.

Sense of Smell

Richard Axel (b. 1946), **Linda B. Buck** (b. 1947)

Animals use their sense of smell to locate food, mark territory, identify their own offspring, and detect the presence and receptivity of potential mates. Of all the animals with a keen sense of smell, bloodhounds—sometimes referred to as "a nose with a dog attached"—stand out. With European ancestors that can be traced back over a thousand years, bloodhounds have been used for their tracking ability to find lost persons, missing children, and escaped prisoners, and are so reliable that their findings are admissible in a court of law. They have been known to follow trails longer than 130 miles (210 kilometers), tracing a mixture of scents emitted by the breath, sweat, and skin that is up to 300 hours old. Humans can distinguish among thousands of different odors. The ability of bloodhounds to differentiate odors is at least 1,000 times greater than humans, whose sense of smell is considered to be relatively simple. The olfactory system is most highly developed in carnivores that hunt.

These remarkable tracking abilities arise from the olfactory organ, located in the nose and consisting of modified nerve cells that have several tiny hairs on their surface. The hairs, which contain olfactory receptors (OR), stick out from the epithelium in a small area on the roof of the nasal cavity into a bed of mucus. These receptors are proteins that detect odorant molecules in the air, located on OR cells. As the animal breathes, odorant molecules in the air dissolve in the mucus and bind to the OR, which triggers an electrical signal sent to the olfactory bulb located in the brain for interpretation. OR cells that are selective for different odorants are distributed throughout the nasal cavity.

In 1991, Linda Buck and Richard Axel at Columbia University investigated the olfactory system of laboratory rats at the molecular level, studies for which they were awarded the 2004 Nobel Prize. They discovered a family of over 1,000 genes (3 percent of all human genes) that encode an equivalent number of olfactory receptors (OR). They demonstrated that each OR cell contains only a single OR that is specialized to recognize only a few odors.

SEE ALSO: Nervous System Communication (1791), Neuron Doctrine (1891), Action Potential (1939), Pheromones (1959).

Bloodhounds, sometimes referred to as "a nose with a dog attached," live up to their remarkable reputation for indefatigability, tracking individuals over tens of miles.

Leptin: The Thinness Hormone

Douglas L. Coleman (b. 1931), **Rudolph L. Leibel** (b. 1942),
Jeffrey M. Friedman (b. 1954)

Mutant obese mice with a voracious appetite appeared at random in a mouse colony at Jackson Laboratory in Bar Harbor, Maine, in 1950. Such animals were found to have a (ob) genetic mutation. During the 1960s, Douglas Coleman found mice with both diabetic (db) and obese (ob) genetic mutations. In 1992, after genetic inbreeding and testing, Coleman, collaborating with Rudolph Leibel, theorized that they had produced obese (ob) mice that were lacking a protein hormone that modulated food intake and body weight and (db) mice that could produce the hormone but lacked the receptor to detect its signal.

Working at Rockefeller University in 1994, Leibel and Jeffrey Friedman discovered the gene and the hormone that could suppress food intake and body weight and named the hormone *leptin* (Greek = "thin"). The obese mice had a genetic mutation that prevented them from producing functional leptin. This hormone is a protein of 167 amino acids that is manufactured primarily in fat cells and acts in a multifaceted manner in the hypothalamus. It blocks neuropeptide Y (NPY), a natural feeding stimulant released by cells in the gut and in the hypothalamus. NPY was earlier found to be a key component in appetite regulation, with small doses stimulating eating, and destruction of NPY nerves causing a loss of appetite. In addition, leptin promoted the synthesis of alpha-melanocyte-stimulating hormone (MSH), a protein hormone produced in the brain that may suppress appetite (but far better established is its role in skin pigmentation).

It has been proposed that leptin is involved in the body's adaption to starvation. Under normal physiological conditions, when body fat is reduced, plasma levels of leptin are reduced, leading to an increase in feeding and a reduction in energy expenditure that continue until normal fat mass is restored.

It was hoped that leptin might provide an answer to treating obese individuals in weight reduction programs. Trials in humans revealed that large doses, at frequent intervals, produced only modest weight loss. Because leptin is a protein, it must be injected and cannot be given by mouth or it faces inactivation by stomach enzymes. The search continues.

SEE ALSO: Human Digestion (1833), Thrifty Gene Hypothesis (1962), Hypothalamic-Pituitary Axis (1968).

One of a series of sumo wrestlers drawn by Utagawa Kuniyoshi (1797–1861), a master of Japanese woodblock prints and painting. Competitors in sumo wrestling have no maximum weight limits and have been known to weigh in excess of 500 pounds (225 kilograms).

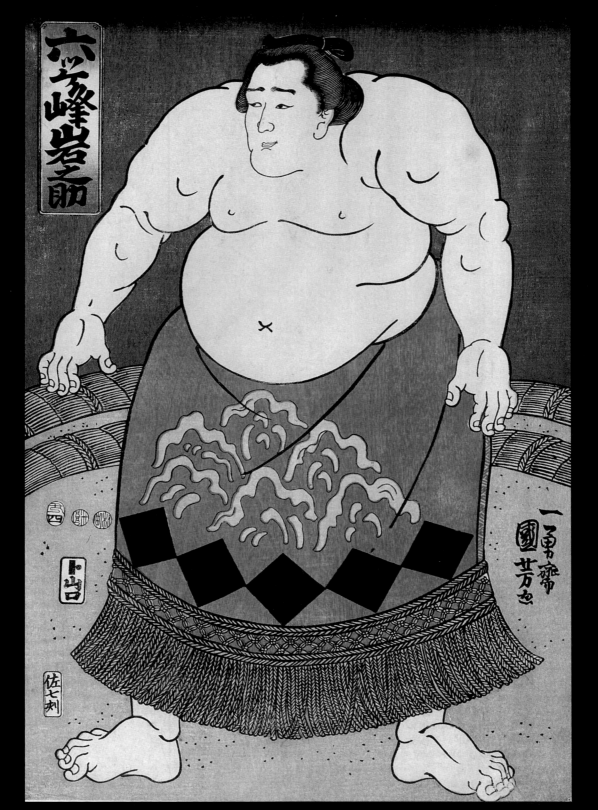

Skin Color

Skin is the body's largest organ, weighing about 6 pounds (13 kilograms), and it's the primary site of interaction with the outside world. Its color has been the source of cultural divisions. Although we tend to focus upon its outward appearance, the skin plays a number of important functions in addition to protecting against mechanical injuries, chemicals, and microbes. It also helps regulate water balance and body temperature, stores fats, and produces hormones and vitamin D_3.

Melanin, primary determinant of skin color in humans and also present in the hair and iris of the eyes, is produced by melanocytes, which are found in the bottom layer of the epidermis. Upon exposure to ultraviolet radiation (UVR), melanin production is increased, causing the skin to tan, and it has been long believed that dark skin pigmentation protects the body against the sun's harmful UVR.

In 2000, anthropologist Nina Jablonski, then at the California Academy of Sciences in San Francisco, and her husband George Chaplin proposed that skin color was an evolutionary adaptation to different levels of UVR to which humans were exposed as they migrated over millennia. They formulated a theory based on their analysis of data generated by NASA's Total Ozone Spectrometer, which, in 1978, measured the UVR reaching the Earth's surface in more than fifty countries around the globe—a level that was progressively weaker farther from the equator. They observed a correlation showing the weaker the UVR, the lighter the skin.

The earliest humans had dark hair covering lightly pigmented skin. By the time they moved to East Africa some 1.2 million years ago and lived closer to the equator, they had become functionally hairless and had acquired dark skin pigmentation. Jablonski and Chaplin hypothesized that as they migrated, skin color changes were required to balance the harmful effects of excessive radiation against the competing need to have sufficient UVR for vitamin D_3 synthesis; this vitamin is required to maintain sufficient blood levels of calcium and phosphorus, which promote bone growth, and is needed for healthy reproduction.

SEE ALSO: Anatomically Modern Humans (c. 200,000 BCE), Depletion of the Ozone Layer (1987).

Upon exposure to ultraviolet radiation (UVR) from sunlight, the skin tans. Over the generations, society has had mixed feelings about the desirability of tanned skin. For light-skinned people, deeply tanned skin is often in vogue, but is also considered an invitation to skin cancer and prematurely aged skin.

Human Genome Project

Thomas Hunt Morgan (1866–1945), **Alfred Sturtevant** (1891–1970),
Francis Crick (1916–2004), **Frederick Sanger** (1918–2013),
James D. Watson (b. 1928)

It was the biological counterpart of putting a human on the moon, the largest biological project ever undertaken. In the late 1980s, the idea was conceived of mapping the human genome, its entire DNA and genes, and in 2003 it was 99 percent completed. The primary objective of the Human Genome Project (HGP) was to find the genetic basis for diseases (such as cancer) and to determine individual variations in our human genetic code that make some of us more susceptible to certain diseases. An understanding of these diseases at a genetic level could potentially lead to the development of highly specific biopharmaceuticals. By 2013, some 1,800 disease-related genes had been reported and 350 biotechnology-based products were in clinical trials.

Jointly funded by the US Department of Energy and the National Institutes of Health, HGP was initiated in 1990 as an international research effort that was projected to be completed in fifteen years. Two years ahead of schedule, in 2003, the project was essentially completed—the human genome had been sequenced—at an approximate cost of $3.8 billion dollars; in 2006, the sequence of the last chromosome was published. Of our 23 chromosomal pairs, 22 are non-sex determining. A human has some 20,000–25,000 genes (about the same number as a mouse), composed of a total of 3.3 billion base pairs. (By contrast, the fruit fly has 13,767 genes.) The DNA of all living organisms consists of the same four base pairs, and it is their specific order that determines whether the organism will be a human, fruit fly, or plant.

The fruits of this effort began almost 100 years earlier. The world's first genetic map was of *Drosophila melanogaster* (fruit fly) and the subject of Alfred Sturtevant's 1911 doctoral dissertation, working under the direction of Thomas Hunt Morgan at Columbia University. In 1953, James Watson and Francis Crick described **the double helix** structure of DNA and the nature of the base pairs of adenine, thymine, guanine, and cytosine. In 1975 Frederick Sanger developed a DNA sequencing technique. Morgan, Watson, Crick, and Sanger were Nobel Prize winners.

SEE ALSO: Deoxyribonucleic Acid (DNA) (1869), Genes on Chromosomes (1910), DNA as Carrier of Genetic Information (1944), The Double Helix (1953), Bioinformatics (1977), Genomics (1986), Human Microbiome Project (2012).

An ultraviolet laser beam passes through a cuvette used for measuring DNA.

Protist Taxonomy

Ernst Haeckel (1834–1919), **Robert H. Whittaker** (1920–1980)

Biologists are classifiers, but as experience has shown over almost two centuries, protists defy a simple classification that has withstood the test of time. To the ancient classification that all living organisms were either plants or animals, unicellular organisms were added as a third kingdom, which in 1866, Ernst Haeckel called *protists*, referring to their primitive forms. In 1959, the American plant ecologist Robert Whitaker proposed a five-kingdom and later a four-kingdom classification, one of which was Protista.

Until recently, protists were classified as one of the four kingdoms within the Domain Eukaryota, which consist of organisms that have membranes enclosing a true nucleus and intracellular organelles. Based on ultrastructure (organelles), biochemistry, and genetics, members of the plant, animal, and fungal kingdoms are considered monophyletic—that is, each group is derived from a single ancestor and all its descendants.

There are more than 200,000 species of protists inhabiting an environment in which water is present some or all the time. They are usually unicellular and differ in size and shape, method of reproduction, motility, and how they obtain their nutrition. But, based on DNA and ultrastructural studies, the protists are even more diverse than previously believed, with some more closely related to other kingdom members than to other protists. Protists do not share a common lineage (they are polyphyletic), and Protista is not really a kingdom but rather a catchall category of eukaryotic organisms that are *not* plants, animals, or fungi. Nevertheless, the name *protist* continues to be used as a shorthand designation for such organisms.

In 2005, the ecologist Sina M. Adl at Dalhousie University in Canada, proposed a classification that forgoes seeking hereditary relatedness and has informally placed all protists into five supergroups, with each subdivided into groups based on how they move and how they obtain their nutrition. In addition, an even simpler and more readily understood classification for protists is based on the following: Protozoa or animal-like, where members ingest food and are motile; Algae or plantlike, including organisms that manufacture their own food by **photosynthesis**; and Fungilike, which digest foods from their environment.

SEE ALSO: Eukaryotes (c. 2 Billion BCE), Leeuwenhoek's Microscopic World (1674), Linnean Classification of Species (1735), Photosynthesis (1845), Endosymbiont Theory (1967), Domains of Life (1990).

A stromatolite reef in Cuatro Ciénegas, Mexico. Stromatolites, among the world's oldest fossils, are formed by the accumulation of multiple layers of prokaryotic cyanobacteria (formerly called blue-green algae) and protists. Before their source was understood, they were referred to as "living rocks."

Induced Pluripotent Stem Cells

James Thomson (b. 1958), **Shinya Yamanaka** (b. 1962)

In recent years, considerable scientific and public interest has been focused on stem cells, both because of their medical potential for tissue and organ transplantation and for the treatment of such diseases as diabetes, Alzheimer's, and Parkinson's. But the use of stem cells involves the destruction of human embryos and the possibility of human **cloning**, which has generated a firestorm of ethical and political controversy derailing further research. The successful development of induced pluripotent stem cells (iPS) was hoped to maintain the intended benefits while abating the criticism.

Stem cells are formed days after the development of the mammalian embryo during the blastocyst stage, which is equivalent to the blastula stage in mammals. Embryonic stem cells are undifferentiated cells capable of dividing to produce additional stem cells; they can differentiate into any of the body's three cell lineages or germ layers (endoderm, mesoderm, ectoderm) and form specialized cells of any cell type. In addition, there are adult stem cells, found primarily in the bone marrow and umbilical cord blood, which can repair and replenish adult cells and tissues.

In 1998, James Thomson, at the University of Wisconsin-Madison, successfully isolated human embryonic stem cells, a considerable but highly contentious scientific achievement because of their source. These issues were largely overcome in 2006, when Shinya Yamanaka, at Kyoto University in Japan, was able to reprogram adult mouse fibroblasts (skin cells) to produce iPS. In 2007, Yamanaka went a step further and produced iPS from human adult skin cells, a feat replicated by Thomson later that year. In this process, multiple transcription factors are added to the skin cells; transcription factors are proteins that control the flow (transcription) of genetic information from DNA to messenger RNA. Yamanaka was a co-recipient of the 2012 Nobel Prize.

The early enthusiasm for the medical possibilities associated with the use of iPS has been tempered because of differences between embryonic stem cells and iPS, and the potential of iPS to cause cancers. To date, they have not been approved for clinical use in the United States.

SEE ALSO: Germ-Layer Theory of Development (1828), The Immortal HeLa Cells (1951), Cloning (Nuclear Transfer) (1952).

Induced pluripotent stem cells are genetically reprogrammed adult cells. They serve as models of disease and can be potentially used for cell-based therapies to treat such conditions as Alzheimer's disease, spinal cord injuries, diabetes, burns, and osteoarthritis.

Viral Mutations and Pandemics

The 1918–1919 influenza pandemic infected 500 million people worldwide—one-fifth of the world's population—and claimed between 20 and 100 million lives; 675,000 deaths occurred in the United States, over ten times the number that died from battle-related injuries during the recently concluded World War I. The same viral strain (H1N1) that caused this Spanish flu reappeared seven decades later as swine flu, infecting almost 20 percent of the world's population, and causing 200,000–300,000 deaths from 2009–2010. The H1N1 viral strain is not your typical seasonal flu.

Viruses have a strong attachment to living organisms, including humans, because they can only reproduce in a host cell. The virus attaches its proteins to the surface of a host cell, injects genetic material (DNA or RNA) into the host, and then pirates the host's cell machinery to make more viruses, before moving on to another host cell. To prevent reinfection by the same viral strain, the host mounts an immune response, producing antibodies that prevent that virus from binding to the host's surface. To survive, viruses mutate, changing their surface proteins, thus evading the host's immune defenses. Antibodies produced in response to a previous flu infection are powerless to protect against a newly mutated strain. Hence, new and different flu vaccines are required each season.

Viruses that replicate using DNA, such as smallpox, carefully check for any errors prior to reproducing its genetic code; therefore, they mutate slowly. By contrast, RNA viruses, such as influenza, bypass the time-consuming proofreading step when copying their genetic code; they mutate very rapidly, far faster than the host's immune response can keep pace with the emergence of new viral strains.

The flu virus can combine pieces of flu from avian, swine, and human sources. Pigs living snout to snout in pens readily pick up the virus from birds and humans and share it with their sty mates. Pigs harbor the mutated virus but are not, themselves, infected, although humans are. Such a mutated strain—the H1N1 Spanish flu and swine flu—was radically different from earlier strains; humans had acquired no immunity and were defenseless against it.

SEE ALSO: Adaptive Immunity (1897), Viruses (1898), DNA as Carrier of Genetic Information (1944).

This 1918 photograph depicts a nurse filling a pitcher from a fire hydrant and wearing a mask as protection against influenza during the Spanish flu pandemic of 1918–1919.

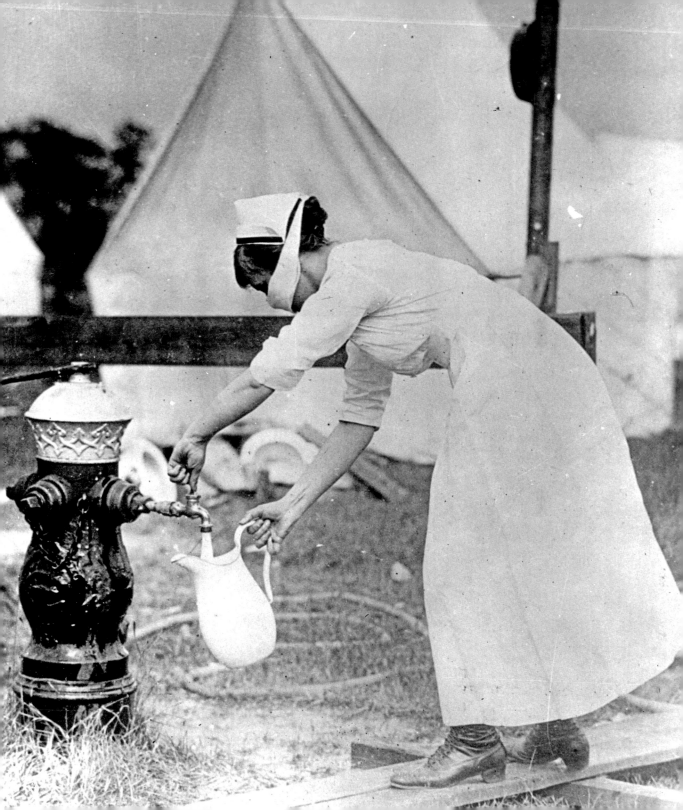

Deepwater Horizon (BP) Oil Spill

The largest accidental marine oil spill in history occurred between April and July 2010 and resulted from an explosion and the sinking of the BP oil rig Deepwater Horizon, causing the deaths of eleven workers. During the eighty-seven days, 4.9 million barrels (210 million gallons) of oil spilled into the Gulf of Mexico causing grave ecologic and economic damage to the Gulf States of Louisiana, Mississippi, Alabama, and Florida, along with its residents and cleanup workers. As of 2013, the criminal and civil settlement will cost BP over $42 billion.

The spilled oil contained 40 percent methane that can potentially suffocate marine life and create "dead zones," where oxygen is depleted from the water. Oil cleanup approaches included physical methods (skimmers, booms or floating barriers, controlled burning), chemical dispersants, and bioremediation with microbes. Some of these were beneficial, while others made a bad situation much worse. The oil dispersant Corexit contained cancer-causing ingredients and was found itself to be toxic to phytoplankton, coral, oysters, and shrimp. It caused mutations in shrimp, crabs, and fish, and produced respiratory and skin irritation, mental health problems, and liver and kidney damage in cleanup workers and residents. The dispersant also made oil sink faster and deeper into beaches. A 2012 study estimated that Corexit increased the toxicity of oil by 52 times. By contrast, bioremediation with the oil-eating microbe *Oceanospirillales* was highly effective. This naturally occurring bacterium appears to have evolved in the Gulf over millions of years and consumes oil from natural seepage. The concern that it would create "dead zones," which would be detrimental to marine plant and animal life, did not occur to the extent feared.

In March 1989, the oil tanker Exxon Valdez ran aground in Prince William Sound, Alaska, spilling 260,000–750,000 barrels (10 million gallons) of oil affecting 1,300 miles of coastline. The oiling of animal fur or feathers caused a loss of insulating capacity, leading to death by hypothermia. Among the mortalities were sea otters (1,000–2,800), harbor seals (300), sea birds (100,000–250,000), and bald eagles (247). Some estimates suggest that environmental recovery will take up to thirty years.

SEE ALSO: Population Growth and Food Supply (1798), Ecological Interactions (1859), Population Ecology (1925), Can Living Organisms Be Patented? (1980).

Oil spills can produce "dead zones" in the water, depleting oxygen and causing suffocation of marine life. In addition to the grave ecological damage are the economic consequences resulting from the absence of tourism on once-inviting beaches.

Translational Biomedical Research

Scientific research is typically classified as basic or applied. Basic research, usually carried out in academic institutions or research institutes, is theoretical and has a long-term vision for success. The rewards for significant discoveries are measured in publications in prestigious journals, academic promotions, and the accolades of the scientific community. Not infrequently, the focus of these basic research scientists is not on applications, and all too often, the research conducted is highly specialized. By contrast, the research conducted in commercial laboratories (biotech, pharmaceutical, agricultural, chemical) is multidisciplinary and has a short-term horizon with a commercial objective and practical utility.

To an increasing extent, the demarcation between basic and applied science is less distinct. Not all basic research is, nor need be, theoretical. With an increasing understanding of humans and microbes at molecular and biochemical levels, new drugs are being developed that target the specific cause of disease, including correcting genetic defects. When the basic research studies of academic scientists are funded by commercial companies, contractual agreements commonly provide the sponsors with priority rights in seeking patents for any discoveries. Moreover, some enlightened and innovative biomedical, chemical, and electronic companies have encouraged their scientists to pursue basic and undirected research, with no immediate commercial objective.

In recent years, there has been increasing governmental emphasis in Europe and the United States on *translational* research, in which basic scientific discoveries generated in the laboratory are directly applied to applications that benefit the health and well-being of society. Nowhere has this impetus for translational focus been greater than in the biomedical community. In December 2011, the National Institutes of Health established the National Center for Advancing Translational Sciences. Directed by the slogan "bench to bedside and back," translational medicine seeks to take potentially fruitful basic research findings—in such areas as **genomics**, transgenic animal models, structural biology, biochemistry, and molecular biology—and apply them as the foundation for clinical studies, which, if successful and refined, can serve as the basis for routine clinical practice.

SEE ALSO: Scientific Method (1620), Blood Pressure (1733), Germ Theory of Disease (1890), Tissue Culture (1902), Antibiotics (1928), Cloning (Nuclear Transfer) (1952), Monoclonal Antibodies (1975), Bioinformatics (1977), Genomics (1986).

The Sick Woman, *a painting by Dutch artist Jan Steen (c. 1625–1679). There were few effective medications available for use before the nineteenth century, with the benefits of many treatments attributable to the presence of a caring physician and the placebo effect.*

Albumin From Rice

Albumin is the most abundant plasma protein in mammals and constitutes 50–55 percent of plasma proteins in humans. Produced in the liver, it serves as a carrier in the blood for hormones, fat-digesting and absorbing bile salts, bilirubin, and **blood clotting** proteins. Albumin's most important function is to regulate blood volume by pulling water into the circulatory system—in particular, the capillaries. It is medically used as a plasma expander for the treatment of shock caused by excessive blood loss and severe burns, as well as in combat settings to stabilize the wounded until whole blood is available. Shock results from an inadequate supply of blood-carried oxygen from reaching the cells, causing them irreversible consequences. It is also used in the production of drugs and vaccines.

Human serum albumin (HSA) is extracted from human blood plasma, the liquid component of blood, but with an annual worldwide demand of some 500 tons (500,000 kilograms), natural sources are in short supply. Difficulties have been encountered producing synthetic or laboratory versions of HSA. More recent attempts to produce HSA have used the tools of genetic engineering. The challenge is to develop a high-yield/low-cost system, producing a product with a minimal risk of causing allergic reactions. Past efforts to grow albumin in potato plants and tobacco leaves have not been satisfactory because of low yields.

The rice genome was sequenced in 2005. In 2011, Daichong Yang and his research colleagues in Wuhan, China, have reported successfully growing HSA in rice (*Oryza sativa*) after using bacteria (*Agrobacterium*) to insert a gene for encoding HSA. The gene is activated during seed production, resulting in the albumin protein being stored in rice grain with nutrients used to nourish the germinating plant embryo. Comparison of the rice- and human-derived HSA revealed that they were chemically and physically identical, with the same 585-amino-acid sequence and three-dimensional shape, in addition to being biologically equivalent in rats. Around 0.1 oz (2.8 grams) of HSA were produced from 2.2 lbs (1 kg) of brown rice, making this process highly cost-effective, with a virtually unlimited supply.

SEE ALSO: Rice Cultivation (c. 7000 BCE), Blood Clotting (1905), Biotechnology (1919), Genetically Modified Crops (1982).

A green terraced rice field in Chiangmai, Thailand.

Human Microbiome Project

Joshua Lederberg (1925–2008)

Microbiome, a term coined by Joshua Lederberg in 2001, refers to the totality of microbes and collective genetic material present in or on the human body. In 2012, the Human Microbiome Project (HMP) found that, by far, the most populous inhabitants of the human body are not human cells, but rather microbes. The human body has ten times as many microbial cells as human cells; they constitute 1–3 percent of a human's total body mass, some 2–6 pounds (0.9–2.7 kilograms). Moreover, whereas the human genome has 22,000 protein-coding genes, bacteria have 360 times more (about eight million).

Completion of the **Human Genome Project** in 2006, which sequenced the entire human genome, provided researchers with a basis to differentiate human and microbial genes. In 2008, the US National Institutes of Health launched the HMP, a five-year study to assess the nature of the microbial population of the healthy human body, establish a reference database, and determine whether changes or differences in these populations predispose individuals to disease. Scientists found some 10,000 microbial species—mostly bacteria, but also protozoa, yeasts, and viruses—and as of June 2012, identified 81–99 percent of these. The largest numbers of microbes are found on or in the skin, the genital area, the mouth, and, in particular, the intestines. Not surprisingly, populations of microbes are most alike in similar body locales. The microbial composition was found to change over time and is influenced by, among other factors, disease and medications—in particular, **antibiotics**.

Earlier identification of microbes involved their labor-intensive isolation and growth on cultures. The HMP utilized DNA sequencing machines and computer analysis of the genome sequences using bacterial ribosomal RNA (16S rRNA), uniquely found in bacteria, and in phylogenetic studies to classify and identify microbes.

It was long believed that the human body maintained normal health independently and that microbes were responsible for infectious disorders. It is now recognized that some microbes play an indispensible role in our digestion and absorption of many nutrients, in addition to being responsible for the synthesis of some vitamins and natural anti-inflammatory substances, and the metabolism of drugs and other foreign chemicals.

SEE ALSO: Ecological Interractions (1859), Probiotics (1907), Antibiotics (1928), Genomics (1986), Human Genome Project (2003).

Enterococci (shown, colorized) are normal residents in the intestines of humans and animals but are also responsible for causing serious infections, particularly in hospital settings. Of greatest concern is that enterococci have natural or acquired resistance to multiple antibiotics.

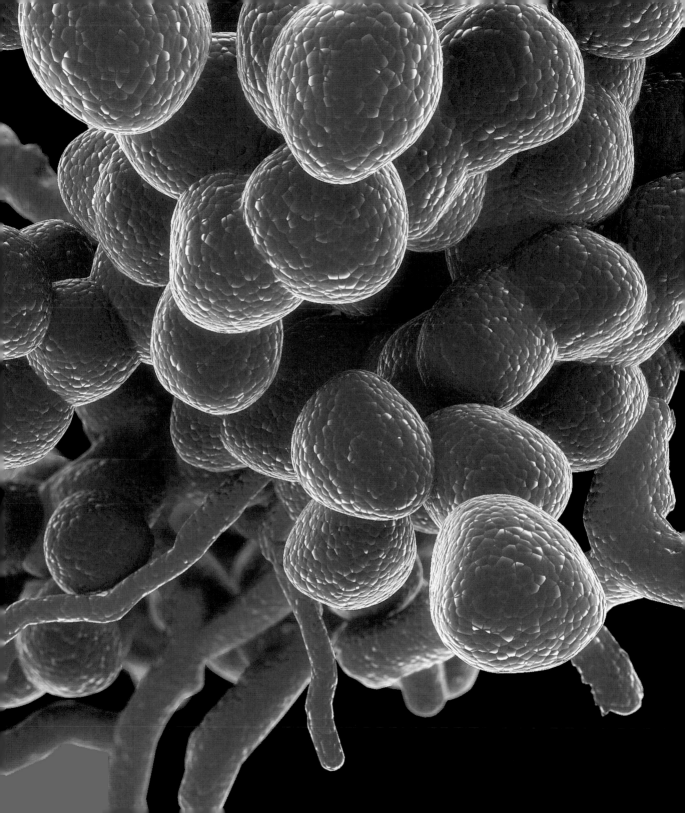

Epigenetics

Jean-Baptiste Lamarck (1744–1829)

Two centuries ago, Jean-Baptiste Lamarck postulated that environmental factors could influence traits, which could be transmitted to offspring. This theory was soundly rejected during his lifetime. Based on animal and human studies, scientists are now having second thoughts.

Almost six decades after the World War II Hongerwinter (hunger winter), survivors exhibited abnormal patterns of methylation that turn on or off genes associated with a number of disorders. The Hongerwinter in Holland began late during 1944, when food supplies were so drastically cut that they provided less than one-quarter of the recommended daily intake. This continued until the liberation of Holland in May 1945; 18,000–22,000 starved to death. The children conceived during this period were small, underweight, and, compared to their siblings conceived before and after the famine, at greater risk of obesity, heart disease, diabetes, and hypertension. Similar results were seen in children after the 1968–1970 Biafra famine in Nigeria. Individuals prenatally exposed to maternal food deprivation during the Great Chinese Famine (1958–1961) showed an increased risk of schizophrenia.

Traits are determined by genes—information carried by DNA. DNA dictates the manufacture of proteins and RNA molecules, which is translated to the manufacture of proteins, the link between our genetic makeup and our outward or physical traits. Epigenetics refers to all changes to genes other than changes in the DNA sequence. Such epigenetic changes commonly include the addition of methyl ($-CH_3$) groups to the DNA backbone, "marking" the DNA, and interfering with its ability to transcribe messages to RNA. Epigenetic marks have also been seen in some cancers.

In 2012, Andrew Feinberg reported differences in DNA methylation patterns in female worker bees who, within a hive, share identical genetic sequences but exhibit different behavioral patterns. Some remain in the hive and nurse the queen, and, upon maturity, leave the hive and forage for pollen. Nursing and foraging bees each have their own distinct DNA methylation patterns. When nurse bees were removed from the hive, the foragers stepped in to replace them, and their methylation patterns changed to that seen with nurse bees. Hence, epigenetic "marks" are reversible and linked to behavior.

SEE ALSO: Larmarckian Inheritance (1809), Mendelian Inheritance (1866), Genetics Rediscovered (1900), Genes on Chromosomes (1910), DNA as Carrier of Genetic Information (1944).

Differences in DNA methylation patterns have been found between genetically identical nursing and foraging female worker bees, suggesting that behavior can alter gene expression.

American Chestnut Tree Blight

In 1900, there were four billion American chestnut trees; they numbered one-quarter of all northeastern hardwoods and were among the tallest and most majestic residents of the eastern forests. Soaring fifty feet before their first branches, they were prized by furniture makers for their hard, straight, rot-resistant wood. In 1904, a ship arrived carrying fungal-infested Japanese chestnut nursery stock, an infestation that accidentally spread to and decimated its iconic American cousin. By 1950, the American chestnuts were virtually all dead, the victims of a fungal blight.

The fungus *Cryphonectria parasitica* enters the tree through wounds or age-related breaks in the bark. It grows and encircles under the bark, forming a canker, and produces oxalic acid that rapidly kills the tree. In 2013, two different approaches were being tested for their ability to resist the chestnut blight: creating a hybrid chestnut tree and, more recently, a genetically modified tree.

Since 1940, the American Chestnut Foundation has been attempting to successfully blend the fungal-resistant Chinese chestnut with the American chestnut. When attacked by the fungus, the Chinese variety is genetically equipped to build a wall around the fungus before it can secrete a sufficient quantity of acid to encircle the tree; the American chestnut responds in a similar manner but reacts far too slowly to save itself. The foundation's object is to produce a hybrid that possesses all the attributes of the American chestnut—its hardness, ability to withstand cold and drought and thrive in its normal range from Maine to Louisiana—with the unique characteristic of fungal resistance.

William A. Powell, forest biotechnologist, and Charles A. Maynard, a geneticist—researchers from the State University of New York College of Environmental Science and Forestry in Syracuse—are seeking a genetic solution. They are testing a genetically modified chestnut, containing a gene obtained from wheat that manufactures an enzyme capable of inactivating oxalic acid, which stops the fungus before it can kill the tree. As with other genetically modified plants, these trees must be grown in a segregated experimental field where there is no potential for their pollen to fertilize other trees. The success of these approaches will take years to evaluate.

SEE ALSO: Fungi (c. 1.4 Billion BCE), Land Plants (c. 450 Million BCE), Gymnosperms (c. 300 Million BCE), Artificial Selection (Selective Breeding) (1760), Biotechnology (1919), Genetically Modified Crops (1982).

One century ago, the American chestnut was among the most common and prized hardwood trees in the northeastern United States. Fifty years later, after a devastating fungal blight, few were to be found alive—a loss that persists today.

De-Extinction

Georges Cuvier (1769–1832)

Extinction is the end of a species, marked by the death of the last member of that species. In 1796, the French naturalist Georges Cuvier presented convincing evidence that established extinction to be a fact, and it is estimated that more than 99 percent of all species that have ever existed are now extinct. Fossil evidence documents that over the past 500 million years, there have been five major extinction events—the most recent occurring during the Cretaceous period 65 million years ago, which eliminated more than half of all marine species as well as many families of terrestrial plants and animals. Its cause is thought to have been an asteroid or comet. In more recent times, extinctions have been attributed to climatic changes, genetic factors, and habitat destruction and pollution. Other identified causes are overexploitation by hunting and fishing, the introduction of **invasive species**, and disease.

In more recent times, extinct species have included the woolly mammoth (3,000–10,000 years ago), passenger pigeon (1914), Tasmanian tiger (1930), and the Pyrenean or Spanish ibex (2,000). But not all scientists accept the premise that "extinct is forever," and there have been active de-extinction efforts to bring back extinct animal and plant species. The most widely proposed method has been by **cloning**, which was popularized in John Brosnan's *Carnosaur* (1984) and Michael Crichton's *Jurassic Park* (1990). In this de-extinction process, a viable DNA sample is taken from a species that has been extinct for no more than thousands of years—not millions of years, as in these novels—and gestated in a host animal.

Thus far, limited de-extinction success has been achieved. In 2003, Spanish researchers used frozen tissue obtained from the last living Pyrenean ibex that had died three years earlier, which was implanted in a goat; this attempt was unsuccessful. In 2009, an ibex clone was born alive but died seven minutes later from an unrelated respiratory ailment. Renewed and enthusiastic interest in de-extinction occurred in 2013, with active discussion and debate, as well as plans by Russian and South Korean scientists to clone a woolly mammoth from well-preserved remains found in Siberia.

SEE ALSO: Devonian Period (c. 417 Million BCE), Dinosaurs (c. 230 Million BCE), Paleontology (1796), Invasive Species (1859), Coelacanth: "The Living Fossil" (1938), Cloning (Nuclear Transfer) (1952), *Silent Spring* (1962), Punctuated Equilibrium (1972), Polymerase Chain Reaction (1983).

This illustration of a Pyrenean ibex appeared in the book Wild Oxen, Sheep & Goats of All Lands: Living and Extinct *(1898) by English naturalist Richard Lydekker (1849–1915).*

Oldest DNA and Human Evolution

Svante Pääbo (b. 1955)

In December 2013, the world's oldest evidence of human development was discovered, a finding that raised a number of evolutionary questions. A femur (thigh bone) fossil was recovered from the "pit of bones," an underground cave in northern Spain, from which scientists had recovered twenty-eight nearly complete human skeletons since the 1970s. From the powdered femur, Matthias Meyer and his colleagues at the Max Planck Institute in Leipzig, Germany, extracted mitochondrial DNA (mtDNA) that dated back some 400,000 years, 300,000 years older than the previous humanoid DNA sample.

Upon preliminary examination, the anatomy of the femur resembled a **Neanderthal**, but a comparison of the DNA evidence showed a much closer relationship to the Denisovans, whose DNA had been previously analyzed from 80,000-year-old remains found in Siberia, 4,000 miles to the east. This finding challenged the narrative of human development based on previously discovered fossil remains and DNA analysis. Humans, Neanderthals, and Denisovans were generally believed to have had a common ancestor in Africa some 500,000 years ago. This ancestor diverged from humans, left Africa, and split once again, 300,000 years ago, into the Neanderthals and Denisovans. The Neanderthals traveled west, toward Europe, and the Denisovans, east. Our human ancestor remained in Africa, giving rise to *Homo sapiens*, who 60,000 years ago migrated to Europe and Asia, where they interbred with the Neanderthals and Denisovans, who became extinct. But the new DNA evidence raises the question: Why are Denisovan fossil remains in Spain?

The new DNA findings were only made possible because of advances in retrieving ancient DNA. When a biological organism dies, its DNA breaks down into small fragments, which in time mix and become contaminated with DNA from other species—in particular, soil bacteria. In 1997, Svante Pääbo, a Swedish biologist specializing in evolutionary genetics, also working at the Max Planck Institute, discovered a new technique for retrieving DNA fragments, which he used to determine the genome sequence of the Neanderthal in 2010 and the femur in Spain. It is possible that advances such as this may rewrite our biological history.

SEE ALSO: Neanderthals (c. 350,000 BCE), Anatomically Modern Humans (c. 200,000 BCE), Lucy (1974), Genomics (1986), Mitochondrial Eve (1987), Human Genome Project (2003).

The jaws of Homo heidelbergensis, *an extinct species that lived in Europe, Africa, and western Asia possibly as far back as 1.3 million years ago. It was the first human species to live in colder climates and may have been the first to bury its dead.*

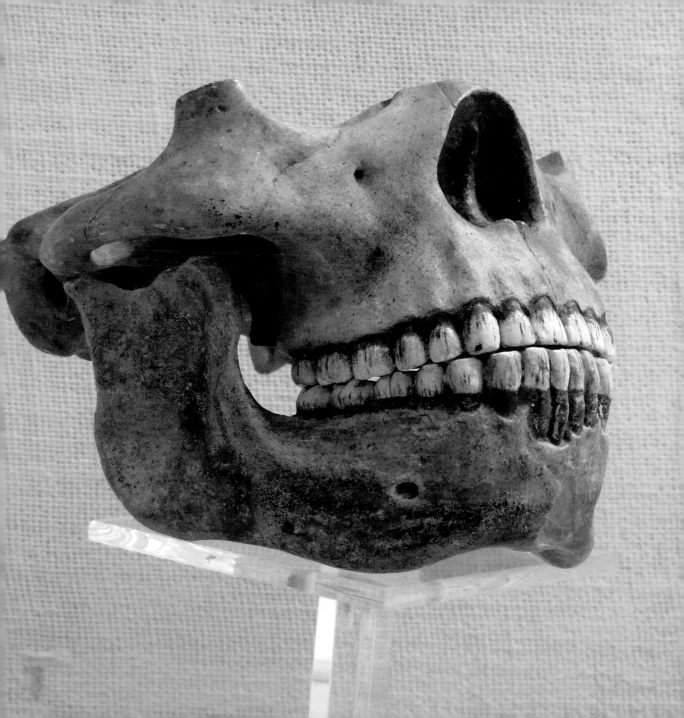

Further Reading

Many sources have been used to write this book, only some of which we have listed. We have also included these readings to provide you with sources for greater in-depth information on the topics discussed. We welcome your comments about any significant or interesting milestones in biology that might be included in future editions of this book at mcgeraldweb@gmail.com.

General Reading and Sources

Alberts, B., et al., *Molecular Biology of the Cell*. New York: Garland Science, 2007.

Alcock, J., *Animal Behavior*. Sunderland, MA: Sinauer Associates, 2013.

Gardner, E. J., *History of Biology*. Minneapolis, MN: Burgess Publishing, 1972.

Kandel, E. R. (ed.), et al., *Principles of Neural Science*. New York: McGraw-Hill Professional, 2012.

Kardong, K., *Vertebrates: Comparative Anatomy, Function, Evolution*. New York: McGraw-Hill, 2011.

Mauseth, J. D., *Botany*. Burlington, MA: Jones & Bartlett, 2012.

Nelson, D. L., et al., *Lehninger Principles of Biochemistry*. New York: W.H. Freeman, 2012.

Pierce, B. A., *Genetics: A Conceptual Approach*. New York: W.H. Freeman, 2010.

Reece, J. B., et al., *Campbell Biology*. San Francisco: Benjamin Cummings, 2013.

Sherwood, L., et al., *Animal Physiology: From Genes to Organisms*. Independence, KY: Cengage, 2012.

Smith, T. M., *Elements of Ecology*. San Francisco: Benjamin Cummings, 2012.

Tortora, G. J., et al., *Microbiology: An Introduction*. San Francisco: Benjamin Cummings, 2012.

c. 4 Billion BCE, Origin of Life

Dyson, F, *Origin of Life*. New York: Cambridge University Press, 1999.

c. 3.9 Billion BCE, Last Universal Common Ancestor

Woese, C., *Proceedings of the National Academy of Sciences USA* 1998 95(11): 9710.

c. 3.9 Billion BCE, Prokaryotes

Lengeler, J. (ed.), et al., *Biology of Prokaryotes*. New York: Wiley-Blackwell, 1999.

White, D., et al., *The Physiology and Biochemistry of Prokaryotes*. New York: Oxford University Press. 2011.

c. 2.5 Billion BCE, Algae

Graham, J. E., et al., *Algae*. San Francisco: Benjamin Cummings, 2008.

c. 2 Billion BCE, Eukaryotes

Keeling, P., et al., Eukaryotes, http://tolweb.org. Tree of Life Web Project.

c. 1.4 Billion BCE, Fungi

Petersen, J. H., *The Kingdom of Fungi*. Princeton, NJ: Princeton University Press, 2013.

c. 570 Million BCE, Arthropods

Minelli, A. (ed.), et al., *Arthropod Biology and Evolution: Molecules, Development, Morphology*. New York: Springer, 2013.

c. 530 Million BCE, Medulla: The Vital Brain

Kandel, E. R. (ed.), et al., *Principles of Neural Science*. New York: McGraw-Hill Professional, 2012.

c. 530 Million BCE, Fish

Bone, Q., *Biology of Fishes*. New York: Taylor & Francis, 2008.

Helfman, G., et al., *The Diversity of Fishes: Biology, Evolution, and Ecology*. New York: Wiley-Blackwell, 2009.

c. 450 Million BCE, Land Plants

Lewis, L. A., et al., *American Journal Botany*. 2004 91(10): 1535.

c. 417 Million BCE, Devonian Period

McGhee, G. R., *When the Invasion of Land Failed: The Legacy of the Devonian Extinctions*. New York: Columbia University Press, 2013.

c. 400 Million BCE, Insects

Daly, H. V., et al., *Introduction to Insect Biology and Diversity*. New York: Oxford University Press, 1998.

Gullan, P. J., *The Insects: An Outline of Entomology*. New York: Wiley-Blackwell, 2010.

c. 400 Million BCE, Plant Defenses against Herbivores

Howe, H. F., *Ecological Relationships of Plants and Animals*. New York: Oxford University Press, 1988.

c. 360 Million BCE, Amphibians

Duellman W. E., et al., *Biology of Amphibians*. Baltimore: Johns Hopkins University Press, 1994.

Wells, K. D., *The Ecology and Behavior of Amphibians*. Chicago: University of Chicago Press, 2007.

c. 350 Million BCE, Seeds of Success

Bewley, J. D., et al., *Seeds: Physiology of Development, Germination and Dormancy*. New York: Springer, 2012.

c. 320 Million BCE, Reptiles

Vitt, L. J., et al., *Herpetology: An Introductory Biology of Amphibians and Reptiles*. New York: Academic Press, 2013

c. 300 Million BCE, Gymnosperms

Bhatnagar, S. P., *Gymnosperms*. New Delhi: New Age International, 1996.

c. 230 Million BCE, Dinosaurs
Fastovsky, D. E., et al., *Dinosaurs: A Concise Natural History*. New York: Cambridge University Press, 2012.
Samson, S. D., *Dinosaur Odyssey: Fossil Threads in the Web of Life*, Berkeley, CA: University of California Press, 2009.

c. 200 Million BCE, Mammals
Attenborough, D., *The Life of Mammals*. Princeton, NJ: Princeton University Press, 2002.
Macdonald, D. W., *The Princeton Encyclopedia of Mammals*. Princeton, NJ: Princeton University Press, 2009.

c. 150 Million BCE, Birds
Cornell Laboratory of Ornithology, *Cornell Laboratory of Ornithology Handbook of Bird Biology*. Princeton, NJ: Princeton University Press, 2004.

c. 125 Million BCE, Angiosperms
Soltis, P.S., et al., *Phylogeny and Evolution of Angiosperms*. Sunderland, MA: Sinauer, 2005.

c. 65 Million BCE, Primates
Fleagle, J. G., *Primate Adaption and Evolution*. New York: Academic Press, 2013.
Nystrom, P., et al., *The Life of Primates*. Cambridge, UK: Pearson, 2008.

c. 55 Million BCE, Amazon Rainforest
Morgan, B., *Rainforest*. London: DK, 2006.

c. 350,000 BCE, Neanderthals
Papagianni, D., et al., *The Neanderthals Rediscovered: How Modern Science is Rewriting Their History*. London: Thames & Hudson, 2013.

c. 200,000 BCE, Anatomically Modern Humans
Lieberman, D., *The Story of the Human Body: Evolution, Health, and Disease*. New York: Pantheon, 2013.
Smith, F. H., et al., *The Origins of Modern Humans: Biology Reconsidered*. New York: Wiley-Blackwell, 2013.

c. 60,000 BCE, Plant-Derived Medicines
Gerald, M. C., *The Drug Book: From Arsenic to Xanax, 250 Milestones in the History of Drugs*. New York: Sterling, 2013.
Pendleton, J., *Plants as Medicine: Healing Compounds Derived from Medicinal Herbs*. Traditional Healing Press, 2013.

c. 11,000 BCE, Wheat: The Staff of Life
Kirby, E. J. M., Food and Agricultural Organization of the United Nations—Botany of the Wheat Plant.

c. 10,000 BCE, Agriculture
Denison, R. F., *Darwinian Agriculture: How Understanding Evolution Can Improve Agriculture*. Princeton, NJ: Princeton University Press, 2012.

c. 10,000 BCE, Domestication of Animals
Manning, A., et al., *Animals and Human Society: Changing Perspectives*. New York: Routledge, 1994.

c. 8000 BCE, Coral Reefs
Sheppard, C. R. C., et al., *The Biology of Coral Reefs (Biology of Habitats)*. New York: Oxford University Press, 2009.

c. 7000 BCE, Rice Cultivation
Tsunoda, S. (ed.), *Biology of Rice*. New York: Elsevier Science (1984).

c. 2600 BCE, Mummification
Aufderheide, A. C., *The Scientific Study of Mummies*. New York: Cambridge University Press, 2011.

c. 2350 BCE, Animal Navigation
Gould, J. L., et al., *Nature's Compass: The Mystery of Animal Navigation*. Princeton, NJ: Princeton University Press, 2012.

c. 400 BCE, Four Humors
paei.wikidot.com/Hippocrates-galen-the-four-humors

c. 330 BCE, Aristotle's *The History of Animals*
Lennox, J. G., *Aristotle's Philosophy of Biology: Studies in the Origins of Life Science*. Cambridge, MA: Cambridge University Press, 2000.

c. 330 BCE, Animal Migration
Milner-Guilland, et al., *Animal Migration: A Synthesis*. New York: Oxford University Press, 2011.
Ueda, H. (ed.), *Physiology and Ecology of Fish Migration*. Boca Raton, FL: CRC Press, 2013.

c. 320 BCE, Botany
Evert, R. F. et al., *Raven Biology of Plants*. New York: W.H. Freeman, 2012.
Taiz, L. et al., *Plant Physiology*. Sunderland, MA: Sinauer Associates, 2010.

77, Pliny's *Natural History*
Galus Plinius Secundus (Pliny the Elder), *Natural History: A Selection*. New York: Penguin Classics, 1991.

c. 180, Skeletal System
Kardong, K., *Vertebrates: Comparative Anatomy, Function, Evolution*. New York: McGraw-Hill Science, 2014.

1242, Pulmonary Circulation
Peacock, A. J., et al., *Pulmonary Circulation*. Boca Raton, FL: CRC Press, 2011.

1489, Leonardo's Human Anatomy
O'Malley C. D., et al., *Leonardo da Vinci on the Human Body: The Anatomical, Physiological, and Embryological Drawings of Leonardo da Vinci*. New York: Gramercy, 2003.

1521, Sense of Hearing
Webster, D. B., et al., *The Evolutionary Biology of Hearing*. New York: Springer, 1991.

1543, Vesalius's *De humani corporis fabrica*
Vesalius, A, et al., *On the Fabric of the Human Body: A Translation of De Humani Corporis Fabrica Libri Septem*. Boston: Jeremy Norman Co., 2003.

1611, Tobacco
Gately, I., *Tobacco: A Cultural History of How an Exotic Plant Seduced Civilization.* New York: Grove Press, 2002.

1614, Metabolism
Salway, J. G., *Metabolism at a Glance.* New York: Wiley-Blackwell, 2004.
Stipanuk, et al., *Biochemical, Physiological, and Molecular Aspects of Human Nutrition.* St. Louis, MO: Saunders, 2012.

1620, Scientific Method
Beveridge, W. I. B., *The Art of Scientific Investigation.* Caldwell, NJ: Blackburn, 2004.
Wilson, Jr., E. B., *An Introduction to Scientific Research.* Mineola, NY: Dover, 1991.

1628, Harvey's *De motu cordis*
Mohman, D., et al., *Cardiovascular Physiology.* New York: McGraw-Hill Professional, 2013.

1637, Mechanical Philosophy of Descartes
Slowik, E., *Descartes' Physics* http://plato.stanford.edu/archives/fall2013/entries/descartes-physics/>

1651, Placenta
Power, M. L., et al., *The Evolution of the Human Placenta.* Baltimore: Johns Hopkins Press, 2012.

1652, Lymphatic System
Buckley, V., *Christina, Queen of Sweden: The Restless Life of a European Eccentric.* New York: Harper Perennial (2005).
Sabin, F. R., *The Origin and Development of the Lymphatic System (Classic reprint).* Forgotten Books, 2012.

1658, Blood Cells
Kaushansky, K., et al., *Williams Hematology.* New York: McGraw-Hill Professional, 2010.

1668, Refuting Spontaneous Generation
Farley, J., *The Spontaneous Generation Controversy From Descartes to Oparin.* Baltimore: The Johns Hopkins Press, 1977.

1669, Phosphorus Cycle
Emsley, J., *The 13th Element: The Sordid Tale of Murder, Fire, and Phosphorus.* New York: Wiley, 20002.

1670, Ergotism and Witchcraft
Kors, A. C. (ed.), et al., *Witchcraft in Europe, 400–1700: A Documentary History.* Philadelphia: University of Pennsylvania Press, 2000.

1674, Leeuwenhoek's Microscopic World
de Kruif, P., *Microbe Hunters.* New York: Mariner Books, 2002.

1677, Spermatozoa
De Jonge, C. J., et al., *The Sperm Cell.* New York: Cambridge University Press. 2006.

1717, Miasma Theory
Litsios, S., *Plague Legends: From the Miasmas of Hippocrates to the Microbes of Pasteur.* Ballwin, MO: Science & Humanities Press, 2001.

1729, Circadian Rhythms
Foster, R. G., *Rhythms of Life: The Biological Clocks that Control the Daily Lives of Every Living Thing.* New Haven, CT: Yale University Press, 2005.

1733, Blood Pressure
Naqvi, N. H., et al., *Blood Pressure Measurement: An Illustrated History.* Boca Raton, FL: CRC Press, 1998.

1735, Linnaean Classification of Species
Blunt, W., et al., *Linnaeus: The Compleat Naturalist.* Princeton, NJ: Princeton University Press, 2002.

c. 1741, Cerebrospinal Fluid
Davson, H., et al., *Physiology of the CSF and Blood-Brain Barriers.* Boca Raton, FL: CRC Press, 1996.

1744, Regeneration
Carlson, B. M. (ed.), *Principles of Regenerative Medicine.* New York: Academic Press, 2007.
Lenhoff, S. G., et al., *Hydra and the Birth of Experimental Biology, 1744: Abraham Trembley's Memoires Concerning the Polyps.* Pacific Grove, CA: Boxwood Press, 1986.

1759, Theories of Germination
Bewley, J. D., et al., *Seeds: Physiology of Development, Germination and Dormancy.* New York: Springer, 2012.

1760, Artificial Selection (Selective Breeding)
Wood, R. J., et al., *Genetic Prehistory in Selective Breeding: A Prelude to Mendel.* New York: Oxford University Press, 2001.

1786, Animal Electricity
Plonsey, R., et al., *Bioelectricity: A Quantitative Approach.* New York: Springer, 2007.

1789, Gas Exchange
Dejours, P. J., *Respiration in Water and Air: Adaptions Regulations Evolution.* New York: Elsevier Science Ltd, 1988.

1791, Nervous System Communication
LeDoux, J., *Synaptic Self: How Our Brains Become Who We Are.* New York: Viking, 2002.
Squire, L. (ed), et al., *Fundamental Neuroscience.* New York: Academic Press, 2012.

1796, Paleontology
Foote, M., et al., *Principles of Paleontology.* New York: W.H. Freeman, 2006.

1798, Population Growth and Food Supply
Malthus, T, *An Essay on the Principle of Population.* New York: Oxford University Press, 2008.
Weeks, J. R., *Population: An Introduction to Concepts and Issues.* Independence, KY: Cengage Learning, 2011.

1809, Lamarckian Inheritance
Honeywell, R., *Lamarck's Evolution: Two Centuries of Genius and Jealousy.* London: Murdock Books, 2008.

1828, Germ-Layer Theory of Development
Gilbert, S., *Developmental Biology.* Sunderland, MA: Sinauer Associates, 2013.
McGeady, T. A., et al., *Veterinary Embryology.* Oxford, UK: Blackwell, 2006.

1831, Cell Nucleus

Misteli, T. et al., *The Nucleus*. Cold Spring Harbor, NY: Cold Spring Harbor Laboratory Press, 2010.

1831, Darwin and the Voyages of the *Beagle*

Darwin, C., *The Voyage of the Beagle* (many editions)

Moorehead, A., *Darwin and the Beagle: Charles Darwin as Naturalist on the HMS Beagle Voyage*. New York: Harper & Row, 1970.

1832, Anatomy Act 1832

Abbott, G., *Grave Disturbances: A History of the Body Snatchers*. Cardiff, Wales: Eric Dobby Publishing, 2006.

Fido, M., *Bodysnatchers: A History of the Resurrectionists*. Chicago: Academy Chicago Pub (1992).

1833, Human Digestion

Chivers, D. J. (ed.), et al., *The Digestive System in Mammals: Food Form and Function*. New York: Cambridge University Press, 1994.

Karlawish, J., *Open Wound: The Tragic Obsession of Dr. William Beaumont*. Ann Arbor, MI: University of Michigan Press, 2011.

1836, Fossil Record and Evolution

Switek, B., *Written in Stone: Evolution, the Fossil Record, and Our Place in Nature*. New York: Bellevue Literary Press, 2010.

Taylor, T. N, et al., *Paleobotany: The Biology and Evolution of Fossil Plants*. New York: Academic Press, 2008.

1837, Nitrogen Cycle and Plant Chemistry

Stevenson, F. J., et al., *Cycles of Soil: Carbon, Nitrogen, Phosphorus, Sulfur, Micronutrients*. New York: Wiley, 1999.

1838, Cell Theory

Alberts, B., et al., *Molecular Biology of the Cell*. New York: Garland Science, 2007.

Karp, G., *Cell and Molecular Biology: Concepts and Experiments*. New York, Wiley, 2013.

1840, Plant Nutrition

Mendel, K., et al., *Principles of Plant Nutrition*. New York, Springer, 2001.

Stevenson, F. J., et al., *Cycles of Soil: Carbon, Nitrogen, Phosphorus, Sulfur, Micronutrients*. New York: Wiley, 1999.

1842, Urine Formation

Eaton, D., *Vanders Renal Physiology*. New York: McGraw-Hill Medical, 2013.

1842, Apoptosis (Programmed Cell Death)

Green, D. R., *Apoptosis: Physiology and Pathology*. New York: Cambridge University Press, 2011.

1843, Venoms

Gjersoe, J., et al., *Venoms: Sources, Toxicity and Therapeutic Uses*. Hauppauge, NY: Nova Science Publishers, 2010.

White, J., et al., *Handbook of Clinical Toxicology of Animal Venoms and Poisons*. Boca Raton, FL: CRC Press, 1995.

1843, Homology versus Analogy

Kardong, K., *Vertebrates: Comparative Anatomy, Function, Evolution*. New York: McGraw-Hill Science, 2014.

1845, Photosynthesis

Eaton-Rye, J. J. (ed), et al., *Photosynthesis: Plastid Biology, Energy Conversion and Respiration*. New York, Springer, 2011.

Hall, D. O., et al., *Photosynthesis (Studies in Biology)*. New York: Cambridge University Press, 1999.

1848, Optical Isomers

chemistry.tutorvista.com>Organic Chemistry nomenclature, Isomers.

1849, Testosterone

Nieschlag, E. (ed.) et al., *Testosterone: Action, Deficiency, Substitution*. New York: Cambridge University Press, 2012.

Smith, L. B., et al., *Testosterone: From Basic Research to Clinical Applications*. New York: Springer, 2013.

1850, Trichromatic Color Vision

Land, M. F., et al., *Animal Eyes*. New York: Oxford University Press, 2012.

Lazareva, O. F., *How Animals See the World: Comparative Behavior, Biology, and Evolution of Vision*. New York: Oxford University Press, 2012.

1854, Homeostasis

Bernard, C., *An Introduction to the Study of Experimental Medicine*. Mineola, NY: Dover Publications, 1957.

Cannon, W. B., *The Wisdom of the Body*. New York: W.W. Norton, 1963.

1856, The Liver and Glucose Metabolism

Berg, J. M., et al., *Biochemistry*. New York: W.H. Freeman, 2010.

Nelson, D. L., et al., *Lehninger Principles of Biochemistry*. New York: W.H. Freeman, 2012.

1857, Microbial Fermentation

El-Mansi, E. M. T. (ed), et al., *Fermentation Microbiology and Biotechnology*. Boca Raton, FL: CRC Press, 2006.

1859, Darwin's Theory of Natural Selection

Darwin, C., *The Origin of Species by Means of Natural Selection* (many editions)

Quammen, D., *The Reluctant Mr. Darwin: An Intimate Portrait of Charles Darwin and the Making of His Theory of Evolution*. New York: W.H. Norton, 2007.

Ridley, M, *Evolution*. New York: Wiley-Blackwell, 2003.

1859, Ecological Interactions

Howe, H. F., *Ecological Relationships of Plants and Animals*. New York: Oxford University Press, 1988.

Schoonhoven, L. M., et al., *Insect-Plant Biology*. New York: Oxford University Press, 2006.

1859, Invasive Species

Elton, C. S., *Ecology of Invasions by Animals and Plants*. Chicago: University of Chicago Press, 2000.

1861, Localization of Cerebral Functions

Pelton, R.W., *The Age Old Science of Phrenology: An Age Old Science Developed by Franz Joseph Gall, M.D. of Vienna*. CreateSpace Independent Publishing Platform, 2012.

1862, Biological Mimicry
Brower, L. P., *Mimicry and the Evolutionary Process*. Chicago: University of Chicago Press, 1989.
Wickler, W., *Mimicry in Plants and Animals*. New York: McGraw-Hill, 1968.

1866, Mendelian Inheritance
Edelson, E., *Gregor Mendel: And the Roots of Genetics*. New York: Oxford University Press, 1999.

1866, Ontogeny Recapitulates Phylogeny
Gould, S. J., *Ontogeny and Phylogeny*. Cambridge, MA: Harvard University Press, 1977.
Haeckel, E., et al., *Art Forms in Nature: The Prints of Ernst Haeckel*. New York: Prestel, 2008.

1866, Hemoglobin and Hemocyanin
Weatherall, D. (ed). et al., *Hemoglobin and its Diseases (Cold Spring Harbor Perspectives in Medicine)*. Cold Spring Harbor, NY: Cold Spring Harbor Laboratory Press, 2013.

1869, Deoxyribonucleic Acid (DNA)
Portugal, F. H., et al., *A Century of DNA: A History of the Discovery of the Structure and Function of the Genetic Substance*. Cambridge, MA: MIT Press, 1977.

1871, Sexual Selection
Campbell, B. (ed.), *Sexual Selection and the Descent of Man: The Darwinian Pivot*. Piscataway, NJ: Transaction Publishers, 2006.
Dixson, A. F., *Sexual Selection and the Origins of Human Mating Systems*. New York: Oxford University Press, 2009.

1873, Coevolution
Thompson, J. N., *The Geographic Mosaic of Coevolution*. Chicago: The University of Chicago Press, 2005.

1874, Nature versus Nurture
Gould, S. J., *The Mismeasure of Man*. New York: W.W. Norton, 1996.

1875, Biosphere
Vernadsky, V. I., *The Biosphere: Complete Annotated Edition*. Göttingen, Germany: Copernicus, 1998.

1876, Meiosis
John, B., *Meiosis (Developmental and Cell Biology Series)*. New York: Cambridge University Press, 2005.

1876, Biogeography
Lomolino, M. V., et al., *Biogeography*. Sunderland, MA: Sinauer Associates, 2010.
Slotten, R.A., *The Heretic in Darwin's Court: The Life of Alfred Russel Wallace*. New York: Columbia University Press, 2004.

1877, Marine Biology
Castro, P., et al., *Marine Biology*. New York: McGraw-Hill Science, 2012.

1878, Enzymes
Berg, J. M., et al., *Biochemistry*. New York: W.H. Freeman, 2010.
Nelson, D. L., et al., *Lehninger Principles of Biochemistry*. New York: W.H. Freeman, 2012.

1880, Phototropism
Davies, P., *Plant Hormones and Their Role in Plant Growth and Development*. Boston: Martinus Nijhoff, 2013.

1882, Mitosis
Murray, A., et al., *The Cell Cycle: An Introduction*. New York: Oxford University Press, 1993.

c. 1882, Thermoreception
Braun, H. A. (ed), et al., *Thermoreception and Temperature Regulation*. New York: Springer, 1990.

1882, Innate immunity
Parham, P., *The Immune System*. New York: Garland Science, 2009.

1883, Germ Plasm Theory of Heredity
Gilbert, S., *Developmental Biology*. Sunderland, MA: Sinauer Associates, 2013.

1883, Eugenics
Kevles, D., *In the Name of Eugenics: Genetics and the Uses of Human Heredity*. New York: Knopf, 2013.

1884, Gram Stain
Tortora, G. J., et al., *Microbiology: An Introduction*. San Francisco: Benjamin Cummings, 2012.

1885, Negative Feedback
Cosentino, C., et al., *Feedback Control in Systems Biology*. Boca Raton, FL: CRC Press, 2011.

1890, Germ Theory of Disease
Waller, J. *The Discovery of the Germ: Twenty Years That Transformed the Way We Think About Disease*. New York: Columbia University Press, 2003.

1890, Animal Coloration
Diamond, J., et al., *Concealing Coloration in Animals*. Cambridge, MA: Belknap Press, 2013.

1891, Neuron Doctrine
Shepherd, G. M., *Foundations of the Neuron Doctrine*. New York: Oxford University Press, 1991.

1892, Endotoxins
Chaudhuri, K., et al., *Cholera Toxins*. New York: Springer, 2009.

1896, Global Warming
Weart, S. R., *The Discovery of Global Warming: Revised and Expanded Edition*. Cambridge, MA: Harvard University Press, 2008.

1897, Adaptive Immunity
Parham, P., *The Immune System*. New York: Garland Science, 2009.

1897, Associative Learning
Gluck, M. A., et al., *Learning and Memory: From Brain to Behavior*. Richmond, UK: Worth, 2007.

1897, Ehrlich's Side-Chain Theory
Silverstein, A. M., *Paul Ehrlich's Receptor Immunology: The Magnificent Obsession*. New York: Academic Press, 2001.

1898, Malaria-Causing Protozoan Parasite
Shah, S., *The Fever: How Malaria Has Ruled Humankind for 500,000 Years*. New York: Sarah Crichton Books, 2010.

1898, Viruses
Scholthof, K.-B. (ed.), et al., *Tobacco Mosaic Virus: One Hundred Years of Contributions to Virology*. St. Paul, MN: Amer Phytopathological Society, 1999.

1899, Ecological Succession
Walker, L. R., (ed.), et al., *Linking Restoration and Ecological Succession*. New York: Springer, 2007.

1899, Animal Locomotion
Alexander, R. M., *Principles of Animal Locomotion*. Princeton, NJ: Princeton University Press, 2002.

1900, Genetics Rediscovered
Stubbe, H., *History of Genetics: From Prehistoric Times to the Rediscovery of Mendel's Laws*. Cambridge, MA: MIT Press, 1973.

1900, Ovaries and Female Reproduction
Haterius, H. O., *Ohio Journal of Science* 1937, 37 (6): 394.

1901, Blood Types
Kaushansky, K., et al., *Williams Hematology*. New York: McGraw-Hill, 2010.

1902, Tissue Culture
Neumann, K-H, et al., *Plant Cell and Tissue Culture—A Tool in Biotechnology: Basics and Application (Principles and Practice)*. New York: Springer, 2009.

1902, Secretin: The First Hormone
Norris, D. O., et al., *Vertebrate Endocrinology*. New York: Academic Press, 2013.

1904, Dendrochronology
Stokes, M. A., et al., *An Introduction to Tree-Ring Dating*. Tucson: University of Arizona Press, 1996.

1905, Blood Clotting
Doolittle, R. F., *The Evolution of Vertebrate Blood Clotting*. University Science Books, 2012.

1907, Radiometric Dating
Macdougall, D., *Nature's Clocks: How Scientists Measure the Age of Almost Everything*. Berkeley, CA: University of California Press, 2008.

1907, Probiotics
Tannock, G. W., *Probiotics and Prebiotics: Scientific Aspects*. Poole, UK: Caister Academic Press, 2005.

1907, Why Does the Heart Beat?
Silverman, M. E., et al., *Circulation*. 2006, 113: 2775.

1908, Hardy-Weinberg Equilibrium
Hardy-Weinberg Equilibrium Model, anthro.palomar.edu/synthetic/synth_2

1910, Genes on Chromosomes
Kohler, R. E., *Lords of the Fly: Drosophila Genetics and the Experimental Life*. Chicago: University of Chicago Press, 1994.
Sturtevant, A. H., *A History of Genetics*. Cold Spring Harbor, NY: Cold Spring Harbor Laboratory Press, 2001.

1911, Cancer-Causing Viruses
Rockefeller University Press, *A notable career in finding out. Peyton Rous, 1879-1970*. New York: Rockefeller University Press, 1971.

1912, Continental Drift
Colbert, E. H., *Wandering Lands and Animals: The Story of Continental Drift and Animal Populations*, Mineola, NY: Dover Publications, 1985.

1912, Vitamins and Beriberi
McDowell, L. R., *Vitamin History, The Early Years*. First Edition Design Publishing, 2013.

1912, Thyroid Gland and Metamorphosis
Ain, K., et al., *The Complete Thyroid Book*. New York: McGraw-Hill, 2010.

1912, X-Ray Crystallography
Authier, A., *Early Days of X-ray Crystallography*. New York: Oxford University Press, 2013.

1917, Bacteriophages
Kuchment, A., *The Forgotten Cure: The Past and Future of Phage Therapy*. New York: Springer, 2011.

1919, Biotechnology
Slater, A., et al., *Plant Biotechnology: The Genetic Manipulation of Plants*. New York: Oxford University Press, 2008.
Thieman, W. J., et al., *Introduction to Biotechnology*. San Francisco: Benjamin Cummings, 2012.

1920, Neurotransmitters
Iverson, L., et al., *Introduction to Neuropharmacology*. New York: Oxford University Press, 2008.

1921, Insulin
Bliss, M., *The Discovery of Insulin: Twenty-fifth Anniversary Edition*. Chicago: University of Chicago Press, 2007.

1923, Inborn Errors of Metabolism
Saudubray, J-M (ed.), et al., *Inborn Metabolic Diseases: Diagnosis and Treatment*. New York: Springer, 2012.

1924, Embryonic Induction
Gilbert, S. F. et al., *Developmental Biology*. Sunderland, MA: Sinauer Associates, 2006.

1924, Timing Fertility
Gilbert, S. F., et al., *Developmental Biology*. Sunderland, MA: Sinauer Associates, 2006.

1925, Mitochondria and Cellular Respiration
Day, D. (ed.), et al., *Plant Mitochondria: From Genome to Function*. New York: Springer, 2004.
Jacobs, J., *Metabolism Basics: A Walkthrough Guide to Fermentation and Cellular Respiration*. Seattle, WA: Amazon Digital Services, 2012.

1925, "The Monkey Trial"
Larson, E. J., *Summer for the Gods: The Scopes Trial and America's Continuing Debate Over Science and Religion*. New York: Basic Books, 1997.

1925, Population Ecology,
Begon, M., et al., *Population Ecology: A Unified Study of Animals and Plants.* New York: Wiley-Blackwell, 1996.

1927, Food Webs
Polis, G. A., et al., *Food Webs.* New York: Springer, 1995.

1927, Insect Dance Language
Stearcy, W. A., et al., *The Evolution of Animal Communication: Reliability and Deception in Signaling Systems.* Princeton, NJ: Princeton University Press, 2005.

1928, Antibiotics
Lax, E., *The Mold in Dr. Florey's Coat: The Story of the Penicillin Miracle.* New York: Henry Holt and Co., 2004.

1929, Progesterone
Avise, J. C., *Evolutionary Perspectives on Pregnancy.* New York: Columbia University Press, 2013.

1930, Osmoregulation in Freshwater and Marine Fish
Bradley, T., *Animal Osmoregulation.* New York: Oxford University Press, 2009.

1931, Electron Microscope
Bozzola, J. J., et al., *Electron Microscopy.* Burlington, MA: Jones & Bartlett, 1998.

1935, Imprinting
Burkhardt, Jr, R. W., *Patterns of Behavior: Konrad Lorenz, Niko Tinbergen, and the Founding of Ethology.* Chicago: University of Chicago Press, 2005.

1935, Factors Affecting Population Growth
Weeks, J. R., *Population: An Introduction to Concepts and Issues.* Independence, KY: Cengage Learning, 2011.

1936, Stress
Seyle, H., *The Stress of Life.* New York: McGraw-Hill, 1978.

1936, Allometry
Reiss, M. J., *The Allometry of Growth and Reproduction.* New York: Cambridge University Press, 1991.

1937, Evolutionary Genetics
Smith, J. M., *Evolutionary Genetics.* New York: Oxford University Press, 1998.

1938, Coelacanth: "The Living Fossil"
Werner, C., et al., *Evolution: The Grand Experiment: Vol. 2- Living Fossils.* Green Fossil, AR: New Leaf Press, 2009.

1939, Action Potential
Kandel, E. R. (ed.), et al., *Principles of Neural Science.* New York: McGraw-Hill Professional, 2012.

1941, One Gene-One Enzyme Hypothesis
Pierce, B. A., *Genetics: A Conceptual Approach.* New York: W.H. Freeman, 2010.

1942, Biological Species Concept and Reproductive Isolation
Grant, P. R., et al., *How and Why Species Multiply: The Radiation of Darwin's Finches.* Princeton, NJ: Princeton University Press, 2011.

1943, Arabidopsis: A Model Plant
Koomneef, M., *Plant Journal*, 2010 61 (6): 909

1944, DNA as Carrier of Genetic Information
Portugal, F. H., et al., *A Century of DNA: A History of the Discovery of the Structure and Function of the Genetic Substance.* Cambridge, MA: MIT Press, 1977.

1945, Green Revolution
Brown, L. R., *Full Planet, Empty Plates: The New Politics of Food Scarcity.* New York: W.W. Norton, 2012.

1946, Bacterial Genetics
Snyder, W. et al., *Molecular Genetics of Bacteria.* Washington, DC: ASM Press, 2013.

1949, Reticular Activating System
Reticular Activating System: reticularactivatingsystem.org

1950, Phylogenetic Systematics
Wiley, E. O.,et al., *Phylogenetics: Theory and Practice of Phylogenetic Systematics.* New York: Wiley-Blackwell, 2011

1951, The Immortal HeLa Cells
Skloot, R. *The Immortal Life of Henrietta Lacks.* New York: Crown, 2010.

1952, Cloning (Nuclear Transfer)
Cibelli, J. (ed.), et al., *Principles of Cloning.* New York: Academic Press, 2013.

1952, Amino Acid Sequence of Insulin
Stretton, A.O.W., *Genetics* 2002 **182**(2): 527.

1952, Pattern Formations in Nature
Ball, P., *The Self-Made Tapestry: Pattern Formation in Nature.* New York: Oxford University Press, 1999.

1952, Plasmids
Pierce, B. A., *Genetics: A Conceptual Approach.* New York: W.H. Freeman, 2010.

1952, Nerve Growth Factor
Habenicht, A. (ed.), *Growth Factors, Differentiation Factors, and Cytokines.* New York: Springer, 2011.

1953, Miller-Urey Experiment
Wikipedia: en.wikipedia.org/wiki/Miller-Urey Experiment

1953, The Double Helix
Judson, H. F., *The Eighth Day of Creation: Makers of the Revolution in Biology.* Cold Spring Harbor, NY: Cold Spring Harbor Laboratory Press, 1996.
Watson, J. D., *The Double Helix: A Personal Account of the Discovery of the Structure of DNA.* New York: Touchstone, 2001.

1953, REM Sleep
Siegel, J. M., *Sleep Medicine Reviews.* 2011 **(15)**3: 139.
National Institute of Neurological Disorders and Stroke: www.ninds.nih.gov/disorders/brain_basics/understanding_sleep/condition_information

1953, Acquired Immunological Tolerance and Organ Transplantation
Hamilton, D., *A History of Organ Transplantation: Ancient Legends to Modern Practice.* Pittsburgh: University of Pittsburgh Press, 2012.

1954, Sliding Filament Theory of Muscle Contraction
MacIntosh, B., et al., *Skeletal Muscle: Form and Function*. Champaign, IL: Human Kinetics, 2005.

1955, Ribosomes
Garrett, R. A., et al., (eds.), *The Ribosome: Structure, Function, Antibiotics, and Cellular Interactions*. Washington, DC: American Society Microbiology 2000.

1955, Lysosomes
British Society for Cell Biology: bscb.org/learning-resources/softcell-e-learning/lysosome/

1956, Prenatal Genetic Testing
Stanford Children's Health: lpch.org/DiseaseHealthInfo/HealthLibrary/pregnant/tests.html

1956, DNA Polymerase
Lehman, I. R., *Journal Biological Chemistry* 2005 **280**: 42477

1956, Second Messengers
Lodish, H., et al., *Molecular Cell Biology*. New York: W.H. Freeman, 2000, Sect. 20.6.

1957, Protein Structures and Folding
Reynaud E., *Nature Education* 2010 **3**(9): 28.

1957, Bioenergetics
Nicholls, D. G., et al., *Bioenergetics*. New York: Academic Press, 2013.

1958, Central Dogma of Molecular Biology
Ridley, M., *Francis Crick: Discoverer of the Genetic Code*. New York: Eminent Lives, 2006.

1958, Bionics and Cyborgs
Fischman, J., *National Geographic Magazine* (1) 2010.

1959, Pheromones
Wyatt, T. D., *Pheromones and Animal Behaviour: Communication by Smell and Taste*, New York: Cambridge University Press, 2003.

1960, Energy Balance
Body Recomposition: bodyrecomposition.com/fat-loss-the-energy-balance-equation.html

1960, Chimpanzee Use of Tools
Sanz, C. (ed.), et al., *Tool Use in Animals: Cognition and Ecology*. New York: Cambridge University Press, 2013.

1961, Cellular Senescence
Burton, D. G. A., *Age* 2009; **31**(1): 1

1961, Cracking the Genetic Code for Protein Biosynthesis
Alberts, B., et al., *Molecular Biology of the Cell*. New York: Garland Science, 2007.

1961, Operon Model of Gene Regulation
Müller-Hill, B., *The Lac Operon: A Short History of a Genetic Paradigm*. Berlin: de Gruyter, 1996.

1962, Thrifty Gene Hypothesis
Speakman, J.R., *International Journal Obesity* 2008 **32**: 1611.

1962, *Silent Spring*
Carson, R., *Silent Spring*. New York: Houghton Mifflin, 1962.

1963, Hybrids and Hybrid Zones
Price, J. (ed.), *Hybrid Zones and the Evolutionary Process*. New York: Oxford University Press, 1993.

1964, Brain Lateralization
Schwartz, J. M., et al., *The Mind and the Brain: Neuroplasticity and the Power of Mental Force*. New York: Regan Books, 2003.

1964, Animal Altruism
Dawkins, R., *The Selfish Gene*. New York: Oxford University Press, 2006.

1966, Optimal Foraging Theory
Stephens, D. W., *Foraging Theory*. Princeton, NJ: Princeton University Press, 1986.

1967, Bacterial Resistance to Antibiotics
Drlica, K. S., et al., *Antibiotic Resistance: Understanding and Responding to an Emerging Crisis*. Upper Saddle River, NJ: FT Press, 2011.

1967, Endosymbiont Theory
Kozo-Polyansky, B. M., et al., *Symbiogenesis: A New Principle of Evolution*. Cambridge, MA: Harvard University Press, 2010.

1968, Multi-Store Model of Memory
Gluck, M. A., et al., *Learning and Memory: From Brain to Behavior*. Richmond, UK: Worth, 2007.
Rudy, J. W., *The Neurobiology of Learning and Memory*. Sunderland, MA: Sinauer Associates, 2013.

1968, Hypothalamic-Pituitary Axis
Norris, D. O., et al., *Vertebrate Endocrinology*. New York: Academic Press, 2013.

1968, Systems Biology
Voit, E., *A First Course in Systems Biology*. New York: Garland Science, 2012.

1969, Cellular Determination
Kenyon College Biology Department: biology.kenyon.edu/courses/biol114/Chap11/Chapter_11.html

1970, Cell Cycle Checkpoints
Murray, A., et al., *The Cell Cycle: An Introduction*. New York: Oxford University Press, 1993.

1972, Punctuated Equilibrium
Gould, S. J., *The Structure of Evolutionary Theory*. Cambridge, MA: Belnap Press, 2002.

1972, Sustainable Development
Rogers, P., et al., *An Introduction to Sustainable Development*. New York: Routledge, 2007.

1972, Parental Investment and Sexual Selection
Low, B. S., *Why Sex Matters: A Darwinian Look at Human Behavior*. Princeton, NJ: Princeton University Press, 2001.

1974, Lucy
Jurmain, R., et al., *Introduction to Physical Anthropology, 2013-2014 Edition*. Independence, KY: Cengage Learning, 2013.

1974, Cholesterol Metabolism
Goldstein, J. L., *Arterioscerosis Thrombosis Vascular Biology.* 2009 **29**(4): 431.

1974, Sense of Taste
Finger, T. E. (ed.), et al., *The Neurobiology of Taste and Smell.* New York: Wiley-Liss, 2009.

1975, Monoclonal Antibodies
Wikipedia: www.wikidoc.org/index.php / Monoclonal_antibodies

1975, Sociobiology
Wilson, E. O., *Sociobiology: The Abridged Edition.* Cambridge, MA: Belknap Press, 1980.

1976, Cancer-Causing Genes
Cooper, G. M., *Oncogenes.* Burlington, MA: Jones & Bartlett Learning, 1995.

1977, Bioinformatics
Haddock, S., *Practical Computing for Biologists.* Sunderland, MA: Sinauer Associates, 2010.

1978, In Vitro Fertilization (IVF)
Elder, K., et al., *In-Vitro Fertilization.* New York: Cambridge University Press, 2011.

1979, Biological Magnification
biologicalmagnification.org/ understanding-biological-magnification

1980, Can Living Organisms Be Patented?
Wikipedia: En.wikipedia.org/wiki/ Biological_patent

1982, Genetically Modified Crops
Halford, N. G., *Genetically Modified Crops.* London: Imperial College Press, 2011.

1983, HIV and AIDS
Kallen, S. A., *The Race to Discover the AIDS Virus: Luc Montagnier Vs Robert Gallo.* Springfield, MO: 21st Century, 2012.

1983, Polymerase Chain Reaction
Rabinow, P., *Making PCR: A Story of Biotechnology.* Chicago: University of Chicago Press, 1996.

1984, DNA Fingerprinting
Rudin, N., et al., *An Introduction to Forensic DNA Analysis.* Boca Raton, FL: CRC Press, 2001.

1986, Genomics
Ridley, M., *Genome: The Autobiography of a Species in 23 Chapters.* New York: Harper Collins, 2006.

1987, Mitochondrial Eve
Sykes, B., *The Seven Daughters of Eve: The Science That Reveals Our Genetic Ancestry.* New York: W.W. Norton, 2002.

1987, Depletion of the Ozone Layer
Parson, E. A., *Protecting the Ozone Layer: Science and Strategy.* New York: Oxford University Press, 2003.

1990, Domains of Life
Margulis, L., et al., *Kingdoms and Domains: An illlustrated Guide to the Phyla of Life on Earth.* New York: Academic Press, 2009.

1991, Sense of Smell
Finger, T. E.(ed.), et al., *The Neurobiology of Taste and Smell.* New York: Wiley-Liss, 2009.

1994, Leptin: The Thinness Hormone
Akaba, S., et al., *Textbook of Obesity, Biological, Psychological and Cultural Influences.* New York: Wiley-Blackwell, 2012.

2000, Skin Color
Jablonski, N. G., *Living Color: The Biological and Social Meaning of Skin Color.* Berkeley, CA: University of California Press, 2012.

2003, Human Genome Project
Ridley, M., *Genome: The Autobiography of the Species in 23 Chapters.* New York: Harper Collins, 2006.

2005, Protist Taxonomy
Reece, J. B., et al., *Campbell Biology.* San Francisco: Benjamin Cummings, 2013.

2006, Induced Pluripotent Stem Cell
Lanza, R. (ed.), et al., *Essentials of Stem Cell Biology.* New York: Academic Press, 2009.

2009, Viral Mutations and Pandemics
Kolata, G., *Flu: The Story of the Great Influenza Pandemic of 1918 and the Search for the Virus that Caused It.* New York: Touchstone, 2001.

2010, Deepwater Horizon (BP) Oil Spill
Steffy, L. C., *Drowning in Oil: BP & the Reckless Pursuit of Profit.* New York: McGraw-Hill 2010.

2011, Translational Biomedical Research
Srivastava, R. (ed.), et al., *Lost in Translation: Barriers to Incentives for Translational Research in Medical Sciences.* London: World Scientific Publishing Co., 2013.

2011, Albumin From Rice
Evans, T. W., *Alimentary Pharmacology & Therapeutics* 2002 **16** (Suppl. 5): 6.
Yang He, et. al,, *Proceedings of the National Academy of Sciences USA* 2011 108(47): 19078.

2012, Human Microbiome Project
Hooper, L. V., *Science* 2001 **292**: 1115.

2012, Epigenetics
Francis, R. C., *Epigenetics: The Ultimate Mystery of Inheritance.* New York: W.W. Norton, 2011.

2013, American Chestnut Tree Blight
Freinkel, S., *American Chestnut: The Life, Death, and Rebirth of a Perfect Tree.* Berkeley, CA: University of California Press, 2007.

2013, De-Extinction
MacLeod, N., *The Great Extinctions: What Caused Them and How They Shape Life.* Richmond Hill, ON: Firefly Books, 2013.

2013, Oldest DNA and Human Evolution
Jurmain, R., et al., *Introduction to Physical Anthropology, 2013–2014 Edition.* Independence, KY: Cengage Learning, 2013.

Index

Photo Credits

"Thus, from the war of nature, from famine and death, the most exalted object which we are capable of conceiving, namely, the production of the higher animals, directly follows. There is grandeur in this view of life, with its several powers, having been originally breathed into a few forms or into one; and that, whilst this planet has gone cycling on according to the fixed law of gravity, from so simple a beginning endless forms most beautiful and most wonderful have been, and are being, evolved."

—Charles Darwin, *Origin of Species* (1859)